Environmental Science and Engineering

The ultimate goal of this series is to contribute to the protection of our environment, which calls for both profound research and the ongoing development of solutions and measurements by experts in the field. Accordingly, the series promotes not only a deeper understanding of environmental processes and the evaluation of management strategies, but also design and technology aimed at improving environmental quality. Books focusing on the former are published in the subseries Environmental Science, those focusing on the latter in the subseries Environmental Engineering.

More information about this series at https://link.springer.com/bookseries/7487

Swapnila Roy · Alok Garg · Shivani Garg ·
Tien Anh Tran
Editors

Advanced Industrial Wastewater Treatment and Reclamation of Water

Comparative Study of Water Pollution Index during Pre-industrial, Industrial Period and Prospect of Wastewater Treatment for Water Resource Conservation

Springer

Editors
Swapnila Roy
Department of Chemistry, School of Applied Sciences
K. K. University
Nalanda, Bihar, India

Shivani Garg
Institute of Environmental Studies
Kurukshetra University
Kurukshetra, Haryana, India

Alok Garg
Department of Chemical Engineering
National Institute of Technology
Hamirpur, Himachal Pradesh, India

Tien Anh Tran
Faculty of Marine Engineering
Vietnam Maritime University
Haiphong, Vietnam

ISSN 1863-5520 ISSN 1863-5539 (electronic)
Environmental Science and Engineering
ISBN 978-3-030-83810-2 ISBN 978-3-030-83811-9 (eBook)
https://doi.org/10.1007/978-3-030-83811-9

This Springer imprint is published by the registered company Springer Nature Switzerland AG
The registered company address is: Gewerbestrasse 11, 6330 Cham, Switzerland

Preface

This book presents recent developments in advanced industrial wastewater treatment technologies with basic principles, applications, merits, demerits of the aforementioned technologies with their limitations and future research needs and reclamation that is attracting increasing attention or that have a high potential for large-scale application in the near future. It explores the Advanced Oxidation Processes (AOPs) for the remediation of organic contaminants in wastewater, the recent advances in thermochemical processes; used to produce activated carbon for water treatment. This book also highlights the recycling and regeneration strategies with challenges and future prospects of activated carbon, so the ecosystem can be protected by the usage of activated carbon in multidimensional direction by purifying the wastewater as well as solid biomass waste management up to a greater extent and the removal of beta lactam drugs from water using AOPs. The effects of various parameters have been discussed in detail with critical analysis of each parameter like concentration effect, water matrix effect, and toxic effects in this book. Bioaugmentation with microbes and their mechanism for wastewater treatment, contamination of water-bodies by heavy metals, their toxic effect along with the mechanism of heavy metal removal using biofiltration process and its efficiency are highlighted in detail in this book. This book focused on the identification of Indian & global aspects and best practices followed in wastewater segments to reuse, retreat, recycle for efficient energy and nutrient recovery and the various AOPs used to remove specific pollutants and COD from petrochemical wastewater to meet discharge standards are explicitly explained. The effect of influential operational parameters on the eradication of EDCs (Endocrine Disrupting Compounds) from aqueous media via various processes has been highlighted. In this book it is highlighted that some modern wastewater treatment solutions to restrict the ocean water source from the maritime industry and shipbuilding. The influence in changes of water pollution index on the environment index and human health quality especially the preindustrial and industrial period is elaborated here. This book provides the in-depth mechanism of heavy metal interactions with the bio-polymeric surface using π-π interactions, dipole-dipole interaction, hydrogen bonding with homopolymer, copolymers, blends, or biopolymer/nanocomposite matrix containing anions or electron-rich group and also the application of graphene/GO-based composites for the elimination of contaminants from wastewater. This book also highlights the idea of adsorption integrated with biological treatment of wastewater using powdered activated carbon membrane bioreactor (PAC-MBR) and Biological Activated Carbon reactor (BAC). Developments in molecular biology techniques over the last twenty years provide insights into the complex microbial diversity found in biological treatment systems. This book highlights these techniques in detail and presents the state of the art in this field and the opportunities these techniques offer to improve process performance.

Nalanda, India — Swapnila Roy
Hamirpur, India — Alok Garg
Kurukshetra, India — Shivani Garg
Haiphong, Vietnam — Tien Anh Tran

Contents

About the Editors

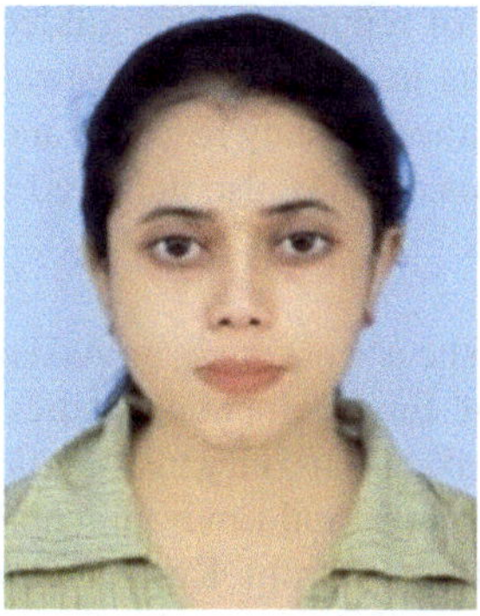

Dr. Swapnila Roy is an Assistant Professor in the Department of Chemistry (School of Applied Sciences), K K University, Nalanda, Bihar for last 2 years. Prior to KKU, she worked as a Lecturer of Regent Institute of Science and Technology, Kolkata, India. She worked as a Scientific Assistant in World Bank Funded Project in West Bengal Pollution Control Board. She worked as a Editor in reputed publishing house. She also worked as Analytical Chemist in different Pharmaceutical and Chemical companies. She received his Ph.D. in Environmental Chemistry from Jadavpur University, Kolkata, India. Her primary areas of research include Defluoridation in wastewater, adsorption method for remediation, design expert tool, application of artificial neural network. She is skilled in handling of sophisticated instruments such as ICP-OES, AAS, FT-IR, HPLC, GCMS, Bomb Calorimeter, Soxhlet Apparatus, and different instruments using for analysis of hazardous waste. She has authored many research and review articles, in SCI journals, presented the papers in National and International Seminars and Conferences and attended faculty development programs. She received a young research excellence award and Researcher Award for her outstanding performances. She has edited successfully the Springer book "The Impact of the COVID-19 Pandemic on Green Societies".

Dr. Alok Garg is working as Assistant Professor at Department of Chemical Engineering, National Institute of Technology, Hamirpursince 2019, prior to this he was Assistant Professor and Lecturer in Department of Chemical Engineering, Thapar Institute of Engineering and Technology, Patiala for 10 years. He has completed his Ph.D. (wastewater treatment) and M. Tech. (Chemical Engineering) from Department of Chemical Engineering, Thapar Institute of Engineering and Technology, Patiala. He is experienced in higher education industry. He is a strong educational professional skilled in teaching and research in the field of chemical engineering, industrial pollution abatement, and wastewater treatment. He has presented many papers in national and international conference of repute. He has guided master's and doctoral students in the area of wastewater treatment. He has also authored many research and review articles, in SCI journals. He has also authored a book chapter in springer book.

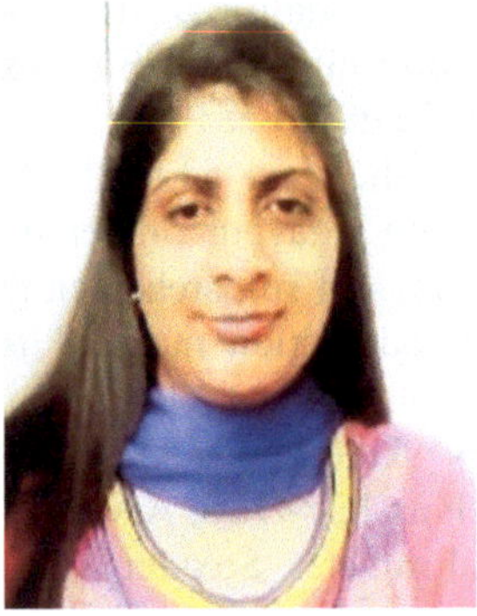

Er. Shivani Garg since 2015 she is working as an Assistant Professor at Institute of Environmental Studies, Kurukshetra University Kurukshetra, Haryana, India. She has completed M.Tech (Energy and Environmental Management) and M.Sc. (Botany) from Kurukshetra University. She is an experienced Assistant Professor with a demonstrated history of working in the higher education industry. She is a strong education professional, skilled in sustainable development, environmental issues, climate change, energy resource management, environmental compliance, water quality, and environmental impact assessment. Her main area of research interest is connected with wastewater treatment, wastewater, energy management, soil science, irrigation and water management, biogas studies, waste management, and environmental pollution. She has also presented the papers in National and International Seminars and Conferences and attended faculty development programmes. She has reviewed many chapters along with book. Her published works are with many renown publishing organizations and some are under publication process. Her most recent publications are "Green Technologies for the Treatment and Utilisation of Dairy

Product Wastes", "Nanocellulose-Based Supercapacitor", "Green Technologies for Restoration of Damaged Ecosystem".

Dr. Tien Anh Tran is a Lecturer at Department of Marine Engineering, Vietnam Maritime University, Haiphong City, Vietnam. He graduated B.Eng. and M.Sc. in Marine Engineering from Vietnam Maritime University, Haiphong City, Vietnam. He received the Ph.D. degree at Wuhan University of Technology, Wuhan City, People's Republic of China in 2018. He has been invited as a speaker, an organization committee member of the international conferences. He is an author/reviewer for the international journals indexed in SCI/SCIE, EI. He is a Leading Guest Editor, Editor Board Member for the journals: Environment, Development and Sustainability (Springer Nature), International Journal of Renewable Energy Technology (InderScience). In 2021, He has edited successfully the Springer book "The Impact of the COVID-19 Pandemic on Green Societies". His current research interest includes Ocean Engineering, Marine Engineering, Environmental Engineering, Machine Learning, Applied Mathematics, Fuzzy Logic Theory, Multi Criteria Decision Making (MCDM), Maritime Safety, Risk Management, Simulation and Optimization, System Control Engineering, Renewable Energy and Fuels, etc.

Chapter 1
Industrial Wastewater: Characteristics, Treatment Techniques and Reclamation of Water

Shivani Garg

Abstract Industries become major consumers of water resources due to the increase in demand for products. The wastewater generated from the industries needs treatment for reuse or discharge in the environment to prevent pollution. Wastewater treatment techniques help in the protection of the health of the public and environment. The treatment techniques vary from one type of industry to other due to the quality of effluent discharge from operations, energy and chemical requirements, process flexibility, and residual disposal options. The chapter focuses on characteristics of wastewater from different industries, their treatment techniques such as preliminary, primary, secondary, and advanced treatment processes, and treatment effectiveness for reuse of the treated water. The chapter also focuses on the comparative study of different wastewater treatment techniques as one type of treatment method selected for the treatment of one type of industry may or may not be suitable for the other type of wastewater from another industry.

Keywords Reclamation of treated water · Reverse osmosis · Ultrafiltration · Nano-filtration

Abbreviations

USEPA	United Nation Environment Protection Agency
SDWA	Safe Drinking Water Act
VOC	Volatile Organic Compound
EPA	Environment Protection Agency
TOC	Total Organic Carbon
CWA	Clean Water Act

S. Garg (✉)
Institute of Environmental Studies, Kurukshetra University, Kurukshetra Haryana 136119, India
e-mail: shivanigarg2015@kuk.ac.in

S. Roy et al. (eds.), *Advanced Industrial Wastewater Treatment and Reclamation of Water*, Environmental Science and Engineering,
https://doi.org/10.1007/978-3-030-83811-9_1

1.1 Introduction

Water is an important part of life and development operations in different industries. The increasing pollution cause increase in higher water demands leads to the requirement of recycling and reuse of wastewater for different purposes (Scheierling et al. 2011). Industrial wastewater is used as the only source for watering the crop plants in some developing countries (Ensink et al. 2004). The freshwater resources are limited for use, so it is required to think and discover the methods of recycling or reusing the wastewater. The water quality parameters are linked with the usage and development of the economy (Chennakrishnan et al. 2008). Water can be contaminated in many ways. Industrial wastewater is a major source of water pollution (Godswill and Godspel 2019). The major contribution to poor water quality is industrial discharge in urban areas (Mathuthu et al. 1997). According to The Food and Agricultural Organization (FAO) the freshwater withdrawal is 3,928 km^2/ year (Raza et al. 2012). The large-scale industries such as food, textile, etc. produce mostly the wastewater that causes water pollution. The untreated wastewater discharged into the environment causes fatal disease through the food chain (Balaji et al. 2005). India is the largest producer of dairy products in the world produces four liters of wastewater by one liter of milk (Chaudhry and Garg 2019). Thus, the requirement for the treatment of wastewater becomes mandatory for recycling (Mara 2000). A large number of industries required water for boiler feed and chilling purposes which can be easily fulfilled by treating wastewater. The treatment process also generates the sludge and other by-products which are rich source of nutrients for agriculture fields. According to the Clean Water Act, 1987, it is required to go through the bioassay testing to ensure that the effluent does not have any toxic effect on the biota after discharge. The wastewater generates from every industry have different physiochemical composition, so it also requires different methods for treatment. The chapter explains the characteristics of wastewater, the guidelines for permissible limits and different traditional and advanced treatment techniques along with modern methods of purification of water for reuse and reclamation.

1.1.1 Guidelines for Wastewater Discharge into the Environment

The standards for the discharged wastewater are decided by different organizations or agencies of different countries. As each country has its own pollution control committee or organization such as in India, Central Pollution Control Board decides the discharge standards. Discharge directly enters into environment or waterbodies. The effluent standard pertained to the quality of the effluent before it discharges into the environment after the treatment. Standards must be followed and compiled by the industries through policies or legislation enforcement so that discharged water has the water quality parameters with the limit sets by the pollution control agency.

The United Nation Environment Protection Agency (USEPA 2012) has classified and enlisted 126 organic and inorganic toxic chemicals that have specific discharge limits. This agency also updates the list of unregulated contaminants that are found in drinking water every five years under the Safe Drinking Water Act (SDWA).

World Health Organization (WHO 1989) set the standards or limits for the quality of wastewater before discharge. This is the world's largest and effective organization for standards for mitigation of pollution. The WHO (2003) is currently revising guidelines for policy creators to permit the use of wastewater safely after treatment.

1.1.2 Types of Industries on the Basis of Wastewater Production

In industries, water is used for washing, drinking, cooling, and production of steams in the boilers, etc. The wastewater generates from these operations has different types of pollutants that cause pollution in the environment and water bodies nearby. The urban wastewater contains both industrial and municipal wastewater. The quality of wastewater depends upon the types of industries. Figure 1.1 shows the types of industries that contribute a large amount of wastewater and use the maximum freshwater for their operations.

In India, the pollution index (PI) for different industrial sectors given by MOEFCC (Ministry of Environment, Forest and Climate Change) decide the standards of categorization of industries. The PI numbers are categorized between the range 0–100 given to the industrial sectors for example pulp paper, sugar, cement, etc. industries are the largest polluting industries have been categorized with PI score 60 or above

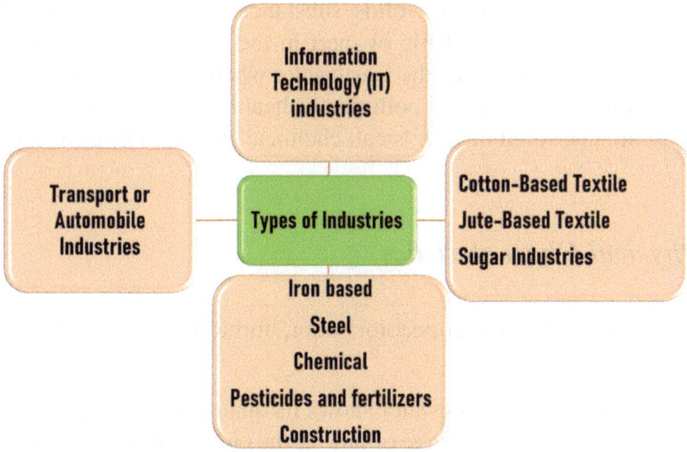

Fig. 1.1 Types of industries all over the world

Table 1.1 The comparative studies of principal pollutant produce by different industries

Name of industries	Major pollutants
Canning	Carbohydrates, Proteins, Suspended Solids
Dairy	Suspended Solids, Milk Residues, Fat, Carbohydrates, Proteins
Tannery	Carbohydrates, Proteins, Dyes
Brewery	Colors, liquors, chemical, carbohydrates
Oil and refinery	Oil, petrochemicals, sludge
Household	Detergents, Soap. Fat, chemical, plant extracts
Mining	Suspended solids, metals, acids and salts
Electronic	Organic chemicals, metals and metal ores, COD

known as Red Category. White category with PI score 20 or below includes bicycles, biofertilizers, etc. Table 1.1 shows the different types of pollutants produced from various industries. These industries produce wastewater during the operation of production of product and wastewater has many types of pollutants, different physiochemical parameters which sometimes exceed the prescribed standards given by different organizations such as WHO, CPCB, etc.

1.2 Characteristics of Industrial Wastewater

Wastewater generates from different industries constitutesa large number of contaminants and nutrients and organic solids substances. These contaminants can be biodegradable or non-biodegradable or inert in the environment and may contain pathogenic microorganisms, thus the wastewater when discharged into the environment and any aquatic body causes pollution or health problems. The characteristics of wastewater are classified into physical, chemical, physiobiological categories.

1.2.1 Physical Characteristics

The physical properties constitutecolor, odor, turbidity, temperature, suspended solids.

- **Color:** The color of the wastewater shows its qualitative character. The wastewater color change with the passage of time as the oldest the wastewater, darkest the color. The change in color occurs due to the decomposition in anaerobic conditions. The dark color generally occurs due to the formation of sulfides during decomposition.

- **Odor:** The odor of the wastewater is a major concern as the general public affected by it. The odor mostly produced due to the decomposition of organic matter in the wastewater in anaerobic conditions. The Sulphur products are mostly unpleasant. The different type of industrial wastewater produces different odors, for example, organic compost, beverages, and most food industries wastewater produces ammonia, Sulphur compounds, and alcohols product during decomposition anaerobically have an unpleasant odor (Brault 1991).
- **Turbidity:** Colloidal material or suspended solids cause turbidity in wastewater. For the removal of turbidity, the process requires coagulation and filtration methods. The photosynthesis and benthic organisms' activity may inhibit due to the blockage or reduction of penetration of light in the water body. Turbidity also causes the death of certain fish populations by clogging their gill with solids in wastewater.
- **Temperature:** The temperature of wastewater is higher than the normal water due to the industrial activity and presence of different pollutants in it. The wastewater temperature also varies from season to season with different regions such in winter or cold region the temperature varies from 7–18 °C and in summer or warmer region vary from 13–24 °C Crites and Tchobanoglous (1998). The measurement of temperature is a required process before the treatment of wastewater as some biological treatment processes are temperature dependent.
- **Solids:** Solids are the soluble or insoluble organic and inorganic matter present in wastewater. The solids in wastewater may be volatile (organic matter such as proteins, carbohydrates, and fats) and fixed solid (non-volatile) (Zainith et al. 2016). The total solids contain suspended solid and soluble compounds dissolved in the wastewater. The suspended solids can be removed with the help of the evaporation, sedimentation process. The average total solids present in wastewater are 40–60%.

1.2.2 Chemical Characteristics

Chemical characteristics are generally defined the inorganic and organic nature (chemical oxygen demand (COD), biochemical oxygen demand (BOD) etc.) of wastewater. The chemical nature of wastewater mostly dependent on the raw material uses, a product produced and the process used in industries.

- ***Inorganic chemicals*** include organic nitrogen, nitrates, and nitrites, organic and inorganic phosphorous. Nitrogen and phosphorus are significant nutrients required for plant growth. Generally, pesticides and fertilizers industries wastewater have large quantities of these nutrients. These promote the growth of plants in aquatic bodies but in higher concentrations cause contamination and pollution in water. Trace element includes heavy metals such as copper, cobalt, zinc, and iron. Heavy

Table 1.2 Some examples of heavy metal in major industries wastewater

Name of the industries	Examples of heavy metals
Fertilizers	Cadmium, Chromium, Zinc, Mercury, Arsenic, Nickel
Pulp and paper	Chromium, Carbon, Lead, Nickel, Zinc
Organic chemical	Arsenic, Mercury, Chromium, Cadmium, lead
Petroleum Refining	Zinc. Arsenic, Cadmium, Lead, Nickel
Textile and Tanning	Chromium, lead
Steel works	Arsenic, Cadmium, Mercury, Nickel, Zinc

metals in large quantities produce toxic effects such as arsenic, cadmium, mercury, chromium, etc. Table 1.2 shows some examples of heavy metal discharge from different industries (Bond and Straub 1974). The determination and treatment of these toxic chemicals are very important. Gases such as oxygen, methane, carbon dioxide, and sulfide are helping to operate the system but high quantity leads to pollution and imbalance in the environment as hydrogen sulfide produces during anaerobic processes is an odorous and toxic gas and also causes corrosion of the long sewer pipes. Dissolved oxygen affects the aerobic biological treatment processes so its measurement is required before treatment.

- ***Organic chemical*** in wastewater determines with BOD, CODand total organic carbon (TOC). The organic chemical may be biodegradable sometimes and leads to toxicity. So, determination and treatment of such chemicalsare required before discharge. The other toxic organic chemical includes pesticides, phenols, and oils. Certain specific organic pollutants are monitored to access the concentration of priority pollutants. Soaps, detergents, and other polyphosphates cause environmental pollution by adding to complex calcium (Munter 2003). In 1966, pollutant polychlorinated biphenyls (PCB) compound was the first pollutant found in the drinking water, fish tissues, sediments, and bird tissues all over the world (Manahan 1994). These pollutants have high thermal, biological, and chemical stability made them highly suitable components in transformers and capacitors as coolants and insulation fluids (Munter 2003).
- ***Volatile organic compounds (VOC)*** most commonly leak from underground storage tanks, improperly discarded solvents, and landfills in built-in industrial and commercial areas such as benzene, xylenes, trichloroethane, dichloromethane, etc. VOC are common soil pollutants in industrial areas.

  Heavy metals such as cadmium, mercury, lead and copper ions combine with cell membranes hinder the transport process throughcell walls. Cadmium, Mercury, and Lead havean affinity for Sulphur and disrupt enzymes. Cadmium is widely used in metal plating may discharge in water from industrial discharge and mining wastes. This is poisonous for human health in water causes kidney failure, high blood pressure, and damage of red blood cells. Inorganic lead discharge from mining sources and leaded gasoline is a major source of atmospheric and land contamination. Its contamination causes severe dysfunction of the kidney, reproductive system, brain, and nervous system in humans. The hardness of water due to calcium and magnesium concentration cause damage to appliances and equipment of household and industries (Hudson and Gilcreas 1976).

1.2.3 Biological Characteristics

The biological properties broadly divided into two sub-categories, water borne disease for public health considerations and biological processes such as bioaccumulation or eutrophication from nutrient additions. The viruses, bacteria, protozoans, and helminths are some examples of causative organisms of waterborne diseases. Some examples of waterborne diseases are hepatitis, giardiasis, and jaundice, etc. The most serious waterborne disease of bacterial origin is cholera which causes high fatalities and morbidities in the United States and in some developing countries in 1800 century (Englande et al. 2015). The water contamination with nutrients such as nitrogen or phosphorus causes excessive biological growth and deterioration of water quality of water resources. It also affects the growth of aquatic fauna by deteriorating the oxygen level in water bodies.

Wastewater pollutants are also categorized into toxic, conventional, or non-conventional. Conventional pollutants constitute BOD, suspended solids, oil and grease, pH, coliforms, etc.

Toxic pollutants include priority pollutants. Non-conventional pollutants constitute COD, TOC, and nutrients. Nanoscale particles such as sub-colloidal particles including viruses and natural organic matter are also found in wastewater (Song et al. 2010). Some examples of nanoscale particles are oxidation compounds of iron, magnesium, and lead (Lytle and Schock 2005).

1.3 Industrial Wastewater Treatment Techniques

Proper treatment of wastewater is required for the safety of the environment. The most important objectives of treating wastewater are to separate as much as solid and innocuous material that will cause adverse effects on the environment and reuse the treated wastewater and residual material as much as possible. There are many physico-chemical as well as biological methods for the treatment of industrial wastewater (Table 1.3). The common treatment techniques, include pre-treatment, primary treatment, secondary treatment, and tertiary treatment (Fig. 1.2). In conventional treatment techniques tertiary treatment methods include only when there is a requirement. Each industrial wastewater required a different approach for the treatment process as the constituents of wastewater are different in each industry. The technique used to treat industrial wastewater is specific and different than municipal

Table 1.3 Overview of wastewater systems main mechanisms for removal of specific pollutants

Pollutants	Subdivision of pollutants	Methods
Organic matter	BOD Particulate (>~1 μm)	Sedimentation, Adsorption, Hydrolysis, Stabilization
	Soluble BOD (<~1 μm)	Absorption, Stabilization
Solids	Coarse solids (>~1 cm)	Screening
	Suspended solids (>~1 μm)	Sedimentation
	Dissolved solids (<~1 μm)	Adsorption
Pathogenic organisms	Bacteria, Fungi, Viruses	Sedimentation, Filtration, Disinfection, UV irradiation
Nitrogen	Organic nitrogen	Ammonification
	Ammonia	Nitrification, Bacterial assimilation
	Nitrate	Denitrification
Phosphorus	Phosphate	Bacterial assimilation, Precipitation, Filtration

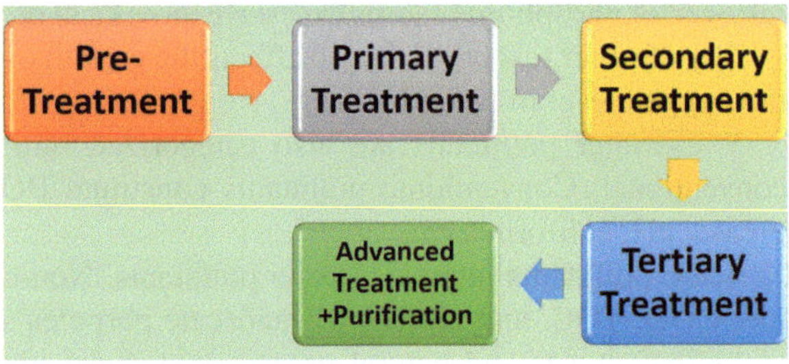

Fig. 1.2 The flow chart of common wastewater treatment processes

wastewater treatment (Woodard 2005). The most common approaches are explained as follow:

- Physical Methods

These methods mostly remove substances by sedimentation, adsorption, screening, deep bed filters, flotation, membranes, and ion exchange, etc. These methods used natural forces such as gravitational, electrical attraction, van der Wall forces,etc. The majority of total suspended solids and contaminants are removed by the segregation process that in turn plays an important role in overall wastewater treatment processes. Membrane methods are advanced methods used to make the water safer and reusable.

- **Chemical Methods**

These methods generally convert the complex substances into a simpler form by using chemicals or certain agents that enhance the process. These methods include removal and conversion of solid into gases, non-biodegradable tobiologically degradable materials, chelating agents, etc. These processes along with advanced oxidation or biological processes can remove difficult pollutants from industrial wastewater. In recent years, some more advanced technologies such as ozonation, photocatalysis, and ultrasound oxidation have been studied by researchers for better results in treatment.

- **Biological Methods**

Biological processes include biological pathways or natural decomposition/ conversion of organic matter into a simpler form. The biological process helps in carbonaceous organic material removal that representing BOD, nitrification, denitrification, phosphorus removal, and stabilization. The management of nitrogen in wastewater treatment is the most required process (Tchobanoglous 2003). Aerobic and anaerobic biological processes are also involved in this stage. Aerobic/anaerobic bioreactors are more commonly used in the treatment of wastewater nowadays as these reactors required small space for setup, economically efficient, and good in COD removal (Chan et al. 2009).

Wastewater treatment processes include two methods for collection and treatment setup: one is an offsite system (wastewater is collected and transported to other treatment plant site) and another is an onsite system (wastewater treatment plant is installed at the same site where it produced and further reuse for different purposes). Onsite system maybe small- or large-scale treatment plants for sewage, municipal and industrial wastewater treatment.

The important step for treatment and protection of the environment and human health is to keep the quantity and pollutant to minimal at the point of origin for design and installation of specific wastewater treatment plant such as the substitution of more environmentally and biodegradable material used during production, recycling of as much as the water at different stages for example the waste steam can reuse in the feeding of the boiler again.

The main goal of wastewater treatment techniques is to achieve the best water quality for reclamation and discharge into the environment. The common treatment processes with advanced treatment techniques are discussed as below:

1.3.1 Pre-treatment Methods

The initial stage of wastewater treatment is the pre-treatment process. In this process used physical methods to remove the large waste components from wastewater such as floatable grit and grease, sticks, wood pieces, rags which may cause operational and maintenance failure of the wastewater treatment plant. The common methods involved in this stage are screening, grit chamber, comminutors, floatation units, skimming tanks, etc.

- **Screening**

First stage of wastewater treatment is screening as in this stage the wastewater is clean and screens for large waste particles. These particles are buried, incinerated, or grind in a garbage grinder. To maintain and protect plant machinery, this stage is followed by grit removal. The homogenization of solids during screening contributes to the organic load in the treatment process. The screens are arranged and set at 45-60° with vertical. The screens are required to clean manually or automatically operated systems.

- **Grit Chamber**

Grit chamber used to remove suspended inorganic material such as sand consisting of gravel, cinder, and other heavy solids gritty substances from wastewater when it passes through it. The removal of grit is necessary to avoid any damage to machinery during treatment. Inorganic solids (suspended) are separated in grit chambers whereas organic suspended solids are removed with the sedimentation process due to the different nature and disposal arrangement of these solids. Grit chambers are cleaned mechanically or manually. Mechanically cleaning is done by remove and scrape blades on the base of the grit chamber. The collected grit after the scraping removed and raised to ground level by various processes such as jet pumps, airlifts, and bucket elevators. Manual cleaning is cleaning with a shovel to scrape the grit and must be done once a week.

- **Oil and Grease Removal**

This tank generally helps in the removal of finely divided suspended solids and oil and grease matter. The floatation tank used to remove fine fibers from screened wastewater from the paper and textile industry and oil and grease waste from oil and automobile industries. It also used to treat wastewater from metal finishing and coating, cold-rolling, pharmaceutical, and tanneries industries. The presence of oil and grease in wastewater prevents the solids to settle during the sedimentation process. The aeration system is used to speed up the separation of suspended matter during the process. The air under pressure injected into the effluent which pressurized the air–water mixture to escape atmospheric pressure by forming tiny air bubbles. Flocs (sludges), suspended solids, and oil globules are floated by these bubbles. This also increases the buoyancy of particles, so the particles float on the surface and easily removed. An API separator or parallel plate or corrugated plate oil separators are used to separate free oil by gravity allowing the lighter than water oil globules to rise to the surface. The API separator removed oil globules of size 0.015 cm or larger and can remove >50 mg/l effluent oil. The corrugated plate separators (CPS) remove oil globules of size 0.01 cm and large and as low as 10 mg/l oil from effluent. The separation depends upon the hydraulic variation. Chemical coagulants are added to the effluent to separate oil emulsions and enhance flocculation and flotation. Sometimes due to the presence of volatile organic carbons, methane or nitrogen gases are filled in the space of API separator and floatation unit which leads to explosion hazards. To prevent these hazards, the floatation units are covered generally.

- **Equalization and Neutralization**

Industrial wastewater possesses high pH and the concentration of contaminants depends upon the operational units of particular industries. To neutralize the large fluctuation in pH and concentration of contaminants, the basin is necessary to install inflow to the treatment plant. Equalization will help in maintaining the relative flow and concentration through the plant. The neutralization process generally follows with equalization process, hence acidic or alkaline flows can be partially neutralized in the equalization tank. By adding lime, caustic, or limestone for acidic and hydrogen peroxide or HCl or flue gas for alkaline wastewater can be neutralized. The pH of the wastewater should be a range from 6.5-8.5 for biological treatment.

1.3.2 Primary Treatment

The primary treatment removes approximately 50–70% of total suspended solids, 25–50% BOD and 65% oil and grease during the process. The process main objective to separate settleable organic and inorganic solids by sedimentation and floatable substances by skimming. Some inorganic substances, organic nitrogen and phosphorus and heavy metal present in solids also removed during primary treatment.

- **Sedimentation**

The settleable solids are removed into a large tank in which solids settle by gravity Cheremisinoff (2001). The supernatant flows to the next stage which is secondary treatment. It removes sand, silt, clay along with biological contaminants from water by gravity. The sedimentation process suggested before the supplication of water purification and disinfection methods. The sedimentation process depends upon the retention time of wastewater in the tank. The longer the undisturbed water in the tank, the greater the settling of solids and pathogens to the bottom of the tank. Coagulants are added to accelerate the process of sedimentation such as aluminum sulfate, liquid alum (poly aluminum chloride or PAC), and ferric sulfate. The solid sludge from the bottom can be removed and transferred to a sludge digestor in secondary treatment or biological treatment. The function of the overflow rate is to enhance the efficiency of primary and secondary clarification of solids. The settling velocity is greater than the overflow rate to increase efficiency.

- **Coagulation/ flocculation**

To settle solids during wastewater treatment in sedimentation tanks some natural or synthetic chemicals are added, the process is known as coagulation Flamant et al. (2004). Coagulants are added to neutralize the charges of particles in wastewater as coagulants generally have an opposite charge to the non-settleable material such as clay or organic substances. Due to neutralization, the particles attached to each other become heavy in turn settle down at the bottom. Rapid mixing and high energy are required to disperse coagulants and promote the formation of microflocs (large particles formed after neutralizing the charged particles which can still not be visible with naked eyes).

Flocculation is the gentle mixing stage of coagulation in which small charged particles attached to opposite charge particles and form flocs. The slow mixing help in attaching the microflocs to each other to form large flocs which settled during sedimentation. Polymers with high molecular weight or coagulant aid are added during mixing to help bind, bridge, and strengthening the floc particles which add weight and increase the settling rate of particles. The physical separation of liquid from solids usually achieved by floatation, filtration, or decantation. The most common mineral and organic coagulants/ flocculation additive are added to the wastewater are iron or aluminum salts, polymers, activated carbon or silica, or talcum. Some heavy metal chelating agents are also used during coagulation methods.

1.3.3 Secondary Treatment

The main objective of this stage in wastewater treatment is to remove residual organic and suspended solids from primary treatment. Aerobic and anaerobic biological treatment methods are used in this treatment. The biological treatment process used organic matter as food or substrate by microorganisms and produced inorganic end products such as carbon dioxide, ammonia, and water. The only requirement of secondary treatment is the presence of an accurate number and type of acclimated microorganisms to metabolize organic matter. The activated sludge and trickling filtration process are the two most commonly used biological treatment processes. The fine suspended, colloidal, and soluble organic material mixed with biologically active sludge maintain in suspension with an aeration system in the tank. Aeration not only maintained the oxygen supply for microorganisms but also maintain maximum contact and turbulence in the tank. The activated sludge process used for absorbing, flocculating, and assimilating wastes. The trickling filtration methods wastewater flows over the surface with microorganisms present in it. The microorganisms present in fixed-film surfaces create a gelatinous, slimy film that used the substrate as food for their growth and transfer the end product into water such as carbon dioxide, nitrates, and sulfates. The energy produced during the processes utilized in reproduction and other metabolic activity of microorganisms. The stabilization time used for industrial wastewater is higher than municipal wastewater. The retention time requires during the processes may be few hours to days. Temperature fluctuation is also a determining factor of the efficiency of treatment in both processes.

- **The Activated Sludge Process**

The maintaining of air supply for microorganisms presents in flocs in tank and contact time of incoming wastewater with these microorganisms are the main requirement for efficient treatment. The rectangular tank with continuous aeration is used to maintain microorganism concentration in the conventional process. Aeration is supplied by diffusers in form of bubbles or by an impeller forturbulent agitation of liquid. The

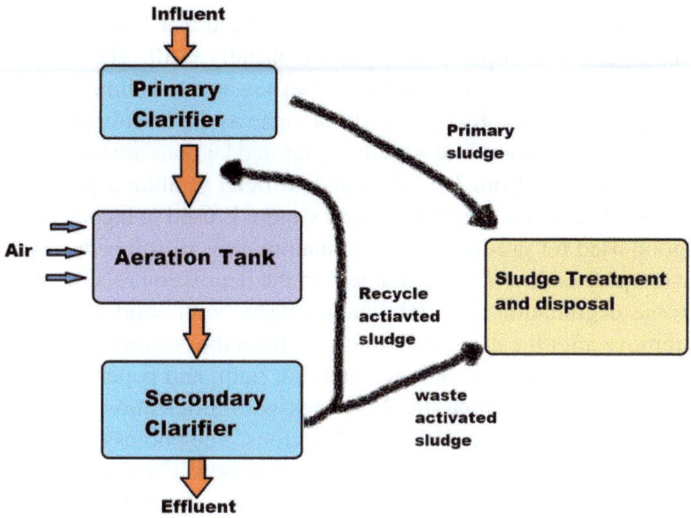

Fig. 1.3 The graphical flow diagram of working process of conventional activated sludge process

concentration of microbes is maintained by mixing a portion of the secondary sediment sludge in the activated sludge tank as the activated sludge process produced new cell microbes which settle in the sedimentation with solids. The rest of the sludge is disposed of after further treatment. Figure 1.3 shows the flow of wastewater through the conventional activated sludge process, sludge recycling to primary sludge, and treatment and disposal of the sludge. The detention time to aeration is given in the process to maintain high solid content or a low food-microorganism ratio or endogenous respiration will occur that leads to the complete oxidation. Extended aeration enhances the removal of priority pollutants to μg/l level and reduces toxicity in industrial wastewater.

Certain modification in the conventional activated sludge systems helps enhance the efficiency of the treatment process such as complete mix activated sludge systems maximize dampening of fluctuations of quality of wastewater. In sequencing batch reactors, a combination of complete mix and plug-in flow operated on an intermittent basis (sedimentation, aeration, and decanting of effluent in the same basin). This system provides high-quality treated effluent and good settling flocs which are rich in nutrients without the need for an external clarifier.

- **Lagoons and Oxidation Ponds**

In many locations in the United States, stabilization ponds are the most common technique to treat wastewater. The utilization and design of the ponds depend upon weather, available land, location, and purpose. The time of detention depending on the type of pond and weather condition and it may vary from 7–180 days. Ponds are classified into facultative, aerated, aerobic, and anaerobic four categories and the most widely used pond is a facultative pond.

Facultative ponds are operated aerobically in the upper layer and anaerobically in the lower layers. The depth of this pond is usually 6-8ft. The detention time in facultative ponds will vary with the nature of wastewater and climate conditions. The efficiency of ponds affected by the growth of algae as the dissolved oxygen concentration becomes low for microbial activities. Aerated lagoons are usually 6–20 ft deep with detention time vary from 3–12 days and artificial aeration is provided mechanically for microbial activity. The three-basin system is used in lagoons systems. First aerated lagoon used for pre-treatment and maintaining all solids in suspension. The second basin operated lower level and help in the deposition of solid at the bottom. The anaerobic degradation and stabilization occur with solid at the bottom. The solid will remove after the anaerobic digestion from the bottom. Aeration basins are used to treat nontoxic wastewater such as food, pulp, and paper effluents. Aerobic ponds are with depths of about 18–36 inches with a detention time of 3–5 days. These ponds are used in warm and sunny locations. The photosynthesis and surface reaeration process fulfill the oxygen demand for ponds. These ponds generate high microbial quality effluent. Anaerobic ponds are used to treat strong agricultural and industrial wastewater as a pre-treatment process. The ponds are usually 8–15 ft deep with detention time 20–50 days. The wastewater treated in these ponds produces a very stinky odor. By adding sodium nitrate, the odor can be removed.

- **Trickling Filtration Process**

The trickling filtration process is similar to the activated sludge process except in this process microorganisms are attached to the fixed bed rather than the suspended solids. The wastewater flows on the bed (contains coarse, rough, and hard material) with rotary nozzles. The organic matter is assimilated with bacteria. The film of the bed becomes thick after sometime due to the deposition of solids. The sludge from these beds is transferred into a secondary settling tank. Many factors such as organic and hydraulic loading rate and biodegradability of wastewater affect the performance of the trickling filtration process. This process designed to remove 50% of BOD from wastewater. The effluent is recirculated and mixed with influent to dilute the organic concentration.

- **Rotating Biological Contactors (RBC)**

These are similar to trickling filtration process consist of larger diameter plastic medium mounted on horizontal rotator which rotates the tank of wastewater. Approximately 40% of the tank is submerged and biological slime is developed on the medium in the tank. Rotation of the tank provides aeration and enhances microbial metabolism. BOD removal, denitrification, and nitrification may be achieved by RBCs. The operating cost of RBC is less than the activated sludge process.

1.3.4 Tertiary Treatment

These treatment methods are required when some specific contaminants cannot remove during secondary treatment. The alternative for conventional activated sludge technologies is membrane bioreactors that do not require primary and secondary treatment before tertiary treatment (Le-clech et al. 2006). Direct raw wastewater is applied through membrane bioreactors for treatment. Some tertiary treatment techniques are explained as follows:

- **Filtration**

Filtration process is the key method for the removal of suspended and colloidal particles including microorganisms. The filtration process most commonly used are surface, depth, and membrane filtration.

Surface Filtration: This method is used as a pre-treatment for disinfection and membrane filtration. Mechanical sieving with thin filter material that is made up of different cloth fabrics, metal fabrics, or synthetic material is used for particulate matter removal. This method can be done at a higher filtration rate to achieve low turbidity.

Depth Filtration: This process is important for the complete disinfection of wastewater. Depth filtration helps in the removal of suspended solid including sorbed contaminants. This is an effective method for the reclamation of water. Mono or dual filter media using sand and anthracite coal are used in depth filtration. The large plants with multiple filters do not use backwash so the turbidity value is elevated in this process.

Membrane Filtration: The use of membrane technologies gives the advantage in wastewater treatment by complete suspended solids removal, high rate, high disinfection capability and high efficiency in organic and nutrient removal, and low carbon footprint (Stephenson et al. 2000). Membrane bioreactors technologies remove a wide range of dissolved contaminants. The membrane filtration techniques are subclassified into different processes such as reverse osmosis, microfiltration, ultrafiltration, and nanofiltration. Reverse osmosis (RO) is the most commonly used membrane filtration process that helps in the removal of large molecules and ions from effluents. RO is most specifically used to remove dissolved organic components and total dissolved solids in addition to turbidity, pathogens removal. The process involves the flow of influent through a dense membrane that is a thin, micro-porous sheet. The membrane used in the process is made up of cellulose acetate most commonly. The membranes retain the pollutants and pass the water molecules through them. The membrane bioreactor (MBR) technology usage limits by the membrane fouling, maintenance and operating cost (van Nieuwenhuijzen et al. 2008). The pre-treatment of raw wastewater is required before feeding it into the RO system to avoid the membrane fouling and blockage of pores. The typical conventional pre-treatment constitutes the addition of chemicals such as coagulants or flocculent, acids to prepare the water for granular media filtration (Sauvet-Goichon 2007).

In microfiltration, the pore size is in the range of 0.08–2 m and for ultrafiltration, the pore size of filters is in the range of 0.005–0.2 m. So, both the process can remove large organic molecules, microorganisms, and colloidal particles. The ultrafiltration process completely removes the *E. coli* and total coliform Bray et al. (2021). Nanofiltration (NF) is also known as low-pressure reverse osmosis technology or membrane softening. The pore size is in the range of 0.5 to 1 nm. NF is inexpensive as compared to the RO system as it used lower operating pressure and higher flow rates. NF can reduce color, odor, hardness, total solids, and heavy metals from water.

1.4 Advanced Treatment and Purification of Treated Water for Reclamation

As the treatment of wastewater goes through different stages, contaminants or pollutants are removed with different methods. But the water after treatment requires more techniques to make it reusable. Many factors affect the selection of reclamation technologies such as wastewater characteristics, compatibility to existing conditions, type of water reuse application, the objective of reclaimed water quality, flexibility, operating and maintenance requirement, disposal of residual, energy, and chemical required (Asano et al. 2007). Some techniques are discussed as follows:

1.4.1 Disinfection

Disinfection methods are generally designed for the reduction of waterborne disease-causing pathogens for example bacteria, viruses, and protozoa. In the United States, the protozoa, bacteria and viruses such as *Cryptosporidium, Giardia, Shigella, Escherichia coli,* and enteroviruses respectively are the most commonly reported waterborne pathogens (Craun et al. 2006). Common chemical agents used for the disinfection of wastewater are chlorine and ultraviolet (UV) irradiation. Conventionally, chlorine and hypochlorite are the most common disinfectants. UV irradiation is more common as compared to chlorine and ozonation in drinking water application recently (Asano et al. 2007). UV disinfection technology ensure high-quality discharge of effluent. The removal of any large particles (down to the level of 15 microns) by large anthracite coal filter to enable the maximum penetration of UV light before UV disinfection technology is required.

The effect of disinfectants on pathogen depends on amount, contact time, and water quality variables. The relationship between residual concentration, contact time, and microbial inactivation may be affected by temperature, pH, and turbidity for example for chlorine is more effective below 7.6 pH (Fair et al. 1948), and when turbidity is <1 turbidity unit, have minimal effect on disinfection (LeChevallier et al. 1981).

1.4.2 Carbon Adsorption

Adsorption process help in the removal of refractory organic substances, some heavy metal, toxins, and volatile organic carbons. It is a surface process that increases the concentration of particular constituents on the surface or interface between two phases (Gupta et al. 2007). The most common absorbent used for this process is activated carbon which is solid and has a very porous, very high surface area to unit weight ratio along with high absorption capacity for the hydrophobic soluble substances in wastewater. The granular activated carbon is used for advanced wastewater treatment plants. As the wastewater flows through the bed, the organics and toxic substances are absorbed by the carbon. The bed after usage can be replaced bya new one or a thermal regeneration process. However, the capacity of regenerated carbon will be reduced as compared to the new one. The performance of the process can be enhanced by added powdered activated carbon to the activated sludge process and various kind of compounds can be removed by >90% and more (Baccar et al. 2012). In recent years, nanoparticles such as carbon composite or nanotubes have been used for the exclusion of heavy metals and organic substances. Some emerging contaminants such as ibuprofen, caffeine, primidone, levofloxacin, etc. can be removed by activated sludge and microfiltration treatment processes (Yang et al. 2011). The origin of activated carbon also affects the absorption process for example activated carbon extracted from wood removed >90% of acetaminophen than activated carbon extracted from other sources (Cabrita et al. 2010). Adsorption process in combination with other treatment processes such as ultrafiltration, coagulation, etc.

1.4.3 Ion Exchange

Ion exchange process used to remove the hardness and certain metal ions from water. This process used the ability of some natural and synthetic material to exchange ions present in material with the ion present in water (Xu 2005) for example if calcium ion present in water pass through ion-exchange material and it prefers calcium ion as compare to sodium ion present in the material, the calcium ions adhere to the material in the place of sodium (Fig. 1.4). The sodium ions pass in the effluent. The working principle of the ion-exchange method is the principle of selectivity which states that a stronger binding ion is displaced by a weakly bound ion in the water. This selectivity helps in the removal of distinct ion and replace with other ions in water. The microporous exchange resin is the main component of ion exchange equipment which is supersaturated with certain solutions. The sulfonated polystyrene beds saturated with sodium ions are used for water softening. Water flows through the bed and ions exchange during the passage from resin bed to water (Skipton et al. 2008). Ion exchange softening or Sodium Zeolite Softening is an ion exchange process used to purify the water by removing the hardness caused by calcium and magnesium. This is commonly used in steam boilers and industrial water treatment.

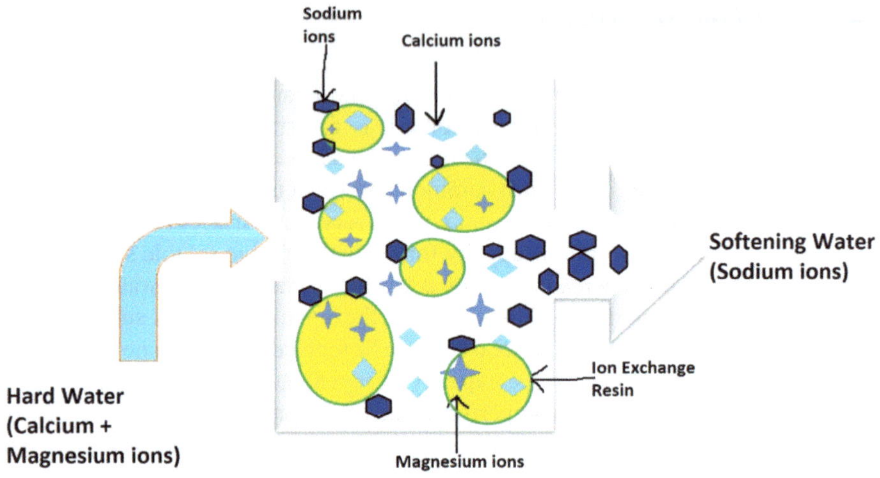

Fig. 1.4 The graphic diagram for Ion-exchange process

The most common and practical ion exchange use is a fixed-bed column system for food and agricultural wastewater treatment (Eckenfelder 2000).

1.4.4 Granular Media Filters

This process is used to reduce turbidity, suspended or colloidal particles, and various size particles from the coarse sediment. The exclusion rating for a different size in granular media filters (GMF) is 10–100 μm. Granular media filters constitute anthracite coal layered bed, sand, finely crushed garnet, and magnetite of particle size 0.8–1.2 mm, 0.5–0.8 mm, 0.4–0.6 mm, and 0.3–0.4 mm respectively. The uppermost layer constitutes lightweight coarsely graded substances such as anthracite coal whereas the lowest layer has heavyweight most finely graded substances such as garnet or magnetite (Fig. 1.5) (Singh 2005). Granular media filters help to avoid membrane fouling in the reverse osmosis filtration system. The regular cleaning of media is required otherwise blockage in the media pore occurs which in turn reduces the efficiency rate of the system.

1.4.5 Bioremediation

Bioremediation is a biological treatment method that uses microorganisms to degrade matter in wastewater. The microorganisms such as bacteria, fungi, and algae used the pollutant and matter as substrate or food and release water and carbon dioxide. The

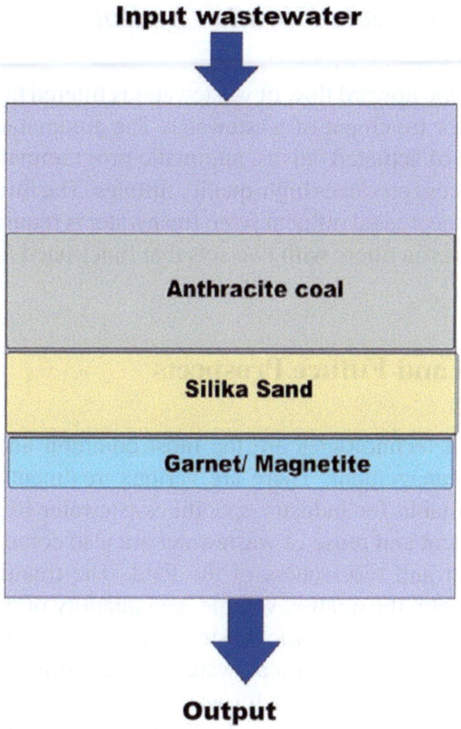

Fig. 1.5 The graphical representation of multimedia layering in granular media filters

metabolism of organic material by microbes formed energy and carbon source that is used in bioremediation process (Fulekar et al. 2009). The technology is well used in some pilot projects by environment agencies such as The National Mission for Clean Ganga for cleaning some part of the river. Some examples of microorganisms used in the bioremediation process are *Rhizoctomia, Pseudomonas, Actinobactor, Bacillins, Flavobacterium, Nocardia, Mycrobacterium, Mycococcus,* etc.

The wastewater from different industries has certain colors such as wastewater from dye and textile industries have various dye compounds that give dark coloration to water. This type of water has been undesirable for industrial or domestic use. Some fungi such as Basidiomycetes and Ascomycetes have been used for wastewater decolorization. An example of fungi used in the decolorization of distillery mill wastewater is *Aspergillus niger* (Patil et al. 2003). The bioremediation and decolorization of distillery effluent have been done with isolated bacteria from different cultures such as *Bacillus thuringiensis, Pseudomonas aeruginosa* (Kumar and Chandra 2006; Mohana et al. 2007).

1.4.6 Automatic Variable Filtration System

In this method setup, the upward flow of wastewater is filtered through the downward flow of filter media for treatment of wastewater. The automatic variable filter technology is constituted of actuated valves, automatic programmable logic controllers, and sensors. This process produces high-quality filtrates. The filter used in the process is further cleaned by the treated influent so no freshwater is required for cleaning. The technologies prefer media filters with two sets that functioned in series or parallelly.

1.5 Conclusion and Future Prospects

Wastewater treatment technologies are the most common and important concept for water resource management. There are various treatment plants and different technologies are available for industry-specific wastewater treatment. The various challenges for treatment and reuse of wastewater are also come into notice in recent years by the scientist and researchers of the field. The treatment plants installed into industries depend on the quality, volume, and quantity of wastewater generates. The treatment techniques for specific wastewater decrease the risk on freshwater resources available along with treatment, water can be disinfected and reclaimed for various purpose. The future prospects for wastewater treatment technologies are to discover advance, safe, cost-effective, and reliable methods. The researches depend upon two major required categories such as health and environmental issues and performances and quality assurance of the plants. The researcher can also address the assessment of human health effects of chemicals or constituents in reclaimed water and also surveillance of waterborne diseases, governmental response infrastructure, and epidemiological research capacity.

1.6 Summary

As challenges with climate change, overpopulation and natural resources are most common for researchers, and scientists' concern. The major concerns for the government and public are safe and healthy drinking water supply with healthy and environmentally sound water resources. Industrialization imposes a major threat to water resources as industries are the largest consumers of freshwater. The different governmental and non-governmental agencies/ organizations give the guidelines for wastewater discharge permissible limit for the safeguarding of water bodies. Different wastewater treatment methods are explained with some advanced treatment technologies in the chapter. The water reclamation concept with the help of various purification and disinfection techniques also discussed in the chapter.

References

Asano T, Burton F, Leverenz H (2007) Water reuse: issues, technologies, and applications. McGraw-Hill Education

Baccar R, Sarrà M, Bouzid J, Feki M, Blánquez P (2012) Removal of pharmaceutical compounds by activated carbon prepared from agricultural by-product. Chem Eng J 211:310–317

Balaji V, Datta S, Bhattacharjee C (2005) Evaluation on biological treatment for industrial wastewaters. Biomed Environ Sci 85(18):320–326

Bond RG, Straub CP (1974) Wastewater, treatment and disposal (Vol. 4). CRC Press

Brault (1991) Water treatment handbook. Volume 1, 6th edition

Bray RT, Jankowska K, Kulbat E, Łuczkiewicz A, Sokołowska A (2021) Ultrafiltration process in disinfection and advanced treatment of tertiary treated wastewater. Membranes 11(3):221

Cabrita I, Ruiz B, Mestre AS, Fonseca IM, Carvalho AP, Ania CO (2010) Removal of an analgesic using activated carbons prepared from urban and industrial residues. Chem Eng J 163(3):249–255

Chan YJ, Chong MF, Law CL, Hassell DG (2009) A review on anaerobic–aerobic treatment of industrial and municipal wastewater. Chem Eng J 155(1–2):1–18

Chaudhry S, Garg S (2019) Industrial wastewater pollution and advanced treatment techniques. In Hussain A, Ahmed S (eds) Advanced treatment techniques for industrial wastewater (pp. 74–97). Hershey, PA: IGI Global

Chennakrishnan C, Stephen A, Manju T, Raveen R (2008) Water quality status of three vulnerable freshwater Lakes of Suburban Chennai, India. Ind J Environ Ecoplan 15(3):591–596

Cheremisinoff NP (2001) Handbook of water and wastewater treatment technologies. Butterworth-Heinemann

Craun MF, Craun GF, Calderon RL, Beach MJ (2006) Waterborne outbreaks reported in the United States. J Water Health 4(S2):19–30

Crites R, Tchobanoglous G (1998) Small and decentralized wastewater management systems (No. 628.3 C934s). Mc Graw Hill

Eckenfelder W (2000) Industrial water pollution control. McGraw-Hill

Englande Jr AJ, Krenkel P, Shamas J (2015) Wastewater treatment & water reclamation. Reference Module in Earth Systems and Environmental Sciences

Ensink JH, Mahmood T, Van der Hoek W, Raschid-Sally L, Amerasinghe FP (2004) A nationwide assessment of wastewater use in Pakistan: an obscure activity or a vitally important one? Water Policy 6(3):197–206

Fair GM, Morris JC, Chang SL, Weil I, Burden RP (1948) The behavior of chlorine as a water disinfectant. J Am Water Works Ass 40(10):1051–1061

Falsanisi D, Liberti L, Notarnicola M (2010) Ultrafiltration (UF) pilot plant for municipal wastewater reuse in agriculture: impact of the operation mode on process performance. Water (basel) 2(4):872–885

Flamant O, Cockx A, Guimet V, Doquang Z (2004) Experimental analysis and simulation of settling process. Proc Safety Environ Protect 82(4):312–318

Fulekar MH, Geetha M, Sharma J (2009) Bioremediation of TrichlorpyrButoxyethyl Ester (TBEE) in bioreactor using adapted Pseudomonas aeruginosa in scale up process technique. Biol Med 1(3):1–6

Godswill AC, Godspel AC (2019) Physiological effects of plastic wastes on the endocrine system (Bisphenol A, Phthalates, Bisphenol S, PBDEs, TBBPA). Int J Bioinform Comput Biol 4(2):11–29

Gupta VK, Jain R, Varshney S (2007) Removal of Reactofix golden yellow 3 RFN from aqueous solution using wheat husk—an agricultural waste. J Hazard Mater 142(1–2):443–448

Hudson HE Jr, Gilcreas FW (1976) Health and economic aspects of water hardness and corrosiveness. J Am Water Works Ass 68(4):201–204

Kumar P, Chandra R (2004) Detoxification of distillery effluent through Bacillus thuringiensis (MTCC 4714) enhanced phytoremediation potential of Spirodelapolyrrhiza (L.) Schliden. Bull Environ Contamin Toxicol 73(5):903–910

Kumar P, Chandra R (2006) Decolourisation and detoxification of synthetic molasses melanoidins by individual and mixed cultures of Bacillus spp. Bioresource Technol 97(16):2096–2102
LeChevallier MW, Evans TM, Seidler RJ (1981) Effect of turbidity on chlorination efficiency and bacterial persistence in drinking water. Appl Environ Microbiol 42(1):159–167
Le-Clech P, Chen V, Fane TA (2006) Fouling in membrane bioreactors used in wastewater treatment. J Membr Sci 284(1–2):17–53
Lytle DA, Schock MR (2005) Formation of Pb (IV) oxides in chlorinated water. J Am Water Works Ass 97(11):102–114
Mara DD (2000) The production of microbiologically safe effluents for wastewater reuse in the Middle East and North Africa. Water Air Soil Pollut 123(1):595–603
Manahan SE (1994) Environmental chemistry. Lewis Publication/CRC Press, USA, pp 161–173
Mathuthu AS, Mwanga K, Simoro A (1997) Impact assessment of industrial and sewage effluents on water quality of the receiving Marimba River in Harare. Lake Chivero: A Polluted Lake. University of Zimbabwe Publications, Harare, Zimbabwe, 43–52
Mohana S, Desai C, Madamwar D (2007) Biodegradation and decolourization of anaerobically treated distillery spent wash by a novel bacterial consortium. Bioresour Technol 98(2):333–339
Mohana S, Shrivastava S, Divecha J, Madamwar D (2008) Response surface methodology for optimization of medium for decolorization of textile dye Direct Black 22 by a novel bacterial consortium. Biores Technol 99(3):562–569
Munter R (2003) Industrial wastewater characteristics. The Baltic University Programme (BUP), Sweden, 185–194
Patil PU, Kapadnis BP, Dhamankar VS (2003) Decolorisation of synthetic melanoidin and biogas effluent by immobilised fungal isolate of Aspergillus niger UM2. Int Sugar J 105(1249):8–13
Raza A, Friedel JK, Bodner G (2012) Improving water use efficiency for sustainable agriculture. Agroecology and strategies for climate change. Springer, Dordrecht, pp 167–211
Sauvet-Goichon B (2007) Ashkelon desalination plant—a successful challenge. Desalination 203(1–3):75–81
Scheierling SM, Bartone CR, Mara DD, Drechsel P (2011) Towards an agenda for improving wastewater use in agriculture. Water Int 36(4):420–440
Singh R (2005) Hybrid membrane systems for water purification. Chapter 2-Water and membrane treatment
Skipton S, Dvorak B, Niemeyer S (2008) Drinking water treatment: water softening (ion exchange). University of Nebraska, Lincoln Extension, Institute of Agriculture and Natural Resources: Lincoln, NE 1491:1–4
Song G, Wang J, Chiu CA, Westerhoff P (2010) Biogenic nanoscale colloids in wastewater effluents. Environ Sci Technol 44(21):8216–8222
Stephenson T, Brindle K, Judd S, Jefferson B (2000) Membrane bioreactors for wastewater treatment. IWA publishing
Tchobanoglous G, Burton FL, Stensel H (2003) Wastewater engineering: treatment and Reuse. 4th edn, 983–1027
USEPA (2012) Guidelines for water reuse EPA/625/R-04/108. USEPA, Washington, DC
Van Nieuwenhuijzen AF, Evenblij H, Uijterlinde CA, Schulting FL (2008) Review on the state of science on membrane bioreactors for municipal wastewater treatment. Water Sci Technol 57(7):979–986
Woodard C (2005) Industrial waste treatment handbook (2nd ed.). Butterworth Heinemann
World Health Organization (1989) Health guidelines for the use of wastewater in agriculture and aquaculture: report of a WHO scientific group, WHO
World Health Organization (2003) Assessing microbial safety of drinking water improving approaches and methods: improving approaches and methods. OECD Publishing
Xu T (2005) Ion exchange membranes: state of their development and perspective. J Membr Sci 263(1–2):1–29

Yang X, Flowers RC, Weinberg HS, Singer PC (2011) Occurrence and removal of pharmaceuticals and personal care products (PPCPs) in an advanced wastewater reclamation plant. Water Res 45(16):5218–5228

Zainith SURABHI, Sandhya S, Saxena GAURAV, Bharagava RN (2016) Microbes: an eco-friendly tool for the treatment of industrial wastewaters. Microbes and Environmental Management, 75–100.

Chapter 2
Advanced Treatment Technologies for Industrial Wastewater

Alok Garg, Pratibha Gautamb, and Darshan Salunkeb

Abstract Exponential growth in industrial development has resulted in the generation of huge quantities of wastewater with a number of organic and inorganic pollutants. As the concentration of these pollutants is continuously increasing with the introduction of new recalcitrant pollutants, environmental statutory bodies are urging industries to move beyond the conventional method of wastewater treatment and orient towards more efficient and sustainable alternatives. Thus, there is an utmost need to shift towards the application of advanced treatment methods of wastewater treatment keeping in mind effluent quality, energy efficiency, economic viability, land availability, operational simplicity, and effluent quality. This chapter delineates advancements in wastewater treatment technologies including various membrane technologies, modifications in advanced oxidation processes, photo-aided treatment technologies, etc. The application of newer methods like forwarding osmosis and nanotechnology for the treatment of industrial wastewater has also been discussed. These technologies are not only promising to provide efficient treatment but also produce minimal after-treatment residue and have been implemented globally to meet environmental-based treatment standards. The chapter delves into detailing of basic principles, applications, merits, and demerits of the aforementioned technologies with their limitations and future research needs.

Keywords Wastewater · Recalcitrant pollutants · Membrane technologies · Forward osmosis · Nanotechnology

A. Garg (✉)
Department of Chemical Engineering, National Institute of Technology, Hamirpur, HP 177005, India

P. Gautamb · D. Salunkeb
Department of Environmental Science & Technology, Shroff S.R. Rotary Institute of Chemical Technology, UPL University of Sustainable Technology, Ankleshwar, Gujarat 393135, India

S. Roy et al. (eds.), *Advanced Industrial Wastewater Treatment and Reclamation of Water*, Environmental Science and Engineering,
https://doi.org/10.1007/978-3-030-83811-9_2

Abbreviations

AOP	Advanced Oxidation Processes
CWAO	Catalytic wet air oxidation
EDR	Electro-dialysis Reversal
ED	Electro-dialysis
FO	Forward Osmosis
MF	Microfiltration
NF	Nanofiltration
ODMP	Osmotically Driven Membrane Process
RO	Reverse Osmosis
UF	Ultrafiltration
WAO	Wet Air Oxidation

2.1 Introduction

Industrial wastewater typically contains a significant number of pollutants. These recalcitrant contaminants in industrial effluent can affect the well-being of human life and ecosystem directly or indirectly.

All pursuit of humankind is dependent on water. As the population of human growing day-by day, quantity of wastewater generation in industrial, commercial, domestic and agricultural sectors is also increased. An increase in water usage demand, do not implies replenishment of over exploited freshwater resources, as per human needs. As a result, a prejudicial supply of freshwater resources takes place amongst these sectors. Thus, a large number of populations in the world face deficiency of pure drinking water, particularly in developing and underdeveloped countries.

Being diverse in nature, characteristics of industrial effluent varies from industry to industry; however, it is also varying within industry. Remediation of industrial effluent is an intricate problem because of the diversified compounds present in it and in varying composition and concentration. To treat this complex wastewater with only conventional method of treatment is not a child's play. Moreover, environmental rules and regulations are getting stringent worldwide, day-by-day and treatment of this complex wastewater with only conventional methods doesn't appear promising and efficient. Thus, there is an utmost need to shift the wastewater treatment technologies from conventional to advanced technologies, which are more efficient, less time and space consuming and capable to treat multiple complex pollutants simultaneously. Many researches are completed; many are still going on at different lab-scale, pilot scale and industrial scale that incorporate treatment of complex industrial effluents, (Nqombolo et al. 2018). Various researches and technologies have already been accepted and are successfully being implemented at industrial level and are being admired as efficient wastewater treatment technologies (Ezugbe and Rathilal

2020). There is no general classification given for these technologies but based on the working principle, these technologies can be classified into several groups as mentioned in Fig. 2.1.

All these processes are further discussed in detail along with their applicability at industrial scale.

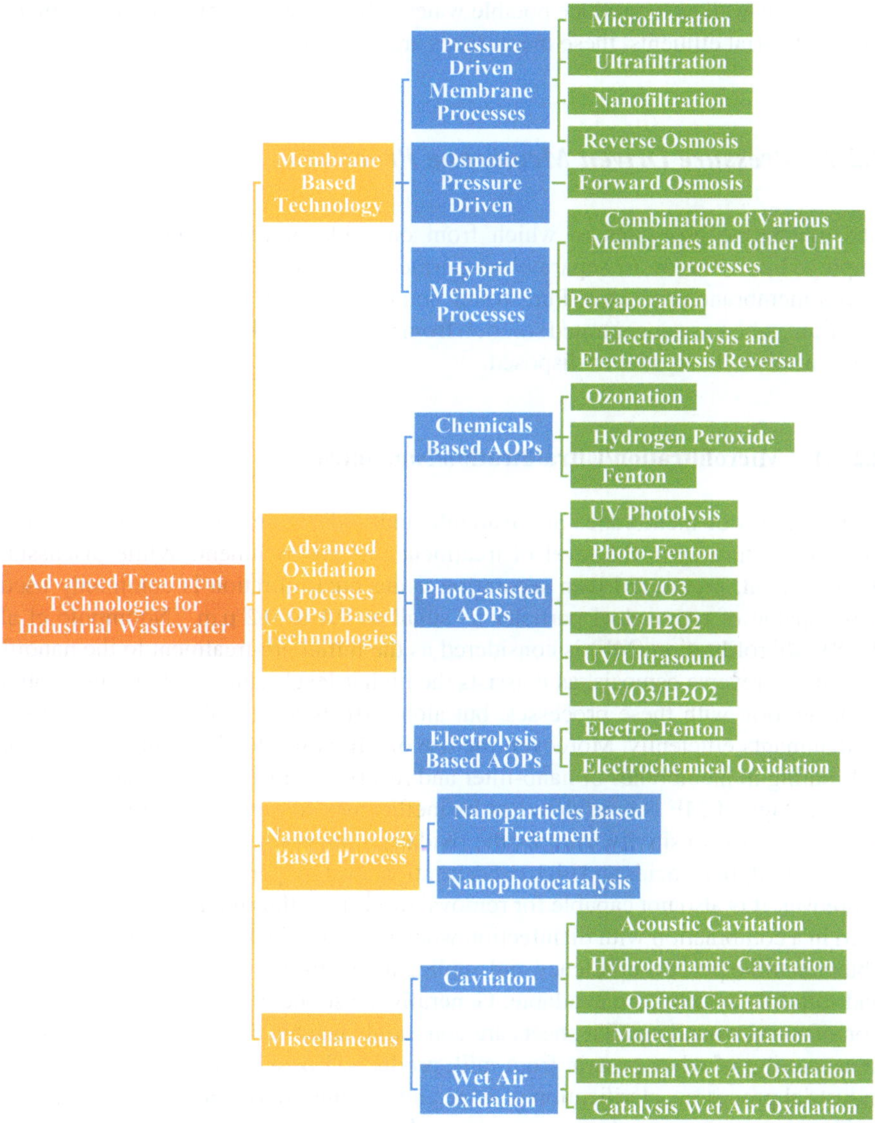

Fig. 2.1 Classification of various advanced wastewater treatment technologies

2.2 Membrane Based Technologies

With a need to deal with complex and high level of water treatment, techniques using membrane separation have grownup. It has evolved from a laboratory tool to a full-fledge commercial industrial application. These days, many membranes play a vital role in large scale desalination of seawater, so that it can be used in industrial processes as well as to produce potable water. Also, to recover precious constituents from industrial effluents, these membranes are admired.

2.2.1 Pressure Driven Membrane Processes

The membrane processes in which from one side of membrane, a pressure is employed on solution to separate out permeate and retentate are known pressure driven membrane processes. Pure water obtained through membrane in known as permeate, whereas concentrated solution from membrane is known as retentate which either needs to be treated or disposed.

2.2.1.1 Microfiltration/Ultrafiltration/Nanofiltration

Various types of membranes are available and widely used for industrial applications depending upon the level of treatment and pre-treatment. While discussing pre-treatment, pressure driven separation using microfiltration is commonly used. It is capable of removal of particles as small as 0.1 to 0.2 μm, (Nqombolo et al. 2018). Microfiltration (MF) is considered as the initial pre-treatment to the nanofiltration and reverse osmosis, as it assists the higher-level removal of organic matter, in integration with these processes, but alone MF is not capable to remove these contaminants efficiently. Moreover, the use of MF is intended to reduce the potential fouling in membranes of nano-filter and reverse osmosis. Possible and foremost disadvantage of MF is considered as its inefficiency to remove dissolved contaminants. Also, its sensitivity towards the oxidative compounds and chemicals mainly nitric and sulphuric acids and high concentration levels of peroxides and persulphates. Moreover, it is also not capable for removal of viruses, thus many a times it is being used in a combination with disinfection whenever microbial control is necessary. To obtain definite process efficiency and application goals, customized MF membranes and filtration systems are available. Generally, these membranes having configuration of spiral wound and flat sheets are commonly used in dairy industrial processes in separation of whey protein from milk products. Likewise, in food and beverage industrial processes, clarification of wine and gelatine, sterilization of beverages.

To separate compounds or contaminants having size in between 0.005 to 10 μm, Ultrafiltration can be applied. Development of ultrafiltration took place in early

Table 2.1 Different membranes treatment process with their pore size and trans membrane pressure

Membranes	Pore size (μm)	Transmembrane pressure (psi)	Species rejected
Microfiltration	0.1–3	1–30	Suspended matter, bacteria
Ultrafiltration	0.01–0.1	1–30	Macromolecules, colour, odour
Nanofiltration	<0.002	75–150	Hardness, small particles, viruses

1860s at lab scale and in mid twentieth century the commercialized use of ultrafiltration membrane took place in chemical and pharmaceutical industries. Whereas for water treatment, the use of ultrafiltration membranes started in 1990s (Li X et al. 2018). Ultrafiltration (UF) membranes are capable of removal of macromolecules, pathogenic microorganisms and suspended matters. Removal of pathogenic bacteria is the major benefit of UF. Energy consumption and required pressure for filtration is lower in UF than Nano filtration and reverse osmosis. Similar to MF, ultrafiltration membranes are also susceptible to acids and oxidative chemicals and compounds. Also, membranes may get damaged at pressure greater than 3 bar.

In water treatment, when the objective is to remove the hardness causing multivalent or divalent ions from water, or for wastewater treatment, when the objective is to recover divalent ions, commercialized use of Nano filtration (NF) comes to existence. Commercialized small scale application of NF is from late 1970s (Bergman 1995). NF serves to remove smaller particles having size range of 0.002 to 0.005 μm somewhat coarser filtration than reverse osmosis. For heavily polluted waters, additional pre-treatment is needed for effective Nano filtration. Demerits of Nano filtration include difficulty in separation of soluble elemental substances and the membrane is also susceptible to free chlorine in water or wastewater. Commercially NF is used for softening of water and organic matter removal from water/wastewater. Table 2.1 provides a brief summary of pore size and trans membrane pressure between MF/UF/NF.

2.2.1.2 Reverse Osmosis

To remove dissolved salts including monovalent ions having particle size smaller than 0.1 nm, the pressure driven technique of Reverse Osmosis (RO) is used. It is only permeable to water molecules. Jean Antoine-Nollet, a French physicist, in 1748 experimentally first observed the reverse osmosis filtration using a pig's bladder as a membrane, (Wiles and Peirtsegaele 2018). To prevail over the natural osmotic pressure of water, enough pressure must be exerted on the concentrate side of the RO membrane to reverse the direction of flow of water molecules. In RO membrane, the pore structure is much tighter that it is practically possible to remove all particles, organic matter, bacteria. But, due to this tighter pore structure, higher pressure requirement is there along with the high level of pre-treatment to avoid fouling of membranes, which in turn makes the RO membrane processes expensive in nature. Application of RO membrane system for industrial purposes including the treatment of water and wastewater is few decades old. (Hrubec et al. 1983) have experimented the pilot scale treatment of municipal wastewater using RO and as a post-treatment with activated carbon. A cellulose acetate membrane was selected and was operated at a pressure of 40 bars. The experiment concluded with a highquality of advanced treated wastewater. Rautenbach and Linn (1996) carried out state of art experimentation to treat leachate at a pilot scale using RO and NF as a post treatment of concentrate of RO to achieve almost a zero-discharge process. RO membrane system showed successful recovery and removal of copper ions from electroplating industrial effluent with an increased flow rate of permeate by increasing the pressure drop across the trans membrane and also by increasing the operating temperature (Chai et al. 1997). Fouling of membrane is the major disadvantage of any available membrane systems, which occurs due the deposition of undesired sediments inside the voids of membrane or on the surface of membranes which in turn reduces the flux permeation and rejection of salts. This fouling can be because of organic fouling (deposition of organic matters), inorganic fouling (due to inorganic scaling), bio fouling (due to deposition of microorganism biofilm beyond the acceptable level) and colloidal fouling (deposition of colloids or suspended matters). Many membrane fouling control methods have been suggested by Jiang et al. (2017) which includes pre-treatment of water and wastewater before RO, real-time monitoring and cleaning to RO membranes, modification of surface of RO membranes for anti-fouling properties by making the surface of membrane as smooth as possible and by use of novel polyamide thin-film composite which improves membrane performance and is having anti-fouling characteristics. Coating RO membranes with azide functionalized graphene oxide which provides smooth and antibacterial nature to the membrane and avoids fouling of membranes. Reuse of RO membranes are seen as an attractive circular economy solution. Veza and Rodriguez-Gonzalez (2003) used $KMnO_4$ to treat outdated reverse osmosis membranes which lessened efficiency of rejection of salt by 2%. The membranes which have reached towards their end-of-life can be used for pre-treatment purposes by recycling them into UF or NF. If furthermore, these membranes can't be used for recycling them into UF or NF, can be applied in removal of low-quality water in membrane biofilm reactors. The RO membranes which are severely spoiled can be reused as a material of support in anion exchange membranes or the contents of polypropylene (PP) can be extracted from them as reported by Lejarazu-Larranaga et al. (2020).

2.2.2 *Osmotically Driven Membrane Processes*

Osmotically driven membrane process (ODMP) is process where naturally produced osmotic pressure is the only source of driving force. ODMP includes forward osmosis membrane process which has attracted its attention in the field of wastewater treatment.

2.2.2.1 Forward Osmosis

Forward Osmosis (FO) is the naturally occurring phenomena where the solvent passes from lower concentration region towards higher concentration region through a permeable membrane. FO is very efficient with low-rate production. Though, rejuvenation of draw solution is very costly for salt removal. Thus, there is a need of NF or RO to regenerate the draw solution. FO consumes low power and is applicable to wide range of feed solutions which is among the merits of it. As compared to pressure-driven membrane processes, osmotically driven membranes are less vulnerable to fouling as they are operated with lesser or no hydraulic pressure which is the major advantage of FO membranes. For effective operation of FO, selection of draw solution plays a vital role. Draw solution is highly concentrated solution which provides a gradient of concentration to drive water molecules from the feed solution through the permeable membrane. It is foremost needed that draw solution is non-poisonous, it should have high stability and high solubility in water so that precipitation can be avoided. Various forms of draw solutes are available mainly organic (EDTA, sucrose, sodium polyacrylate, glucose, etc.) inorganic (sodium chloride, magnesium chloride, sodium sulphate, potassium chloride, potassium nitrate) gases and volatile compounds (NH_3 and carbon dioxide) and nanoparticles having magnetic properties like Fe_2O_4 (Ezugbe and Rathilal 2020). Figure 2.2 describes the schematic diagram of working of FO membrane. The gradient which is provided because of draw solution is the osmotic pressure difference which draw water molecule from feed solution to the draw solution until the equilibrium is reached. Using this gradient, treatment

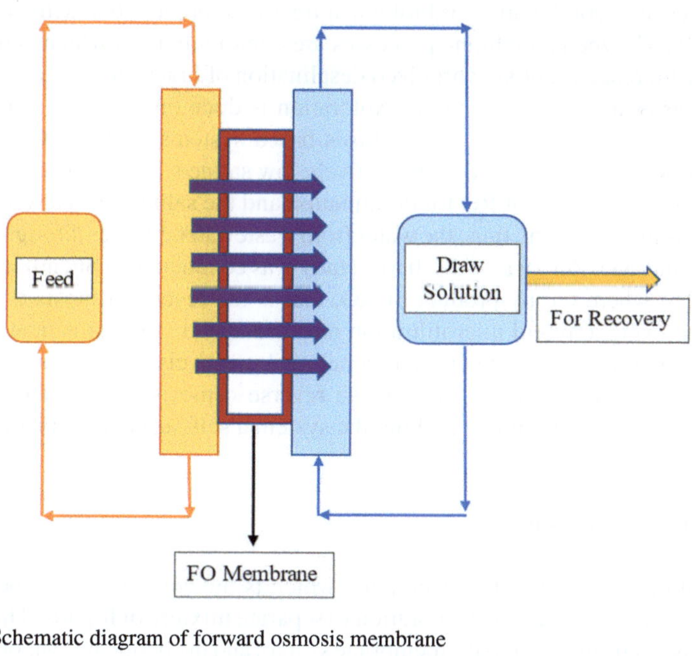

Fig. 2.2 Schematic diagram of forward osmosis membrane

of various wastewater streams can be applied. FO had been applied on management of landfill leachate by using draw solution as salt of sodium chloride and reverse osmosis was applied to recover this salt, the operation had an efficiency of 95% permeate recovery (York et al. 1999). Ferrari et al. (2019) applied FO on raw municipal wastewater treatment using NaCl and $MgCl_2$ as draw solution and had obtained 70% recovery of water. Zhao et al. (2017) obtained 50% decrease in volume of water from gas field process water while investigating the application of hollow fiber FO and by using draw solution as 1 M NaCl solution. The capability of concentrating feed solution having low osmotic pressure with the help of the concentrated draw solution makes this technology capable of treating wide variety of water and wastewater.

2.2.3 Hybrid Membrane Processes

Amalgamation of a membrane process with other membranes or with any other unit processes for example adsorption, coagulation-flocculation or ion exchange processes to provide improved efficiency is identified as hybrid membrane processes. In this process, each individual process tries to balance the demerits of the other process hence to make sure the enhanced quality of treated water. As the discharge standards are getting stringent day-by-day, more exploration of hybrid membrane processes is in demand. For drinking purpose, treatment of contaminated ground water takes place with an application of MF or UF in combination of activated carbon adsorption. For the purpose of wastewater treatment or recovery of resource from wastewater, combination of biological treatment process like activated sludge and MF/UF/NF typed membrane processes are commonly used which is praised as membrane bioreactor. For synchronized desalination of brackish seawater and treatment of wastewater simultaneously, exploration is done on hybrid systems particularly on forward osmosis-reverse osmosis-based systems. In this FO-RO hybrid system, the sea water is diluted with relatively low saline wastewater which reduces the pressure requirement of RO for desalination and the salinity of sea water creates the concentration gradient, thus, the water from wastewater diffuses through semipermeable membrane. Zero waste discharge study was conducted by Shanmuganathana et al. (2016) where the by-product of RO, that is the reverse osmosis concentrate was treated using a hybrid microfiltration membrane and granular activated carbon adsorption treatment. The experiment concluded positively which can be installed at wastewater treatment plant to treat the reverse osmosis concentrate before its discharge into the environment making the system a self-sustainable process.

2.2.3.1 Pervaporation

Pervaporation is the separation technique which is the combination of permeation membrane and the process of evaporation to separate mixture of liquids. The mixture of liquids is fed from one end of a membrane system and the permeate evaporates from the other side. The permeate is sorbed in the upstream side during this process. Under the effect of concentration gradient, the constitute of the mixture are diffused through the membrane. The species diffusing is subsequently evaporated at membrane's other side. To recover liquid, vapours are sent to condenser. This is known as solution-diffusion model where the mass transportation takes place through the membrane. Figure 2.3 demonstrate the schematic diagram of pervaporation membrane. The technology is predominantly applied in separation of ethanol and water. Nevertheless, exploration of this technology in the field of wastewater treatment is also of keen interest. Wijmans et al. (1990) conducted pilot scale experiments using organophilic membranes for the removal of organic solvents from mixture of wastewater streams. Kujawa et al. (2015) investigated experiments with hydrophobic ceramic membranes to remove hazardous VOCs from effluent with the help of vacuum pervaporation. Around 95% removal from aqueous solution of methyl tert-butyl-ether was obtained. Singha and Ray (2012) carried the experiments on pervaporation process by using natural rubber crosslinked membrane to remove pyridine from effluent. Zhang et al. (2012) studied pervaporation by using a membrane made up of polyurethane urea to effectively remove isopropyl acetate from effluent. Similar study was also carried out by Das et al. (2008) for separation of phenols and chlorophenols from wastewater.

Pervaporation membranes are designed specially depending upon the compound's higher affinity. Thus, the nature of chemical and membrane structure are important for attaining the required separation intention. Rate of feed flow, concentration of feed, partial pressure of vapours and temperature are also important factors. Pervaporation is recognized as eco-friendly and energy saving technology apart from its capability to separate liquid mixtures. Demerits of this technology are its operating conditions which are highly sensitive; hence its large-scale industrial applications are still a

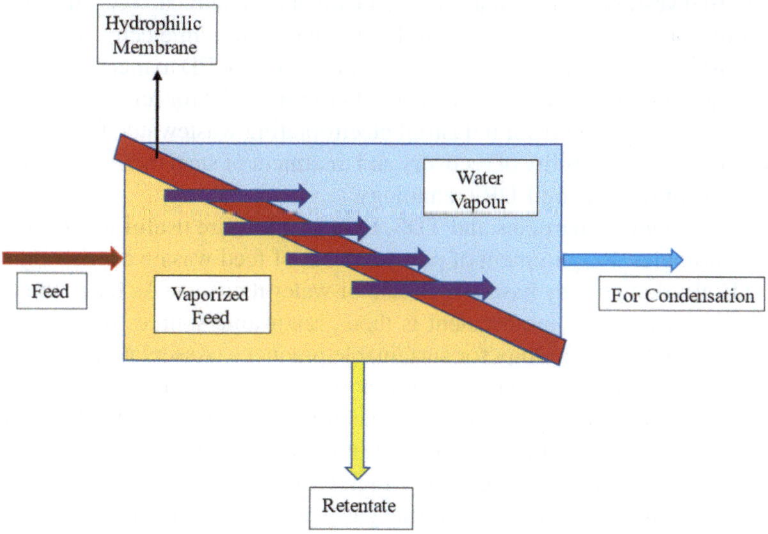

Fig. 2.3 Schematic diagram of pervaporation

point of research. Also, due to lack of specialized membranes and their expensive nature, pervaporation can hardly be applied beyond dehydration.

2.2.3.2 Electro Dialysis and Electro Dialysis Reversal

Electro-dialysis (ED) and electro-dialysis reversal (EDR) are the technologies to separate ions dissolved in water with a combination of electricity and membranes with ion permeability. In these processes, ions from dilute solution are transferred through ion-permeable membrane towards concentrated solution by using electric potential supplied to these solutions. In ED, 2 sets of ion exchange membranes are used, anion-exchange membrane and cation exchange membrane. Anion-exchange membranes are permeable to negatively charged ions and retain positive ions whereas Cation-exchange membranes are permeable to positively charged ions and retain negative ions. Thus, the ions from feed streams are exhausted and the outlet concentrate stream becomes abundant with ions. In 1954, at Arabian Desert, the first commercialized ED plant was installed (Katz 1979).

In EDR, reversing of ion movement takes place due to periodic reversal of electrodes of membrane. Thus, the fouling of ion permeable membrane is reduced as the concentrated stream is periodically converted to diluted stream and diluted stream to concentrated stream. Application of ED and EDR has been done in many ways in recent decades. Valero et al. (2015) recovered 94% of water during electrodialysis treatment of wastewater from almond industry. Abou-Shady (2017) recycled drainage wastewater for the purpose of agriculture using ED and has achieved 99% removal efficiency of sodium ions and other heavy metals. Albornoz et al. (2019) applied ED to recover water from the treatment of university sewage. Deghles and Kurt (2016) investigated more than 90% removal of Cr, chemical oxygen demand and removal of color from tannery effluent by its treatment with electro dialysis. Gally et al. (2018) treated municipal wastewater tertiary through ED to meet discharge standards and ion removal. Sivakumar et al. (2014) applied ED for removal of Cadmium (74.8%) and tin (64.5%) from industrial electroplating wastewater. Chao and Liang (2008) studied the feasibility of recovery and treatment of steel industry's wastewater treatment plant from using EDR technology.

To remove ionized particles and TDS, ED and EDR are useful wastewater treatment technologies. Requirement of pre-treatment of feed wastewater is very little in ED and EDR, though they have a high rate of water recovery. As far as the demand of electricity and power requirement is there, renewable sources of energy can be combined with this technology for sustainable practices (Association AWW, 1995). The major disadvantage of these processes is non-removal of non-ionized chemicals and substances along with harmful bacteria and viruses. Thus, there will be a requirement of post-treatment. Also, to treat highly saline wastewater using ED is very expensive to operate as the energy requirement increases. Moreover, problem of corrosion is also major due to chlorine gas generation at anode.

2.3 Advanced Oxidation Processes

For treatment of organic pollutants in wastewater, Advanced Oxidation Processes (AOPs) are gaining the popularity these days, where either the strong oxidants (chemical based or radiation based) are externally added or oxidative species (hydroxyl radical) are internally (in situ) produced. More enhanced wastewater treatment can be achieved by combination of AOPs with ultraviolet irradiation, ozone or with other highly active catalysts. Establishment of these technologies for treatment of potable water and for reuse of water have been already started at a full scale. Additionally, several researches are also going on for AOPs in combination with microwave treatment, treatment with electrochemical, plasma, ultrasound and electron beam (Ghime and Ghosh 2020). To bridge the gap between physicochemical, secondary treatment and strict environmental regulations, AOPs can be considered as a key tool.

In AOPs, basic principle involved is the production of hydroxyl radical (HO•) which can be produced from strong oxidants. This hydroxyl radical is responsible for taking away the hydrogen atom from an organic pollutant and is responsible in generation of organic radical as mentioned in Eq. (2.1). Further, this organic radical undergoes many chemical reactions to form various products and by-products. Ideally, AOPs, should completely degrade organic matter to CO_2 and H_2O as shown in Fig. 2.4.

$$R - H + HO \rightarrow H_2O + R \tag{2.1}$$

Various types of AOPs are used for the application of degradation of pollutants. In ozone based AOPs, radical formation takes place through a complex reaction series when addition of ozone in water is done. At higher pH concentration, ozone disintegrates at a faster rate. Also, ozone decomposition is increased with the addition of H_2O_2, the reaction of the same is as shown in Eq. (2.2). The performance of the process is dependent on the time of contact, alkalinity of water and ozone dosage. In comparison with all types of AOPs, ozonation based AOPs have more industrial application for treatment of wastewater. Wastewater pollutants from petroleum refineries, electroplating industries, distilleries, textile industries and also in food industries, ozone based AOPs are effectively applied.

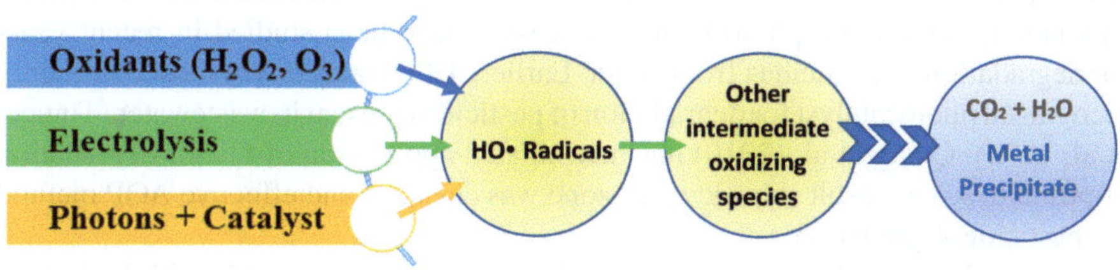

Fig. 2.4 AOP principle

$$2O_3 + H_2O_2 \rightarrow 2HO\bullet + 3O_2 \tag{2.2}$$

Formation of HO• can be achieved by dissociation of H_2O_2 photo-catalytically by UV irradiation at 254 nm wavelength as shown in Eq. (2.3). Hydrogen peroxide-UV is one of the valuable AOP methods because of high rate of reaction and designing of reactors is flexible. H_2O_2/UV based AOP is the cheapest treatment due to easy availability and low cost of H_2O_2. However, there are some demerits of this process such as H_2O_2 has poor absorption characteristic of UV radiations and hindrance in the process due to bromate formation.

$$H_2O_2 + h\upsilon \rightarrow 2\,HO\bullet\ (h\upsilon : \lambda = 254\ nm) \tag{2.3}$$

AOPs using Fenton reagent produces hydroxyl radicals when the H_2O_2 is reacted with iron (II) catalyst. Equation (2.4) shows the formation of hydroxyl radical by Fenton reagent. This is the simplest method for generation of hydroxyl radical as the process does not require any special chemicals or apparatus and the reaction takes place at ambient temperature and pressure. Due to easy availability, environmentally safe and easy handling of salts of iron and H_2O_2, this is an attractive oxidation method. The demerit of this method is iron sludge production however; to overcome this chemical coagulation can be applied as suggested by Lin and Leu (1999).

$$Fe^{2+} + H_2O_2 \rightarrow Fe^{3+} + OH^- + HO\bullet \tag{2.4}$$

As a cohesive solution to treat organic matter like benzene, phenol, toluene, naphthalene, xylene etc., investigation of various AOPs such as Fenton-aided process, photocatalysis, ozonation have been done. Contreras et al. (2002) investigated comparative study to treat wastewater using numerous AOPs like ozone, UV, ozone-hydrogen peroxide, ozone-UV, hydrogen peroxide-UV, photocatalysis process, combination of ozone-UV-hydrogen peroxide and Fenton-hydrogen peroxide. These experimentations were conducted for oxidation of phenol in wastewater and among these, fastest results for removal of phenol from wastewater was shown by Fenton process. Whereas, ozonation was found to be most cost effective.

Chelme-Ayala et al. (2010) studied various photochemical degradation oxidation treatments using ultraviolet-ozone and ultraviolet-peroxide for degradation of recalcitrant pollutants. Also, photocatalysis based AOPs like titanium dioxide-ultraviolet irradiation, Fenton and photo-Fenton processes have been studied in recent years for degradation of pesticides from water. Earlier, TiO_2 catalyst was most commonly recognized photocatalyst for degradation of pesticides present in wastewater (Tinucci et al. 1993). Qiang et al. (2010) has investigated application of O_3 for degradation of pesticides. The result showed that ozone was cheapest and effective AOP method for pesticide degradation.

Koh et al. (2004) have positively combined ultraviolet-peroxide with biological nitrification and denitrification which was then followed by biological oxidation stage to reduce the concentration of COD and BOD_5 within the permissible limits of wastewater discharge standards. Wang et al. (2009) studied the Fenton oxidation process by providing coagulation pre-treatment and post-treatment as biological aerated filtration on landfill leachate to reduce its COD. These studies show effective and promising treatment of secondary wastewater by a combination of AOPs and biological treatment.

2.4 Nanotechnology Based Processes

Technologies based on treatment of wastewater with the aid of particles having their dimensions in nanometers. As these particles have various properties and have high reactivity towards pollutants.

2.4.1 Nanoparticles Based Treatment

Nanoparticles are particles having at least one dimension in the range of 1 to 100 nm (Stark et al. 2015). Nanoparticles have properties like electrical, optical and magnetic that are very unique from the materials used conventionally. Apart from these, Nanoparticles have high reactivity, great catalytic activities and they have high adsorption (Khan et al. 2017). Many researches have been carried out in last few decades, on application of nanoparticles in the field of catalyst, biology, chemistry, medicine and many more. Additionally, widespread application of nanoparticles has been done in the field of wastewater treatment due to its high adsorption and large surface area (Mauter et al. 2018). Nanoparticles have been worldwide reported to treat numerous contaminants like organic matter, inorganic anions, and bacteria etc., (Varjani et al. 2017). Nanomaterials are capable technology for its application in wastewater treatment including zerovalent nanomaterial like zinc, silver, iron nanoparticles, metal oxide nanomaterial like titanium dioxide, zinc oxide, iron oxide, carbon nanotubes (CNTs), Quantum dots and Nano composites (Singh et al. 2019).

2.4.2 Nano-Photo Catalysis

Nano-photo catalysis (nanoparticles activated by photons) generally applied for the elimination of trace pollutants present in wastewater. Nano-photo catalysis is more efficient because of large surface area-to-volume ratio (Martínez et al. 2011) and they have unvarying composition and structured size of particle. Nano-photo catalysis is a class of redox reactions applying light of appropriate wavelength to irradiate the semi-conducting nanomaterial. Since nanoparticles have good adsorbing property, photocatalyst is firstly mixed with polluted wastewater and the mixture is thoroughly agitated in absence of irradiation to achieve an ample adsorption–desorption equilibrium of pollutant son the surface of catalyst, before irradiation by light source. Efficiency of nano-photo catalysts is affected by pH level. Optimum condition of pH is considered when nanoparticles being positively charged are promptly attracted with molecules of pollutant which is charged negatively.

Photo catalytic degradation of a variety of environmental contaminants have been attempted by following catalysts: TiO2, ZnO, FeO, Fe_2O_3, ZnS. Catalyst titanium dioxide is found steady in ultraviolet light and water. Because of its strong oxidizing power, non-toxicity and long-term photo stability it is most preferred photo catalyst (Kwon et al. 2008). Due to high photocatalytic efficiency, cost effectiveness and environmentally safe nature of zinc oxide nanoparticles have been widely applied in wastewater treatment (Eskizeybek et al. 2012). Due to high adsorption abilities of carbonaceous material and removal of pollutants, application of nanostructures of carbon such as Carbon Nano-Fibres (CNFs) and Carbon Nano-Tubes (CNTs) have gained great interest. According to the studies of Banerjee et al. (2012), application of nanoparticles for removal of heavy metals is very effective and most admired technology. The study investigated increased removal of Cr^{6+} ions with gradual increase in dosage of photo catalyst and with decrease in pH due to accumulation of precipitates of Cr $(OH)_3$ on the surface of titanium dioxide.

2.5 Miscellaneous Wastewater Treatment Technologies

2.5.1 Cavitation

As a result of local pressure pulsation in a liquid, formation, collapse or growth of bubbles or cavity takes place. Vapour bubbles in liquid experiences reduction in pressure which in turn increases the size of bubble (growth) and ultimately at critical magnitude of pressure it collapses (implodes). Here the critical pressure becomes equivalent or lower than liquid specific vapour pressure. Thus, the bubble implosion goes together with destruction of chemical bond and hydraulic shock. The implosion generates high temperature of roughly 4725 °C and high pressure of 60,000 kPa in the trapped vapour for a short period of time (Gong and Hart 1998). This intense condition is suitable for breaking the organic structure and biological structure and for dissociating molecules of water into H and HO• radicals. HO• radical generated is an exceptional chain reaction initiator and one of the most powerful oxidants. This HO• radical has ability to degrade unsaturated hydrocarbon spontaneously. According to Mehrvar et al. (2001), during cavitation OH• radical degrades organic pollutant at 3 different places: bulk liquid phase, at the interface of liquid and gas and inside the bubble. Two most commonly applied cavitation technology which has large-scale wastewater treatment applications are Acoustic Cavitation and Hydrodynamic Cavitation. Acoustic Cavitation is produced by acoustic (ultrasound) waves

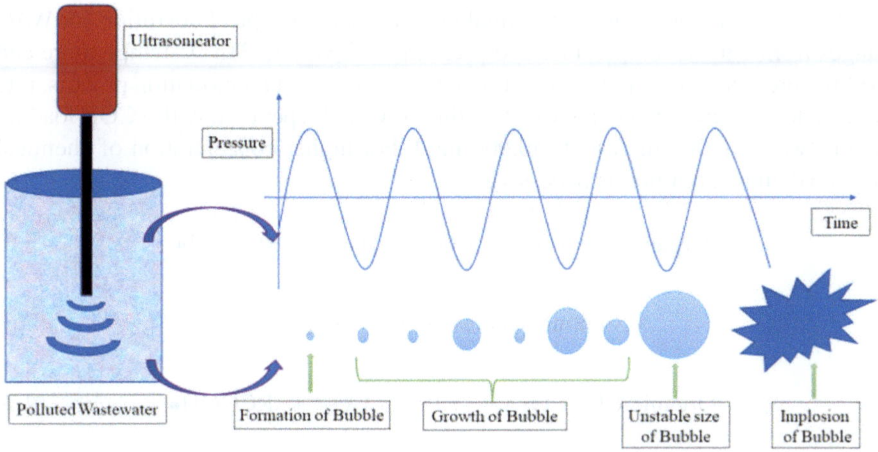

Fig. 2.5 Acoustic cavitation phenomena

whereas; liquid's hydrodynamic phenomena are responsible for hydrodynamic cavitation. Other types of cavitation are Optical cavitation which are produced by strong laser beam and molecular cavitation which are produced due to elementary particles like protons. Commonly, cavitation is applied as a post treatment method after biological treatment processes or in combination with AOPs. Figure 2.5 represents phenomena of acoustic cavitation. The combination of cavitation with various types of AOPs for treatment of wastewater is to recompense the challenges of each of the method with regard to the mineralization of Persistent organic pollutants (Badmus et al. 2017).

2.5.2 Wet Air Oxidation

Wet air oxidation (WAO) also known as wet oxidation is a novel process of oxidizing suspended or dissolved material in wastewater with dissolved oxidation at higher temperature. This type of treatment is meant for wastewater having much higher concentration which can't be treated with biological treatment and too lower for incineration treatment.

The thermal wet air oxidation is operated at high temperature and high-pressure air as oxidant. For high organic process wastewater from chemical or pharmaceutical industries, WAO treatment is a good choice. In wet air oxidation process, the chemistry involved for treatment of waste water, is the radical chain reaction which is formed from organic and inorganic compounds present in the effluent. Thus, complex organic compounds are completely converted to simpler end products and/or inorganic substances are transformed into sulphates or sulphides with higher oxidation value. Equation (2.6, 2.7, 2.8, 2.9, and 2.10) shows conversion of organic species and Sulphur compounds into their final end products. A typical condition for WAO ranges from 180 °C temperature and pressure of 2 MPa to 315 °C temperature and 20 MPa pressure. For typical industrial application of wet air oxidation process, rate of feed flow is between the range of 1 m^3/hr. to 50 m^3/hr. per unit, with a COD loading rate between 10,000 mg/l to 1, 50,000 mg/l. For higher concentration of chemical oxygen demand, dilution is necessary.

$$\text{Organics} + O_2 \rightarrow CO_2 + H_2O + \text{RCOOH}\bullet \tag{2.5}$$

$$\text{Sulphur Species} + O_2 \rightarrow SO_4^{2-} \tag{2.6}$$

$$\text{Organic Cl} + O_2 \rightarrow Cl^- + CO_2 + \text{RCOOH}\bullet \tag{2.7}$$

$$\text{Organic N} + O_2 \rightarrow NH_3 + CO_2 + \text{RCOOH}\bullet \tag{2.8}$$

$$\text{Phosphorus} + O_2 \rightarrow PO_4^{3-} \tag{2.9}$$

Catalytic wet air oxidation (CWAO) is an alternative process to carry out wet air oxidation process at milder operating conditions with lower temperature and pressure. To significantly improve reaction rate, soluble transition metal salts such as salts of copper and iron can be applied for treatment of wastewater, however post-treatment to separate and recycle these salts is much needed. Barbier et al. (1998) reported that mixture of metal oxide catalyst such as copper, zinc, cobalt, manganese and bismuth have shown better treatment efficiency but these catalysts were detected being leaching in many experiments. Further researches for large scale application of catalytic wet air oxidation process are still required along with viability of these catalysts for continuous process treatment needs to be done.

2.6 Conclusion and Future Prospects

Needs of human kind are expanding endlessly, leading to exponential industrial development, exploitation of natural resources (especially water) and generation of huge quantities of waste water. Hence a technology up gradation is also needed simultaneously for treatment and recycling of wastewater, so that this wastewater can again be made available for use and suitable for industrial and municipal purposes and to cease the exploitation of natural water resources and preventing them from being polluted and of course meeting the stringent environmental standards. This chapter delivers a comprehensive study of various technologies which are successfully being implemented for treatment of industrial wastewater. Many technologies as particular or in combination with other process can be used for industrial application.

Various membrane technologies predominantly are being applied in industrial applications particularly in wastewater treatment and are proven to be a successful choice. Although, the sensitivity and fouling of membranes are also a serious industrial issue. However, to overcome this problem, nanomaterial for example graphene oxide, can be used. Additionally, advanced oxidation processes itself and along with biological treatment proves to have a potential industrial application to remove recalcitrant pollutants. Further research scope for AOPs is still open for property enhancement of heterogeneous catalysts. Technologies like nano-photocatalysis and cavitation are also commonly used as lab-scale treatment technologies; however, the application of these technologies at industrial scale can be vied as a part of future research.

References

Abou-Shady A (2017) Recycling of polluted wastewater for agriculture purpose using electrodialysis: Perspective for large scale application. Chem Eng J 323:1–18. https://doi.org/10.1016/j.cej.2017.04.083

Albornoz LL, Marder L, Benvenuti T, Bernardes AM (2019) Electrodialysis applied to the treatment of a university sewage for water recovery. J Environ Chem Engin 7:102982. https://doi.org/10.1016/j.jece.2019.102982

Association AWW (1995) Electrodialysis and Electrodialysis Reversal: M38; American Water Works Association: Denver, CO, USA, Vol 38

Badmus KO, Tijani JO, Massima E, Petrik L (2017) Treatment of persistent organic pollutants in wastewater using hydrodynamic cavitation in synergy with advanced oxidation process. Environ Sci Pollut Res. https://doi.org/10.1007/s11356-017-1171-z

Banerjee P, Chakrabarti S, Maitra S, Dutta BK (2012) Zinc oxide nano-particles—sonochemical synthesis, characterization and application for photo-remediation of heavy metal. Ultras Sonochem 19:85–93. https://doi.org/10.1016/j.ultsonch.2011.05.007

Barbier J Jr, Delanoë F, Jabouille F, Duprez D, Blanchard G, Isnard P (1998) Total oxidation of acetic acid in aqueous solutions over noble metal catalysts. J Catal 177:378–385. https://doi.org/10.1006/jcat.1998.2113

Bergman RA (1995) Membrane softening versus lime softening in Florida: a cost comparison update. Desalination 102:11–24. https://doi.org/10.1016/0011-9164(95)00036-2

Chai X, Chen G, Yue PL, Mi Y (1997) Pilot scale membrane separation of electroplating waste water by reverse osmosis. J Memb Sci 123:235–242. https://doi.org/10.1016/s0376-7388(96)00217-7

Chao YM, Liang TM (2008) A feasibility study of industrial wastewater recovery using electrodialysis reversal. Desalination 221:433–439. https://doi.org/10.1016/j.desal.2007.04.065

Chelme-Ayala P, Gamal El-Din M, Smith DW (2010) Degradation of bromoxynil trifluralin in natural water by direct photolysis and UV plus H2O2 advanced oxidation processes. Water Res 44:2221–2228. https://doi.org/10.1016/j.watres.2009.12.045

Contreras S, Pascual E, Esplugas S, Gim J, Rodri M (2002) Comparison of different advanced oxidation processes for phenol degradation. Water Res 36:1034–1042. https://doi.org/10.1016/s0043-1354(01)00301-3

Das S, Banthia AK, Adhikari B (2008) Porous polyurethane urea membranes for pervaporation separation of phenol and chlorophenols from water. Chem Eng J 138:215–223. https://doi.org/10.1016/j.cej.2007.06.030

Deghles A, Kurt U (2016) Treatment of tannery wastewater by a hybrid electrocoagulation/electrodialysis process. Chem Eng Proces Proc Intensif 104:43–50. https://doi.org/10.1016/j.cep.2016.02.009

Eskizeybek V, Sarı F, Gu¨lce H, Gu¨lce A, Avcı A (2012) Preparation of the new polyaniline/ZnO nanocomposite and its photocatalytic activity for degradation of methylene blue and malachite green dyes under UV and natural sun lights irradiations. Appl Catal B: Environ 119–120:197–206. https://doi.org/10.1016/j.apcatb.2012.02.034

Ezugbe EO, Rathilal S (2020) Membrane technologies in wastewater treatment: a review. Membranes 10:89. https://doi.org/10.3390/membranes10050089

Ferrari F, Pijuan M, Rodriguez-Roda I, Blandin G (2019) Exploring submerged forward osmosis for water recovery and pre-concentration of wastewater before anaerobic digestion: a pilot scale study. Membranes 9:97. https://doi.org/10.3390/membranes9080097

Gally CR, Benvenuti T, da Trindade CDM, Rodrigues MAS, Zoppas-Ferreira J, Pérez-Herranz V, Bernardes AM (2018) Electrodialysis for the tertiary treatment of municipal wastewater: efficiency of ion removal and ageing of ion exchange membranes. J Environ Chem Engin 6: 5855–5869. https://doi.org/10.1016/j.jece.2018.07.052

Ghime D, Ghosh P (2020) 'Advanced oxidation processes: a powerful treatment option for the removal of recalcitrant organic compounds' advanced oxidation processes—applications, trends, and prospects. Edited by Ciro Bustillo-Lecompte, London, pp 3–14. https://doi.org/10.5772/intechopen.90192

Gong C, Hart DP (1998) Ultrasound induced cavitation and sonochemical yields. J Acoust Soc Am 104:2675–2682. https://doi.org/10.1121/1.423851

Hrubec J, van Kreijl CF, Morra CFH, Slooff W (1983) Treatment of municipal waste water by reverse osmosis and activated-carbon-removal of organic micropollutants and reduction of toxicity. Sci Total Environ 27:71–88. https://doi.org/10.1016/0048-9697(83)90125-0

Jiang S, Li Y, Ladewig BL (2017) A review of reverse osmosis membrane fouling and control strategies. Sci Total Environ 595:567–583. https://doi:https://doi.org/10.1016/j.scitotenv.2017.03.235

Katz WE (1979) The electrodialysis reversal (EDR) process. Desalination 28(1):31–40. https://doi.org/10.1016/S0011-9164(00)88124-2

Khan I, Saeed K, Khan I (2017) Nanoparticles: properties, applications and toxicities. Arab J Chem. https://doi.org/10.1016/j.arabjc.2017.05.011

Koh I-O, Chen-Hamacher X, Hicke K, Thiemann W (2004) Leachate treatment by the combination of photochemical oxidation with biological process. J Photochem Photobiol 162:261–271. https://doi.org/10.1016/j.nainr.2003.08.011

Kujawa J, Cerneaux S, Kujawski W (2015) Removal of hazardous volatile organic compounds from water by vacuum pervaporation with hydrophobic ceramic membranes. J Membrane Sci 474:11–19. https://doi.org/10.1016/j.memsci.2014.08.054

Kwon S, Fan M, Cooper AT, Yang H (2008) Photocatalytic applications of micro- and nano-TiO2 in environmental engineering. Crit Rev Environ Sci Technol 38:197–226. https://doi.org/10.1080/10643380701628933

Lejarazu-Larranaga A, Molina S, Ortiz JM, Navarro R, García-Calvo E (2020) Circular economy in membrane technology: using end-of-life reverse osmosis modules for preparation of recycled anion exchange membranes and validation in electrodialysis. J Membrane Sci 593:117423. https://doi.org/10.1016/j.memsci.2019.117423

Li X, Jiang L, Li H (2018) Application of ultrafiltration technology in water treatment. IOP Conf Series: Earth Environ Sci, 186. https://doi.org/10.1088/1755-1315/186/3/012009

Lin SH, Leu HG (1999) Operating characteristics and kinetic studies of surfactant wastewater treatment by Fenton oxidation. Water Res 33:1735–41. https://doi.org/10.1016/s0043-1354(98)00403-5

Martínez C, Canle LM, Ferna´ndez MI, Santaballa JA, Faria J (2011) Aqueous degradation of diclofenac by heterogeneous photocatalysis using nanostructured materials. Appl Catal B: Environ 107:110–118. https://doi.org/10.1016/j.apcatb.2011.07.003

Mauter MS, Zucker I, Perreault F, Werber JR, Kim JH, Elimelech M (2018) The role of nanotechnology in tackling global water challenges. Nat Sustainab 1:166–175. https://doi.org/10.1038/s41893-018-0046-8

Mehrvar M, Anderson WA, Moo-young M (2001) Photocatalytic degradation of aqueous organic solvents in the presence of hydroxyl radical scavengers. Int J Photoenergy 3:187–191. https://doi.org/10.1155/s1110662x01000241

Nqombolo A, Mpupa A, Moutloali RM, Nomngongo PN (2018) Wastewater treatment using membrane technology. Wastewater Water Qual, 29–40. https://doi.org/10.5772/intechopen.76624

Qiang Z, Liu C, Dong B, Zhang Y (2010) Degradation mechanism of alachlor during direct ozonation and O3/H2O2 advanced oxidation process. Chemosphere 78:517–526. https://doi.org/10.1016/j.chemosphere.2009.11.037

Rautenbach R, Linn Th (1996) High-pressure reverse osmosis and nanofiltration, a "zero discharge" process combination for the treatment of waste water with severe fouling/scaling potential. Desalination 105:63–70. https://doi.org/10.1016/0011-9164(96)00059-8

Shanmuganathana S, Johira MAH, Listowskib A, Vigneswarana S, Kandasamya J (2016) Sustainable processes for treatment of waste water reverse osmosis concentrate to achieve zero waste discharge: a detailed study in water reclamation plant. Proc Environ Sci 35:930–937. https://doi.org/10.1016/j.proenv.2016.07.076

Singh S, Kumar V, Romero R, Sharma K, Singh J (2019) Applications of nanoparticles in wastewater treatment. Nanobiot Bioformul 395–418. https://doi.org/10.1007/978-3-030-17061-5_17

Singha N, Ray S (2012) Removal of pyridine from water by pervaporation using crosslinked and filled natural rubber membranes. J Appl Polymer Sci 124:E99–E107. https://doi.org/10.1002/app.35659

Sivakumar D, Shankar D, Gomathi V, Nandakumaar A (2014) Application of electro-dialysis on removal of heavy metals. Pollut Res 33:627–637

Stark WJ, Stoessel PR, Wohlleben W, Hafner A (2015) Industrial applications of nanoparticles. Chem Soc Rev 44:5793–5805. https://doi.org/10.1039/C4CS00362D

Tinucci T, Borgarello E, Minero C, Pelizzetti E (1993) Photocatalysis purification and treatment of water and air. In: Ollis DF, AL Ekabi H (eds). Elsevier, 585–594

Valero D, García-García V, Expósito E, Aldaz A, Montiel V (2015) Application of electrodialysis for the treatment of almond industry wastewater. J Memb Sci 476:580–589. https://doi.org/10.1016/j.memsci.2014.11.007

Varjani SJ, Gnansounou E, Pandey A (2017) Comprehensive review on toxicity of persistent organic pollutants from petroleum refinery waste and their degradation by microorganisms. Chemosphere 188:280–291. https://doi.org/10.1016/j.chemosphere.2017.09.005

Veza JM, Rodriguez-Gonzalez JJ (2003) Second use for old reverse osmosis membranes: wastewater treatment. Desalination 157:65–72. https://doi.org/10.1016/s0011-9164(03)00384-9

Wang X, Chen S, Gu X, Wang K (2009) Pilot study on the advanced treatment of landfill leachate using a combined coagulation, Fenton oxidation and biological aerated filter process. Waste Manag 29:1354–8. https://doi.org/10.1016/j.wasman.2008.10.006

Wijmans JG, Kaschemekat J, Davidson JE, Baker RW (1990) Treatment of organic-contaminated wastewater streams by pervaporation. Environ Prog 9:262–268. https://doi.org/10.1002/ep.670090421

Wiles L, Peirtsegaele E (2018) Reverse Osmosis: a history and explanation of the technology and how it became so important for desalination. 79th International Water Conference (Nov. 4–8, 2018 in Scottsdale, AZ). www.eswp.com/water

York RJ, Thiel RS, Beaudry EG (1999) Full-scale experience of direct osmosis concentration applied to leachate management. Cagliari: Sardinia, Italy, 4–8

Zhang C, Yang L, Bai Y, Gu J, Sun Y (2012) ZSM-5 filled polyurethaneurea membranes for pervaporation separation isopropyl acetate from aqueous solution. Separat Purif Technol 85:8–16. https://doi.org/10.1016/j.seppur.2011.07.008

Zhao S, Minier-Matar J, Chou S, Wang R, Fane AG, Adham S (2017) Gas field produced/process water treatment using forward osmosis hollow fiber membrane: Membrane fouling and chemical cleaning. Desalination 402:143–151. https://doi.org/10.1016/j.desal.2016.10.006

Chapter 3
Advances & Trends in Advance Oxidation Processes and Their Applications

Pratibha Gautam, Amishi Popat, and Snehal Lokhandwala

Abstract Exponential growth in industrialization, urbanization, and commercialization activities in last three decades have contributed to the increasing number of bio-refractory contaminants in the wastewater and this continuous admittance of toxic and persistent organic pollutants in the wastewater are imparting hazardous impacts on the environment. The conventional physical and biological treatment technologies are not very efficient in treating the wastewater having toxic and refractory pollutants and researchers worldwide are working to develop more efficient and sustainable technologies for the same. In recent years, advanced oxidation processes (AOPs) have emerged to be efficient, promising and environmental-friendly methods for the treatment of wastewater having high degree of pollution in comparison to conventional wastewater treatment methods. AOPs have appeared as an important avenue of technologies involving treatment of recalcitrant pollutants through different pathways such as "Enhanced Electrolysis", "Ultraviolet radiation", "Ozonation", "Sonolysis", etc. Many AOPs are based on the in-situ generation of strong oxidants like hydroxyl radicals which can completely mineralize or degrade the organic pollutants into harmless products. This chapter reports the details about various AOPs including theirbasic principle, mechanism, advantages, and limitations of each method to gain better scientific understanding of the most feasible approach to treat industrial wastewater. The authors have tried to summarize some recent AOPs too, which are being practiced seeing the disadvantages of the conventional AOPs. The contents will help to understand the future sustainable challenges for wastewater treatment (WWT).

Keywords Bio-refractory contaminants · Advanced oxidation processes · Enhanced Electrolysis · Ultraviolet radiation · Ozonation · Sonolysis · Hydrodynamic Cavitation

---

P. Gautam (✉) · A. Popat · S. Lokhandwala
Department of Environmental Science & Technology, Shroff S R Rotary Institute of Chemical Technology, UPL University of Sustainable Technology, Ankleshwar, Gujarat 393135, India
e-mail: pratibha.gautam@srict.in

S. Roy et al. (eds.), *Advanced Industrial Wastewater Treatment and Reclamation of Water*, Environmental Science and Engineering,
https://doi.org/10.1007/978-3-030-83811-9_3

3.1 Introduction

Optimized treatment of industrial wastewater is considered a core subject for researchers across the globe, as it poses a severe risk to air, soil, and aquatic life, specifically to animal and human health (Boczkaj et al. 2017). The industrial wastewater having high BOD, COD, color, recalcitrant/toxic compounds, heavy metals, etc. is discharged from various industrial sectors like dyes, textile, distilleries, paper and pulp, tanning, leather manufacturing, and breweries (Chandra and Kumar 2015, 2018), wherein the conventional treatment fails to remove all the pollutants from the effluent due to less efficiency, high operation costs and feasibility of treatment (Scott and Ollis 1995). The characteristics and composition of the industrial effluent vary depending upon the type of industry and there is no universal method/technology for its feasible treatment as it primarily relies on the concentration and nature of pollutants present in the effluent. Since the last decade, specifically, the government has embedded legislations that limits and prescribes the emissions of pollutants which has resulted in a notable increase in the number of research and businesses that are concerned with treating industrial waste water. Nowadays, researchers have more focus on the harmful effect of emerging contaminants on water resources. For environmental researchers, the addition of emerging contaminants like sunscreens, personal care products, pesticides, hormones, pharmaceutical products, flame retardants, nanomaterial, drugs of various compositions, plasticizers, etc. have become a serious issue of concern (Gogoi et al. 2018; Laxma et al. 2017; Reddy et al. 2017). Because of the severe and potential effect on the ecosystem and human health, these contaminants are accepted as the new classes of water contaminants (Barroso et al. 2019). It is estimated that around 700 such different contaminants disposed into the environment; are not completely metabolized and enter into the sewage/drainage system making their way to wastewater treatment plants (Lamastra et al. 2016; Ronquillo et al. 2017). The contaminants present in the effluent cannot be eliminated by biological process or physical separation because of their recalcitrant nature. Also, chemical oxidation with hydrogen peroxide, ozone, and chlorine is being used in recent times to treat the pollutants but in some reactions, the intermediate products generated remains more toxic when compared with initial products. Therefore, the main aim of the twenty-first century will be to quest newer, effective and economical treatment technologies for WWT.

Advanced Oxidation Processes (AOPs) are proving to be the cleaner technologies that effectively degrade persistent/toxic organic contaminants like artificial sweeteners, personal care products, coloring matter, pesticides and pharmaceuticals present in the aqueous medium (Calza et al. 2013; Wang et al. 2014). The processes are known as "Advanced" because the overall rate of oxidation for the contaminant to be removed is faster than that of the simpler compounds. AOPs are chemical-based oxidation processes that tend to eliminate recalcitrant organic compounds through various oxidation reactions with the help of non-selective, strong, and powerful $^{\bullet}OH$ that completely mineralize/degrades the compounds which fail to degrade by conventional methods and it also helps to increase the biodegradability of the effluent. For many years, AOPs are widely applied to target aromatic ring containing organic compounds, emerging contaminants, pharmaceutical drugs, waste discharged from clinical hospitals, personal care products, etc. emitting from different industries (Khetan and Terrence 2007). Moreover, AOPs have been fruitfully applied as a pre-treatment step to decrease the toxicity of the persistent organic compounds prior to biological treatment where microbes are not sufficient to remove such compounds present in the effluent (Stasinakis 2008). With the advances in the field of research, some newer treatment technologies like Ozone based, Fenton based, Electrochemical based, Photo-assisted based, Sonolysis is based and UV assisted based oxidation processes were established to treat the contaminated industrial wastewater (Fig. 3.1). These 'newer' technologies are termed as cleaner technologies as they prefer very less usage of chemicals resulting in a lower amount of sludge generation are energy-saving, eco-friendly, less waste-producing, efficient and sustainable technologies when compared to other biological and chemical processes. As a result of which the concept of Cleaner Technologies has grown interest in recent times. The main advantage of applying AOPs to treat industrial wastewater is that no additional separation is required and also, the pollutants are completely converted into CO_2 and H_2O (Ozdemir et al. 2011; Dalrymple et al. 2007).

The AOPs and hybrid-AOP are considered as recent advancement in the field of treating industrial effluent as complete mineralization of the pollutants can be achieved without generating toxic by-products in the system. So, in that context, the chapter focuses on the conventional AOPs that is being applied and the recent advancements in the field of AOPs to handle the recalcitrant pollutants present in industrial wastewater.

3.2 Advanced Oxidation Processes (AOPs)

Initially, AOPs were put forward to deal with the treatment related to potable water, but in recent times they have developed as efficient treatment technologies in removing refractory, low biodegradable, highly chemically stable, and inhibitory pollutants from wastewater (Kumar et al. 2021). The AOPs work on the mechanism of in situ generations of oxidant species like hydroxyl radicals ($^{\bullet}OH$) which is counted as very reactive in the reaction mechanism. The $^{\bullet}OH$ radical is non-selective, ubiquitous in nature, powerful and has an oxidation potential of 2.8 V. The radicals present in the system degrades the organic pollutants in four ways and they are the abstraction of hydrogen, transfer of electron, addition of radical and the combination of radicals. The $^{\bullet}OH$ radical generated is helpful to mineralize and degrade organic pollutants and also transforms them into CO_2, H_2O and some number of inorganic ions (Eq. 3.1) (Kumar et al. 2021).

$$\text{Organic species} + {}^{\bullet}OH \rightarrow CO_2 + H_2O + \text{inorganic ions} \tag{3.1}$$

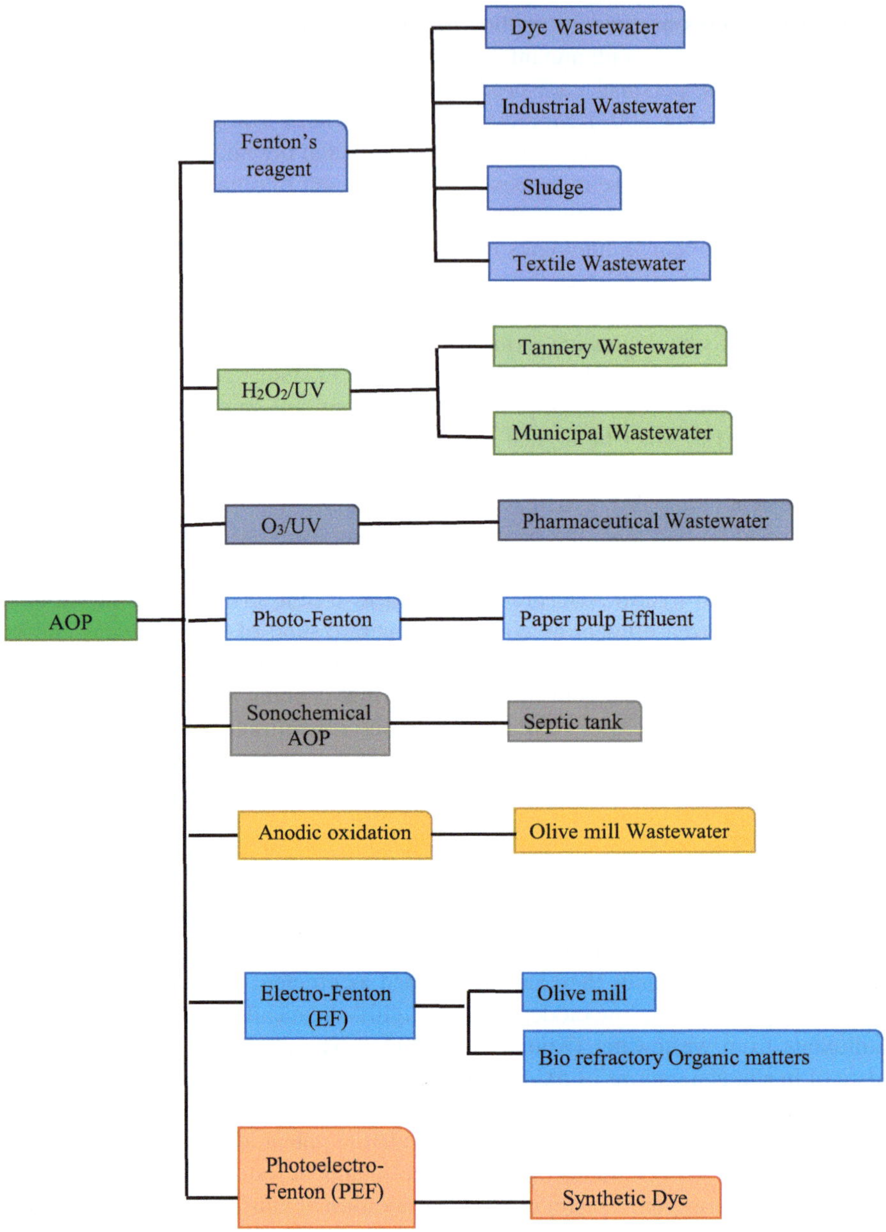

Fig. 3.1 Various AOPs treating different industrial wastewater

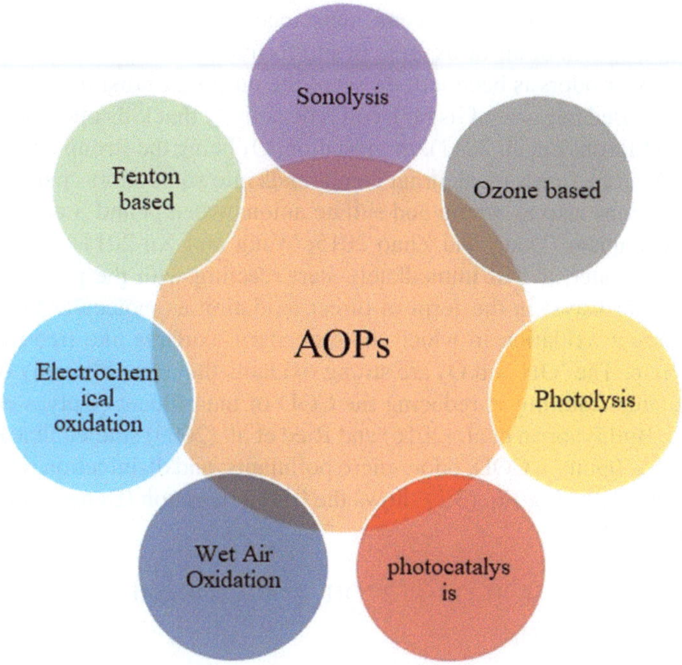

Fig. 3.2 Classification of a major class of AOPs used to treat complex wastewater

However, $^{\bullet}OH$ has a very short lifetime ($t \leq 10^{-3}$ s) and once generated in the electrochemical cell forms many elementary reactions. The sulfate radicals ($SO_4^{-}\bullet$), organic peroxyl radicals (ROO•), hydroperoxyl radicals ($HO_2^{-}\bullet$) and superoxide radicals ($O_2^{-}\bullet$) are other oxygen-containing species and radicals that are generated during AOPs (Manickavachagam et al. 2014; Oturan and Aaron 2014). The radicals are produced from less reactive oxidants like ozone and hydrogen peroxide. The reactive agents can be produced with the help of different processes like Fenton based oxidation, photocatalysis, sonolysis, ozone-based oxidation and also by combining various processes like peroxone (O_3/H_2O_2), peroxone coupled with ultraviolet light ($O_3/ H_2O_2/UV$), Fenton (H_2O_2/Fe^{2+}), $O_3/TiO_2/ H_2O_2$, etc. (Fig. 3.2).

3.2.1 Ozone Based AOPs

3.2.1.1 Ozonation

In recent times, ozonation has been a widely used advanced oxidation process especially in the field of research to deal with industrial wastewater. Because of its high reactivity, ozonation has been applied to treat different recalcitrant compounds like dyes, detergents, pesticides, estrogenic substances, and PPCPs. Due to the electric discharge in a stream of air or oxygen, in situ formation of O_3 takes place where no residual tastes or odors is been left. To serve the purpose of disinfection, ozonation is extensively used because of its high cell lytic activity that kills microbes present in the effluent (Martınez et al. 2011). In ozonation, O_3 being the strong oxidizing agent ($E^0 = 2.07$ V) transforms recalcitrant compounds into some short-chain compounds or and sometimes also to nitrate and sulfate anions when N and S are found in the reaction mechanism (Deng and Zhao 2015; Wang and Xu 2011). As soon as O_3 dissociates in water, it will immediately start reacting with the pollutants present in two different ways: in the form of direct oxidation as molecular O_3 and in the form of indirect oxidation in which the secondary oxidants like free radicals i.e., $^{\bullet}OH$. will help. The $^{\bullet}OH$ and O_3 are strong oxidants that can efficiently oxidize the toxic pollutants resulting in reducing the COD of the effluent (Gulyas et al. 1995; Rice 1995). Buthiyappan et al. (2016) and Ried et al. (2009) studied that at different ozone dosages (g/cm^3), COD, odor, micro pollutants, and disinfection removal were obtained. The following Eq. (3.2) shows the $^{\bullet}OH$ generation (Deng and Zhao 2015; Wang and Xu 2011):

$$O_3 + H_2O \rightarrow 2\,^{\bullet}OH + 4O_2 \text{ (Ozonation)} \tag{3.2}$$

3.2.1.2 Ozonation with H_2O_2

When other oxidants are present, the yield of $^{\bullet}OH$ can be improved. In the peroxone process (O_3/H_2O_2), through H_2O_2 decomposition, $^{\bullet}OH$ generation and O_3 decomposition gets enhanced with the help of hydroperoxide (HO_2^{-}). Further, H_2O_2 generates HO_2^- in basic conditions while reacting with O_3 and produces free radicals (Eqs. 3.3 and 3.4)

$$H_2O_2 \rightarrow HO_2^- + H^+ \tag{3.3}$$

$$HO_2^- + O_3 \rightarrow {}^{\bullet}OH + O_2^- + O_2 \tag{3.4}$$

3.2.1.3 Ozonation with UV

In O_3/UV irradiation, photolysis of ozone takes place, generating an additional primary oxidant such as H_2O_2 that results in the formation of $^{\bullet}OH$. Further H_2O_2 generates more $^{\bullet}OH$ which is shown by the below reactions (Eqs. 3.5, 3.6, and 3.7) (Deng and Zhao 2015).

$$O_3 + H_2O + h\nu \rightarrow H_2O_2 + O_2 \tag{3.5}$$

$$H_2O_2 + h\nu \rightarrow 2\,{}^{\bullet}OH \text{ (Photolysis of } H_2O_2) \quad (3.6)$$

$$H_2O_2 + {}^{\bullet}OH \rightarrow HO_2^- + H_2O \text{ (Photolysis of } H_2O_2) \quad (3.7)$$

The disadvantage of this process is that it is energy-intensive because to lower the COD value, a higher dose of ozone needs to be applied.

3.2.2 Fenton Based AOPs

3.2.2.1 Classical Fenton Process

To treat the complex wastewater, Fenton oxidation is thought to be an efficient, effective, and advanced treatment process (Hadavifar et al. 2009). The Fenton process aims to remove color, toxicity, and remnant COD of the wastewater, and also the effluent biodegradability can be increased by this process (Bhoite and Vaidya 2020). In this process, the mixture containing H_2O_2 and ferrous iron (Fe^{2+})- Fenton's reagent, produces a very reactive hydroxyl radical involving complex intermediate reactions and ultimately removing organic pollutants from effluent (Zhang et al. 2019). In acidic conditions, Fe^{2+} catalyses and initiates the decomposition of H_2O_2 which results in the production of $^{\bullet}OH$ radical in the electrolytic cell. The below are the reactions (3.8–3.11) involved in the Fenton oxidation process:

$$Fe^{2+} + H_2O_2 \rightarrow Fe^{3+} + OH^- + {}^{\bullet}OH \quad (3.8)$$

$$Fe^{2+} + {}^{\bullet}OH \rightarrow Fe^{3+} + OH^- \quad (3.9)$$

$${}^{\bullet}OH + RH \rightarrow H_2O + R^{\bullet} \quad (3.10)$$

$$R^{\bullet} + Fe^{3+} \rightarrow R^+ + Fe^{2+} \quad (3.11)$$

When the Fenton process is applied practically to treat the wastewater, the main factors hindering the process are temperature, Fe: H_2O_2 ratio and pH (Pliego et al. 2015). When the temperature is increased up to 100 °C, higher efficiencies are obtained in terms of the oxidation rate and also in the mineralization degree (Zazo et al. 2010). In the Fenton process, Fe: H_2O_2 the ratio should be optimum. If the ratio is high, it will lead to the enhancement in the catalytic conversion of H_2O_2 and not the mineralization degree as an excess amount of $^{\bullet}OH$ radicals produced will lead to scavenging reactions (Eq. 3.9) (Christensen et al. 1982). If the ratio is low, surplus hydrogen peroxide generated in the reaction cell will act as a scavenger for $^{\bullet}OH$ generated (Gogate and Pandit 2004). The Fenton process is a strongly pH dependent process and in a majority of the cases, pH = 3 is thought as optimum.

The main benefits of the Fenton process are (Li et al. 2010; Lu et al. 2005):

- It requires a short reaction time among all AOPs
- There is no energy required to produce H_2O_2
- The operation and maintenance of this process is comparatively easy

The drawbacks of the process are (Deng et al. 2008; Zhang et al. 2006):

- It is operated at acidic pH (2–3)
- Higher sludge generation
- Ferrous ion consumption is more rapid than its regeneration

However, in small-scale industries, the conventional Fenton process is extensively used to remove special pollutants present in the effluent.

3.2.2.2 Photo-Fenton Process

Different processes have been established to overcome the shortcomings of the Fenton oxidation process. Among other processes, the Photo-Fenton oxidation process is considered the first process that couples the light energy in the form of UV sources. In this process, $^{\bullet}OH$ is formed by photolysis of H_2O_2 in the presence of UV. With the help of UV irradiation, photocatalytically produced Fe^{3+} in Eq. (3.12) gets transformed to Fe^{2+} generating an additional $^{\bullet}OH$ in the system.

$$Fe^{3+} + H_2O + h\nu \rightarrow Fe^{2+} + H^+ + {}^{\bullet}OH \quad (3.12)$$

This process has many benefits when compared to the Fenton oxidation process. The rate of reaction in the photo-Fenton process is a rapid and also higher amount of mineralization of pollutants is achieved. The limitations of the process are the requirement of UV energy for a longer time, high energy consumption, and cost-intensive. Derakhshan and Fazeli (2018) examined that when complex wastewater was treated with a photo-Fenton oxidation process, an increase in biodegradability, detoxification, and COD removal was obtained.

3.2.2.3 Electro-Fenton (EF) Process

To abate persistent organic compounds, the EF process is been extensively used since last decade apart from many other processes. EF is considered to be an attractive treatment technique and it's mainly because of the fast pollutant removal rate, higher degradation rate for persistent organic pollutant removal and due to its compatibility with the environment (Oturan et al. 2000; Brillas et al. 2009). Under this process, hydroxyl radical is formed by the electro-generation of Fenton's reagent in the electrochemical cell.

As shown in Fig. 3.3, when reduction of O_2 electron occurs in the acidic medium, on the surface of cathode there is the generation of hydrogen peroxide (Oturan et al. 2008a; Nidheesh and Gandhimathi 2012). A continuous oxygen source is required near by cathode to generate hydrogen peroxide in the electrochemical cell (Eq. 3.13). When ferrous ions in the form of the catalyst are added to the solution it leads to the formation of $^{\bullet}OH$ according to (Eq. 3.14).

$$O_2 + 2H^+ + 2e^- \rightarrow H_2O_2 \tag{3.13}$$

$$Fe^{2+} + H_2O_2 \rightarrow Fe^{3+} + OH^- + {}^{\bullet}OH \tag{3.14}$$

Main Pros of the EF process are as follows (Pliego et al. 2015):

- Complete mineralization and degradation of pollutants can be achieved.
- No transportation cost for H_2O_2

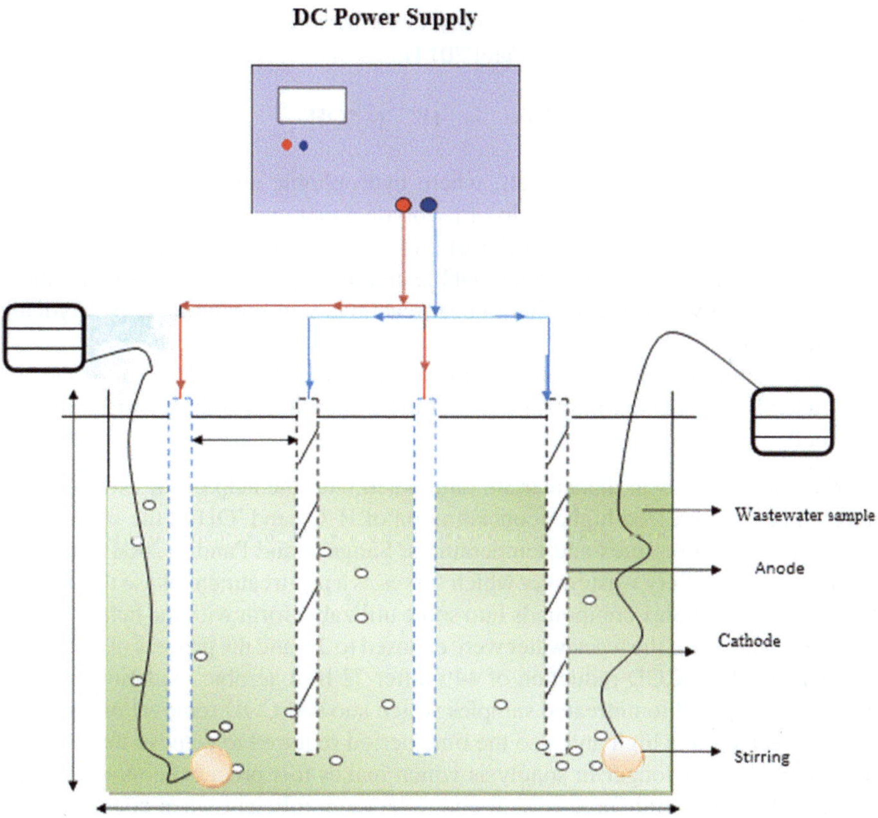

Fig. 3.3 Electro-Fenton process

- OH· is produced in the electrochemical cell by in-situ electro regeneration of H_2O_2 and Fe^{2+}

  The Cons of the EF process are as follows (Pliego et al. 2015):

- The process is operated in acidic conditions (pH 2 to 3)
- Higher operational and maintenance cost is required
- The EF process is linked with the concentration polarization in the electrodes

3.2.3 Sonolysis

To eliminate the pollutants, present in the effluent, sewage sludge, and water, the potential of the ultrasound process has been increased. Sonochemical oxidation uses ultrasound irradiation to generate $^{\bullet}OH$ radicals which results in the phenomenon of cavitation. Sonochemical reactions are dependent on acoustic irradiation of liquids with high frequencies that lead to forming cavitation (20–1000 kHz) (Kumar et al. 2021). In ultrasonically irradiated liquids, there are three sites that are potential for the chemical reactions (Nair and Patel 2014):

$$H_2O \rightarrow H^+ + {}^{\bullet}OH \tag{3.15}$$

1. In the cavitation bubble itself, where hydrophobic and volatile species gets degraded rapidly with the help of pyrolytic equations as well as the equations involving the $^{\bullet}OH$ radical participation by water Sonolysis (Eq. 3.15).
2. The interface region between a bubble and the liquid where hydroxyl radicals are located which proves that radical reactions will predominate pyrolytically reactions.
3. The solution bulk where the Sonolysis activity takes place because the free radicals present have now run away from the interface and shifted to the solution bulk.

The contaminants in the water are eliminated with the help of ultrasound irradiation since the localized higher concentration of H_2O_2 and $^{\bullet}OH$ in the solution and localized higher pressures and temperatures. Sangave and Pandit (2004) examined sonication of distillery wastewater which serves as a pre-treatment phase that helps to transform the complex compounds into some utilizable form with the help of cavitation. The samples of the wastewater were exposed to 2 h and the process of ultrasound had displayed the COD reduction of 44% after 72 h of aerobic oxidation that took place, as compared to untreated samples which showed COD removal of 25%. The power dissipation is high and also the time period required to achieve the complete degradation seems longer in sonolysis which makes this process economically not feasible. Therefore, this process is found to be more efficient when combined with other conventional processes rather than an individual process.

The advantages of the process are that the toxic waste is being treated in mild conditions; as only electricity is used as a reactant it is environmentally friendly process and also the effluents having low conductivity can be easily treated (Nair and Patel 2014). The limitations of the process are that when the process is applied to complex effluent, the installation cost is high and also, low removal of COD is obtained (Nair and Patel 2014).

3.2.4 Electro-Chemical Oxidation

Sustainability and energy are the pillars behind the success of any treatment technology. Keeping that in mind, Electrochemical based AOPs have gained attraction nowadays for degrading the recalcitrant pollutants thereby producing the reactive species in the electrochemical by the usage of electricity. Production of no secondary waste nor chemicals is observed at the end making this process more sustainable (Garcia-Segura et al. 2018). The EAOPs are further divided into two types (Panizza and Cerisola 2009):

1. Direct oxidation where directly the transfer of charge takes place between pollutant and anode surface. Also, electrodes and supply of power are required in direct oxidation.
2. Indirect oxidation is initiated by the in-situ generation of oxygen species being reactive towards anode surface. Also, reagents in the form of the catalyst are required in order to produce a specific type of oxidizing agent in the system.

To treat pharmaceutical wastewater, electrochemical oxidation is considered an effective method of oxidation (Perez et al. 2017).

The main befits of Electrochemical oxidation are as follows:

- The process is easy to handle as equipment used are simple
- Higher pollutant removal rate is obtained
- The process is energy efficient and flexible with automation

The drawbacks of the electrochemical oxidation process are as follows:

- Acidic conditions are required
- Sludge is generated
- Chemical consumption is high in the process

3.2.4.1 Electrocoagulation Process

It is an in-situ process where the dissolution of metal coagulants are produced at anode, whereas, $^{\bullet}OH$ and H_2 gas is released from the cathode which indeed destabilizes the contaminants or the pollutants (Bouhezila et al. 2011). The industrial-scale application of this technique is not feasible which makes this treatment technology confounding (Moreno-Casillas et al. 2007). EC involves three major processes: electrochemistry, coagulation, and floatation. The generalized steps involved in the EC process given by Garg and Prasad (2016) are defined by four steps: first being the electrolytic oxidation process on the sacrificial anode, second being the addition of coagulant, third being an amalgamation of colloids on coagulant by destabilization and adsorption of pollutant on it and fourth being the removal of agglomerated pollutants by sedimentation and floatation.

The main reactions (Eqs. 3.16–3.19) occurring at anode and cathode in the EC cell are as follows:

At anode

$$M_{(Solid)} \rightarrow M^{n+}_{(aqueous)} + ne^- \quad (3.16)$$

$$2H_2O_{(liquid)} \rightarrow 4H^+_{(aqueous)} + O_2(g) + e^- \quad (3.17)$$

At Cathode

$$M^{n+}_{(aqueous)}\ ne^- + \rightarrow M_{(solid)} \quad (3.18)$$

$$2H_2O_{(liquid)} + 2e^- \rightarrow 2OH^- + H_{2(gas)} \quad (3.19)$$

The main advantages of EC are as below (Gautam et al. 2019):

- As no additional chemicals are required, EC is a green technology.
- In EC, treatment cost is low as no mixing pools and sedimentation tank is required.
- Low maintenance cost and simple design.
- The quantity of sludge generated is low in the EC process and also it is of better quality as compared to conventional coagulation processes.

The main drawbacks of the EC are as follows:

- Periodic replacement of electrodes is required.
- Toxicity arises because of the use of chlorinated compounds such as NaCl as an electrolyte which needs to be replaced by non-chlorinated compounds by using sodium sulfate as an electrolyte.

3.2.4.2 Anodic Oxidation

In Anodic Oxidation (AO) process, the hydroxyl radicals are physiosorbed on the anode surface and it is the most widely used EAOPs (Panizza and Cerisola 2009). In some cases, they are also chemisorbed when Pt or DSA are used as anodes. When Boron doped diamond (BDD) anodes are used in the system, the $^{\bullet}OH$ is generated faster as this anode has a higher O_2 evolution ability. There is a two-step mechanism involved in the AO process and it is as follows (Comninellis 184 and Battisti 1996):

1. Electrochemical conversion in which the persistent organic contaminants get converted into short-chain biodegradable products.

2. Electrochemical incineration or combustion in which complete degradation of persistent organic contaminants into some CO_2, H_2O, and inorganic ions occurs.

The main advantages of the AO process are easy to operate, degradation of pollutants occurs rapidly and a higher pollutant removal rate is obtained (Chen et al. 2003). In the AO process, the efficiency depends upon the proper choice of the anode as it is responsible for the oxidation process. To enhance the process, AO is coupled with electro-Fenton, Solar electro-Fenton H_2O_2, etc.

3.2.5 *Photocatalysis*

The wastewater having a low BOD/COD ratio has persistent organic pollutants present in it having complex composition. There are many chemical treatment processes and biological methods available to treat such pollutants but their complete decomposition and mineralization are not possible (Catalkaya and Sengul 2006). When the discovery of photoelectron chemical water splitting reactions with the help of semiconductors was conducted, advanced oxidation processes namely photocatalysis emerged rapidly to deal with the contaminants present in the effluent. Photocatalysis has proved to be an efficient and sustainable treatment technology that helps to increase the biodegradability of non-biodegradable and hazardous pollutants like persistent organic pollutants (Mazuba et al. 2017). The photocatalytic process involves the use of different oxidizing agents in combination with suitable light and or catalyst. At the time of the photocatalysis process, the photons are triggered from the solid semiconductor by the absorption of electromagnetic radiation. In near UV-irradiation, the appropriate semiconductor gets excited by photons which are having sufficient magnitude to get valence band holes (h^+) and conduction band electrons (e^-). This e^- and $h+$ migrating to their bands on the surface of semiconductors reacts in a way with the redox species present in the environment resulting in the splitting of H_2O generating O^{2-} and $^{\bullet}OH$ radicals. The radicals are efficient in degrading the persistent organic contaminants and it also detoxifies the heavy metals ions like Cr^{+6} and As^{+3}. According to Poulopoulos et al. (2019), photocatalysts have played a vital role in degrading the persistent organic contaminants from the effluent through light absorption in the water. Mostly ZnO and TiO_2 are widely used as photocatalysts in the process.

The advantage of the process is that it serves as a substitute for the methods which are energy-intensive and also the process uses pollution-free and renewable sources of energy i.e., solar energy. The shortcomings of the process are that it requires low light absorption abilities; requires wide band gap energies and fast recombination rates of electrons and holes inhibit its real field application.

3.2.6 Photolysis

In UV direct photolysis, the removal of pollutants takes place when incident radiation gets absorbed from UV light. UV photolysis helps reduce few organic pollutants but alone this treatment is not feasible to eliminate the pollutants. With the help of UV photolysis, the oxidation potential can be enhanced for some processes. When it is coupled with O_3, due to more $^{\bullet}OH$ radical generation because of photolysis of H_2O_2, the degradation rate of the pollutants to be removed increases. The $^{\bullet}OH$ radical helps in eliminating the compounds which makes this process successful. It was studied that with the help of UV photolysis process, degradation of 90% was achieved in the case of trifluralin and bromoxynil at pH 2 and UV dosage of 302 and 439 (Chelme-Ayala et al. 2010). Yazdanbakhsh et al. (2018) studied that in UV photolysis process (irradiance = 87 mWcm^{-2} results in degradation of 73% at 80 min for 2, 4, 6- trichlorophenol (C_0 = 250 mgL^{-1}). Therefore, when UV photolysis is applied as pre-treatment and then subsequent biological treatment, more efficiency can be obtained. The main drawback of this process is its high operational cost.

3.2.7 Wet Air Oxidation

Oxidation is evidently not possible without the presence of air or water at the source and their presence governs the oxidation process by elevating the temperature and pressure at the time of reaction, oxidizing the organic and inorganic constituents (Kim et al. 2011). Such reaction is said to be known as Wet Oxidation/Wet air Oxidation (WAO) which is possible in both suspended and solution form. Biological treatments are showing potentially poor results in degrading organic matters and/or hazardous materials present in the effluents in higher measures, whereas WAO technology shows contrast results in the same case (Kim et al. 2011). Oxidation with the help of free radical at the time of the WAO process allows the pollutants to either partially oxidize itself into biodegradable intermediates or completely mineralize itself to water, carbon dioxide, and harmless end-products, amid the temperature ranging between 125 °C and 320 °C and the pressure between 0.5 and 20.0 MPa in the aqueous phase (Kim et al. 2011).

However, WAO having the ability to degrade compounds with higher organic and/hazardous constituents, it is limited to aliphatic and aromatic compounds without including the halogenated group compounds. On the other hand, a hefty amount of Capital Expenditure (CAPEX) and Operational Expenditure (OPEX) of the WAO Plants becomes an obstacle due to excessive temperatures & pressure maintenance, while the additional expenditures raise when the refractory organic pollutants are spoiling the effluent quality (Li et al. 2007; Stuber et al. 2005).

3.3 Recent Trends in AOPs

3.3.1 Cavitation

Cavitation is the latest technology that has found significant advantages in wastewater treatment (Cintas and Luche 1999). Cavitation occurs when vapor cavities are formed inside a homogeneous medium of liquid and it results in different situations (Leighton 1995). Energy is locally recovered to large extent by production, growth, and the subsequent collapse of cavities formed during the process creating a strong oxidizing condition to produce hydroxyl radicals and hydrogen peroxide (Cintas and Luche 1999). The reactors that are designed for the process of cavitation can transform toxic organic compounds like aromatic compounds, pesticides, chlorinated hydrocarbons, dyes, esters, phenolic compounds, and pharmaceuticals into some short-chain compounds, water, carbon dioxide, and also some inorganic ions (Adewuyi 2001). Based on the mode of generation, there are four different types of cavitation, including optic, acoustic, particle, and hydrodynamic. In various chemical processing applications, only two types of cavitation i.e., hydrodynamic and acoustic cavitation is thought to be an effective and efficient treatment option (Mishra and Gogate 2010).

- In acoustic cavitation, cavities are generated with the help of ultrasound waves having a frequency that ranges from 16 kHz to 2 MHz (Gogate 2008a). The drawbacks of the process are high operational cost; lack of scale-of potential and lower energy efficiency (Mishra and Gogate 2010).
- Cavities are produced based on pressure vibration which is introduced in the liquid by bypassing the solution from venture or orifice (Gogate 2008b), and this process is called Hydrodynamic cavitation (HC). The violent breakdown of cavities produces $^{\bullet}OH$ radicals and also forms intense turbulence which intensifies the chemical and physical changes (Barik and Gogate 2018). As per many researchers, complete mineralization of the effluent cannot be obtained with the help of the hydrodynamic cavitation process alone because of the limited formation of oxidation radicals which results in combining the hydrodynamic cavitation process with the other AOPs to obtain better removal efficiency of the wastewater (Bagal and Gogate 2014a; Mishra and Gogate 2010).

By coupling HC with oxidizing agents like ozone, H_2O_2, and catalysts, enhancement in the degradation efficiency of pollutants are observed (Wang et al. 2017; Saharan et al. 2011). By combining various treatment methods with HC, mineralization of various pollutants like dyes (Saharan et al. 2011; Rajoriya et al. 2017), pesticides (Thanekar et al. 2018; Raut-Jadhav et al. 2013), phenolic compounds (Barik et al. 2016, 2018), pharmaceuticals (Wang et al. 2017; Thanekar et al. 2018), etc. is obtained.

3.3.1.1 Hydrodynamic Cavitation with O_3

Ozonation is considered to be an effective treatment as the oxidation potential is 2.8 V and efficient degradation of pollutants like pesticides and dyes take place (Pillai et al. 2009; Wang et al. 2003; Maldonado et al. 2006). Coupling of HC with O_3 is efficient under the effect of cavitation, as ozone dissociates into molecular O_2 and nascent oxygen forming the hydroxyl radicals when reacted with water. The below are the reactions that take place when HC is combined with ozone:

$$H_2O \rightarrow H + {}^{\bullet}OH \tag{3.20}$$

$$O_3 \rightarrow O_2 + O \tag{3.21}$$

$$O + H_2O \rightarrow + 2{}^{\bullet}OH \tag{3.22}$$

$$\text{Pollutant} + OH \rightarrow CO_2 + H_2O + \text{degradation intermediates} \tag{3.23}$$

The combined treatment of HC with ozone was examined by Gore et al. (2014) for the degradation of orange 4 dye with the help of a circular venturi, in which the ozone flow rate was varied between 1 to 8 g/h. When the ozone flow rate was increased from 1 to 3 g/h, the mineralization of pollutants got increased and with the further increase in the ozone flow rate above 3 g/h, the mineralization rate shows constant result. With the ozone flow rate of 3 g/h, 76.25% reduction in TOC was observed when HC combined with ozone keeping the rate constant of 2.06×10^{-2} min $^{-1}$.

3.3.1.2 Hydrodynamic Cavitation with H_2O_2

In HC, the pollutant degradation mechanism occurs through the formation of hydroxyl radicals in the system. H_2O_2 acts like the additional source which intensifies the degradation process with HC treatment. Moreover, at a higher loading rate, H_2O_2 acts like a scavenger decreasing the pollutant removal efficiency (Saharan et al. 2012). Because of extreme conditions, the dissociation of H_2O_2 into OH. Radicalsin the HC reactor cell take place. The following are the reactions that occur when HC is combined with H_2O_2 (Rajoriya et al. 2016):

$$H_2O_2 \rightarrow {}^{\bullet}OH + {}^{\bullet}OH \tag{3.24}$$

$$^{\bullet}OH + H_2O_2 \rightarrow HO_2 + H_2O \tag{3.25}$$

$$^{\bullet}OH + HO_2 \rightarrow H_2O + O_2 \tag{3.26}$$

$$HO_2 + H_2O_2 \rightarrow {}^{\bullet}OH + H_2O + O_2 \tag{3.27}$$

$$\text{Pollutant} +^{\bullet} OH \rightarrow CO_2 + H_2O + \text{degradation intermediates} \tag{3.28}$$

3.3.1.3 Hydrodynamic Cavitation with Fenton

The hybrid method combining HC with Fenton oxidation is considered the most efficient method for the elimination of contaminants present in the wastewater (Bagal and Gogate 2014a). The coupling of hydrodynamic cavitation and the Fenton oxidation process leads to higher degradation efficiency as high amounts of free radicals are generated in the presence of Fe^{2+} ions. Higher degradation of dichlorvos as 91.5% at a loading of $FeSO_4$: H_2O_2 as 3:1 was notified by Barik and Gogate (2016). Their study uses orifice in the cavitation process where Fenton reagent and HC were coupled together. As per their observation, results in higher measures were observed with combined experiments rather than HC alone. The combined HC/Fenton process set-up is represented in Fig. 3.4.

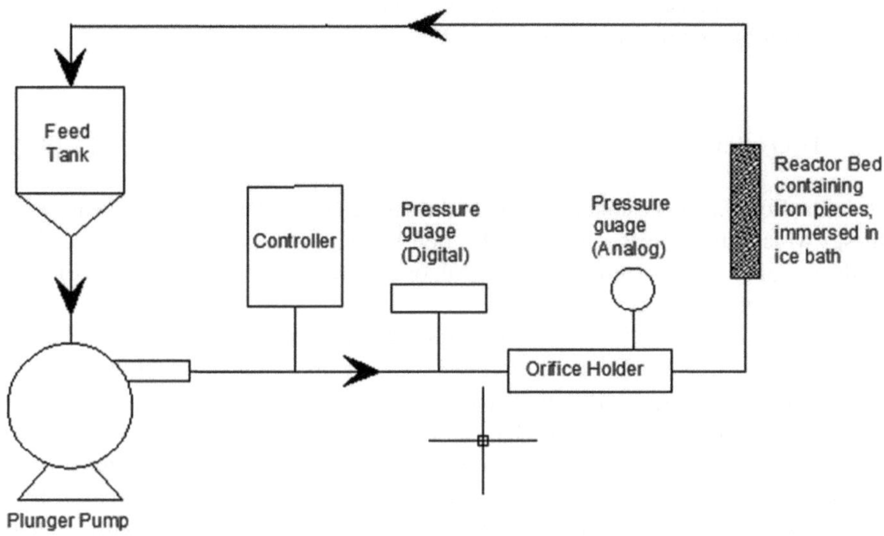

Fig. 3.4. Schematic representation of combined hydrodynamic cavitation with Fenton oxidation process

3.3.1.4 Hydrodynamic Cavitation with UV Photolysis

Apart from the individual process of HC or UV, the combined process of HC and UV has resulted in greater production of •OH radicals and a higher pollutant removal rate. The cavitation effect has resulted in producing •OH radicals and at the same time, UV irradiations have initiated the photolysis of water, ultimately removing the pollutant. The researchers have examined the efficacy of the coupled process of HC with UV to degrade diclofenac sodium in the HC reactor by keeping a 250 W UV lamp. Only 43% and 26.8% degradation was achieved when UV and HC processes were applied individually, while 65.5% degradation was observed in the case of the coupled processes of HC/UV (Bagal and Gogate 2014b).

3.3.2 Ionizing Radiation

Ionizing radiation (IR) is found to be a recent and emerging treatment technology for drinking water and wastewater (Wang and Chu 2016). IR is shown by gamma rays (γ) and electron beam irradiations. Two of them, γ-rays and electron beam are categorized by radiolysis of water, resulting in the generation of •OH, $e_{aq}^{-}$and H·(Eq. 3.29) (Khan et al. 2015) Therefore, oxidation/reduction of the pollutants can be obtained by IR process (Khan et al. 2015).

$$H_2O - - -\wedge -\wedge - - > HO(0.29),\ e_{aq}^{-}\ (0.28)H(0.06),\ H_3O^{+}\ (0.27),$$
$$H_2O_2\ (0.07),\ H_2\ (0.047) \qquad (3.29)$$

In the Eq. (3.29), the concentration of the species is denoted by numbers in parentheses which are expressed in µmol/J, often known as G-value. G-value is defined as the µmol of species (molecules, radical, or ions) formed or decomposed per joule of the absorbed energy.

The $e_{aq}-$ in Eq. (3.25) is a potentially strong reducing agent present in aqueous solution ($E^0 = 2.9$ V)(Buxton et al. 1988). $E_{aq}-$with its higher reactivity towards organic substances especially halogenated organic compounds (HOCs) shows prominent results in comparison with OH. radical. Ionization radiation is said to be the most evolving advanced oxidation/reduction technologies where the generation of oxidizing/reducing species occur simultaneously to degrade the contaminants. IR is a process where the contaminants in water are degraded in an eco-friendly manner eliminating the chemical addition followed by the conventional methods. The chances of pollutant getting fully removed from the effluent are more if IR process is applied widely (Shah et al. 2014).

3.3.3 Hybrid Methods

Recently the studies have revealed that by combining different AOPs, better results for water decontamination can be obtained (Chakma et al. 2014). Consequently, hybrid AOPs like sonophotolysis (Xu et al. 2014), gamma rays/PMS (Wang et al. 2018), cavitation-based hybrid AOPs (Thanekar et al. 2018), EF process, etc. were used to remove persistent organic contaminants from the effluent. Xu et al. (2014) examined the efficacy of the photolysis (UV), sonophotolysis (USUV) and cavitation processes to degrade atrazine. The studies revealed that combined processes like USUV had shown more efficiency as higher degradation of the intermediates has occurred in the combined process when compared with US and UV as an individual process. Wang and Wang (2018) observed the mineralization efficiency of carbamazepine with the help of gamma irradiation process, PMS process, and Fe^{2+}/PMS process but greater efficiency was observed in the case of the combined process of the gamma rays-assisted PMS system. Various hybrid AOPs have been by Horakova et al. (2014) to eliminate persistent organic compounds, like surfactants, PPCPs, pesticides, and dyes. The study revealed that degradation of compounds was enhanced when hybrid methods like UV radiation, TiO_2 photocatalyst and inorganic oxidants were used. Recent advancement in the field of water treatment with the help of metal–organic frameworks (MOFs)-catalyzed AOPs were studied by Sharma and Feng (2017) and it was also observed that by using MOFs based AOPs, removal efficiency got enhanced. Gagol et al. (2018) reviewed AOPs based on cavitation and showed that maximum degradation of the pollutants like phenols, pesticides, etc. was obtained by using hybrid methods consisting of acoustic or hydrodynamic cavitation with Fenton reagent, H_2O_2, UV irradiation, persulfate, or O_3. By combining flocculation-ozonation-Fenton seems an efficient hybrid method for the treatment of distillery wastewater (Martins et al. 2011).

3.4 Conclusion and Way Forward

One of the challenges that today's world is facing are the proper disposal and efficient treatment of industrial wastewater although widespread research has been done to prevent the entire ecosystem. However, AOPs and their various combinations have proved to be an efficient treatment technology to deal with the recalcitrant properties of the effluent generating from different industries. They have been applied in pilot and bench scales. But their commercial and long-term implementation is difficult as these technologies are energy-consuming and costly considering their maintenance and operation. So, efforts are being raised to evaluate the advantages and limitations of its real scale implementation in various industries. AOPs are not being applied in the industries either because of lack of awareness regarding different AOPs or maybe the knowledge regarding the recently advanced oxidation processes to deal with specific pollutants is not sufficient. However, the future for the real scale implementation of

AOPs is brighter as its combination with other conventional processes is feasible and reliable. The future scope of the AOPs will depend on the collaboration of engineers, electrochemists, and chemists to make sure regarding the field scale application and utilization of these processes.

References

Adewuyi YG (2001) Sonochemistry: environmental science and engineering applications. Ind Eng Chem Res 40:4681–4715. https://doi.org/10.1021/ie010096l

Bagal MV, Gogate PR (2014a) Wastewater treatment using hybrid treatment schemes based on cavitation and Fenton chemistry: a review. Ultrason Sonochem 21(1):1–14. https://doi.org/10.1016/j.ultsonch.2013.07.009

Bagal MV, Gogate PR (2014b) Degradation of diclofenac sodium using combined processes based on hydrodynamic cavitation and heterogeneous photocatalysis. Ultrason Sonochem 21:1035–1043. https://doi.org/10.1016/j.ultsonch.2013.10.020

Barik AJ, Gogate PR (2018) Hybrid treatment strategies for 2,4,6-trichlorophenol degradation based on combination of hydrodynamic cavitation and AOPs. Ultrason Sonochem 40(A):383–394. https://doi.org/10.1016/j.ultsonch.2017.07.029

Barik AJ, Gogate PR (2016) Degradation of 4-chloro 2-aminophenol using a novel combined process based on hydrodynamic cavitation, UV photolysis and ozone. Ultrason Sonochem 30:70–78. https://doi.org/10.1016/j.ultsonch.2015.11.007

Barroso PJ, Santos JL, Martın J, Aparicio I, Alonso E (2019) Emerging contaminants in the atmosphere: analysis, occurrence and future challenges. Crit Rev Environ Sci Technol 49(2):104–171. https://doi.org/10.1080/10643389.2018.1540761

Bhoite GM, Vaidya PD (2020) Fenton oxidation and adsorption pretreatment for superior biogas recovery from biomethanated spent wash. Chem Eng Commun 207(10):1347–1357. https://doi.org/10.1080/00986445.2019.1650031

Boczkaj G, Fernandes A (2017) Wastewater treatment by means of advanced oxidation processes at basic pH conditions: a review. Chem Eng J 320:608–633. https://doi.org/10.1016/j.cej.2017.03.084

Brillas E, Sires I, Oturan MA (2009) Electro-Fenton process and related electrochemical technologies based on Fenton's reaction chemistry. Chem Rev 109(12):6570–6631. https://doi.org/10.1021/cr900136g

Buthiyappan A, Aziz AR, Daud WM (2016) Recent advances and prospects of catalytic advanced oxidation process in treating textile effluents. Rev Chem Eng 32(1):1–47. https://doi.org/10.1515/revce-2015-0034

Buxton GV, Greenstock CL, Helman WP, Ross AB (1988) Critical review of rate constants for reactions of hydrated electrons, hydrogen atoms and hydroxyl radicals (OH/O) in aqueous solution. J Phys Chem Ref Data 17(2):513–886. https://doi.org/10.1063/1.555805

Calza P, Sakkas VA, Medana C, Vlachou AD, Dal Bello F, Albanis TA (2013) Chemometric assessment and investigation of mechanism involved in photo-Fenton and TiO2 photocatalytic degradation of the artificial sweetener sucralose in aqueous media. Appl Catal B 129:71–79. https://doi.org/10.1016/j.apcatb.2012.08.043

Catalkaya EC, Sengul F (2006) Application of Box-Wilson experimental design method for the photodegradation of bakery's yeast industry with UV/H2O2 and UV/H2O2/Fe(II) process. J Hazard Mater 128(2–3):201–207. https://doi.org/ https://doi.org/10.1016/j.jhazmat.2005.07.052

Chakma S, Moholkar VS (2014) Investigations in synergism of hybrid advanced oxidation processes with combinations of sonolysis + Fenton process + UV for degradation of bisphenol A. Ind Eng Chem Res 53(16):6855–6865. https://doi.org/10.1021/ie500474f

Chandra R, Kumar V (2015) Biotransformation and biodegradation of organophosphates and organohalides. In: Chandra R (ed) Environmental waste management. CRC Press, Boca Raton, pp 475–524

Chandra R, Kumar V (2018) Phytoremediation: a green sustainable technology for industrial waste management. In: Chandra R, Dubey NK, Kumar V (eds) Phytoremediation of environmental pollutants. CRC Press, Boca Raton, pp 1–42

Chelme-Ayala P, El-Din MG, Smith DW (2010) Degradation of bromoxynil and trifluralin in natural water by direct photolysis and UV plus H2O2 advanced oxidation process. Water Res 44(7):2221–2228. https://doi.org/ https://doi.org/10.1016/j.watres.2009.12.045

Chen X, Chen G, Yue PL (2003) Anodic oxidation of dyes at novel Ti/B-diamond electrodes. Chem Eng Sci 58(3–6):995–1001. https://doi.org/10.1016/S0009-2509(02)00640-1

Christensen H, Sehested K, Corfitzen H (1982) Reactions of hydroxyl radicals with hydrogen peroxide at ambient and elevated temperatures. J Phys Chem 86:1588–1590. https://doi.org/10.1021/j100206a023

Cintas P, Luche J-L (1999) Green chemistry. A sonochemical approach. Green Chem 1:115–125. https://doi.org/10.1039/A900593E

Comninellis C, De Battisti A (1996) Electrocatalysis in anodic oxidation of organics with simultaneous oxygen evolution. Journal de Chimie Physique et de Physico-ChimieBiologique 93:673–679. https://doi.org/10.1051/jcp/1996930673

Dalrymple OK, Yeh DH, Trotz MA (2007) Removing pharmaceuticals and endocrine-disrupting compounds from wastewater by photocatalysis. J Chem Technol Biotech 82(2):121–134. https://doi.org/10.1002/jctb.1657

Deng J, Jiang J, Zhang Y, Lin X, Du C, Xiong Y (2008) FeVO4 as a highly active heterogeneous Fenton-like catalyst towards the degradation of Orange II. Appl Catal B 84(3–4):468–473. https://doi.org/10.1016/j.apcatb.2008.04.029

Deng Y, Zhao R (2015) Advanced oxidation processes (AOPs) in wastewater treatment. Curr Poll Reports 1(3):167–176. https://doi.org/10.1007/s40726-015-0015-z

Derakhshan M, Fazeli M (2018) Improved biodegradability of hardly decomposable wastewaters from petrochemical industry through photo-Fenton method and determination of optimum operational conditions by response surface methodology. J Biol Eng 12:10. https://doi.org/10.1186/s13036-018-0104-9

Gagol M, Przyjazny A, Boczkaj G (2018) Wastewater treatment by means of advanced oxidation processes based on cavitation—a review. Chem Eng J 338:599–627. https://doi.org/10.1016/j.cej.2018.01.049

Garcı´a-Espinoza JD, Mijaylova-Nacheva P, Aviles-Flores M (2018) Electrochemical carbamazepine degradation: effect of the generated active chlorine, transformation pathways and toxicity. Chemosphere 192:142–151. https://doi.org/10.1016/j.chemosphere.2017.10.147

Garg KK, Prasad B (2016) Treatment of a multicomponent aqueous solution of purified terephthalic acid wastewater by electrocoagulation process: optimization of process and analysis of sludge. J Taiwan Inst Chem Eng 60:383–393. https://doi.org/10.1016/j.jtice.2015.10.038

Gautam P, Kumar S, Lokhandwala S (2019) Advanced oxidation processes for treatment of leachate from hazardous waste landfill: a critical review. J Clean Prod 237: 117639. https://doi.org/10.1016/j.jclepro.2019.117639

Gogate PR, Pandit AB (2004) A review of imperative technologies for wastewater treatment I: oxidation technologies at ambient conditions. Adv Environ Res 8(3–4):501–551. https://doi.org/10.1016/S1093-0191(03)00032-7

Gogate PR (2008a) Treatment of wastewater streams containing phenolic compounds using hybrid techniques based on cavitation: a review of the current status and the way forward. Ultrasonic Sonochem 15(1):1–15. https://doi.org/10.1016/j.ultsonch.2007.04.007

Gogate PR (2008b) Cavitational reactors for process intensification of chemical processing applications: a critical review. Chem Eng Process 47:515–527. https://doi.org/10.1016/j.cep.2007.09.014

Gogoi A, Mazumder P, Tyagi VK, Tushara Chaminda GG, An AK, Kumar M (2018) Occurrence and fate of emerging contaminants in water environment: a review. Groundw Sustain Dev 6:169–180. https://doi.org/10.1016/j.gsd.2017.12.009

Gore MM, Saharan VK, Pinjari DV, Chavan PV, Pandit AB (2014) Degradation of reactive orange 4 dye using hydrodynamic cavitation based hybridtechniques. Ultrason Sonochem 21:1075–1082. https://doi.org/10.1016/j.ultsonch.2013.11.015

Gulyas H, von Bismarck R, Hemmerling L (1995) Treatment of industrial wastewaters with ozone/hydrogen peroxide. Water Sci Technol 32(7):127–134. https://doi.org/10.2166/wst.1995.0217

Hadavifar M, Zinatizadeh AA, Younesi H, Galehdar M (2009) Fenton and photo-Fenton treatment of distillery effluent and optimization of treatment conditions with response surface methodology. Asian-Pacific J Chem Eng 5(3):454–464. https://doi.org/10.1002/apj.313

Horakova M, Klementova S, Krız P, Balakrishna SK, Spatenka P, Golovko O (2014) The synergistic effect of advanced oxidation processes to eliminate resistant chemical compounds. Surface Coat Technol 241:154–158

Khan JA, Shah NS, Nawaz S. Ismail, M. Rehman, F, Khan HM (2015) Role of eaq, OH and H in radiolytic degradation of atrazine: a kinetic and mechanistic approach. J Hazard Mater 288:147–157. https://www.cheric.org/research/tech/periodicals/doi.php?art_seq=1384028

Khetan SK, Terrence JC (2007) Human pharmaceuticals in the aquatic environment: a challenge to green chemistry. Chem Rev 107(6):2319–2364. https://doi.org/10.1021/cr020441w

Kim K-H, Ihm S-K (2011) Heterogeneous catalytic wet air oxidation of refractory organic pollutants in industrial wastewaters: a review. J Hazard Mater 186(1):16–34. https://doi.org/10.1016/j.jhazmat.2010.11.011

Kumar V, Singh K, Shah MP (2021) Advanced oxidation processes for complex wastewater treatment. In Maulin P. Shah (ed) Advanced oxidation processes for effluent treatment plants. Elsevier, pp 1–31. ISBN 9780128210116, https://doi.org/10.1016/B978-0-12-821011-6.00001-3

Lamastra L, Balderacchi M, Trevisan M (2016) Inclusion of emerging organic contaminants in groundwater monitoring plans. MethodsX 3(16):459–476. https://doi.org/10.1016/j.mex.2016.05.008

Laxma Reddy PV, Kavitha B, Kumar Reddy PA, Kim K-H (2017) TiO2-based Photocatalytic disinfection of microbes in aqueous media: a review. Environ Res 154:296–303. https://doi.org/10.1016/j.envres.2017.01.018

Leighton TG (1995) Bubble population phenomena in acoustic cavitation. Ultrason Sonochem 2(2):S123–S136. https://doi.org/10.1016/1350-4177(95)00021-W

Li N, Descorme C, Besson M (2007) Catalytic wet air oxidation of chlorophenols over supported ruthenium catalysts. J Hazard Mater 146(3):602–609

Li W, Zhou Q, Hua T (2010) Removal of organic matter from landfill leachate by advanced oxidation processes: a review. Int J Chem Eng 270532. https://doi.org/10.1155/2010/270532

Lu MC, Zhang H, Huang YY, Wang SY (2005) Influence of inorganic ions on the mineralization of 2, 4-dinitrophenol by the Fenton reaction. Fresenius Environ Bull 14:101–104

Mabuza J, Otieno B, Apollo S, Matshediso B, Ochieng A (2017) Investigating the synergy of integrated anaerobic digestion and photodegradation using hybrid photocatalyst for molasses wastewater treatment. Euro-Mediter J Environ Integ 2:17. https://doi.org/10.1007/s41207-017-0029-6

Maldonado MI, Malato S, Perez-Estrada LA, Gernjak W, Oller I, Domenech X, Peral J (2006) Partial degradation of five pesticides and an industrial pollutant by ozonation in a pilot-plant scale reactor. J Hazard Mater 138:363–369. https://doi.org/10.1016/j.jhazmat.2006.05.058

Manickavachagam M, Sillanpaa M, Swaminathan M, Ahmmad B (2014) Advanced oxidation processes for wastewater treatment. Int J Photoenergy 2014(682767). https://doi.org/10.1155/2014/682767

Martı´nez SB, Perez-Parra J, Suay R (2011) Use of ozone in wastewater treatment to produce water suitable for irrigation. Water Res Manag 25(9):2109–2124. https://doi.org/10.1007/s11269-011-9798-x

Martins RC, Pinto FL, Castro-Silva S, Quinta-Ferreira RM (2011) Flocculation, ozonation, and fenton's process in the treatment of distillery effluents. J Environ Eng 139:110–116

Mishra KP, Gogate PR (2010) Intensification of degradation of Rhodamine B using hydrodynamic cavitation in the presence of additives. Sep Purif Technol 75:385–391. https://doi.org/10.1016/j.seppur.2010.09.008

Moreno-Casillas H, Cocke D, Gomes J, Morkovsky P, Parga J, Peterson E (2007) Electrocoagulation mechanism for COD removal. Separ Purif Technol 56:204–211. https://doi.org/10.1016/j.seppur.2007.01.031

Nair R, Patel R (2014) Treatment of dye wastewater by sonolysis process. Int J Res Mod Eng Emerg Technol 2(1)

Nidheesh PV, Gandhimathi R (2012) Trends in electro-Fenton process for water and wastewater treatment: an overview. Desalination 299:1–15. https://doi.org/10.1016/j.desal.2012.05.011

Oturan MA, Aaron J (2014) Advanced oxidation processes in water/wastewater treatment: principles and applications. Rev Crit Rev Environ Sci Technol 44(23):2577–2641. https://doi.org/10.1080/10643389.2013.829765

Oturan MA, Guivarch E, Oturan N, Sirés I (2008a) Oxidation pathways of malachite green by Fe3+-catalyzed electro-Fenton process. Appl Catal B Environ 82(3–4):244–254. https://doi.or/https://doi.org/10.1016/j.apcatb.2008.01.016

Oturan MA, Peiroten J, Chartrin P, Acher AJ (2000) Complete destruction of p-nitrophenol in aqueous medium by electro-Fenton method. Environ Sci Technol 34(16):3474–3479. htps://doi.org/https://doi.org/10.1021/es990901b

Ozdemir C, Oden MK, Şahinkaya S, Kalipc IE (2011) Color removal from synthetic textile wastewater by sono-Fenton process. Clean: Soil, Air, Water 39(1):60–67. https://doi.org/10.1002/clen.201000263

Panizza M, Cerisola G (2009) Direct and mediated anodic oxidation of organic pollutants. Chem Rev 109(12):6541–6569. https://doi.org/10.1021/cr9001319

Perez AJ, Jacquet Q, Batuk D, Iadecola A, Saubane're M, Rousse G, Larcher D, Vezin H, Doublet ML, Tarascon JM (2017) Approaching the limits of cationic and anionic electrochemical activity with the Li-rich layered rocksalt Li3IrO4. Nat Energy 2(12):954–962

Pillai KC, Kwon TO, Moon IS (2009) Degradation of wastewater from terephthalic acid manufacturing process by ozonation catalyzed with Fe2+, H2O2 and UV light: Direct versus indirect ozonation reactions. Appl Catal B- Environ 91:319–328. https://doi.org/10.1016/J.APCATB.2009.05.040

Pliegoa G, Juan A. Zazoa JA, Garcia-Muñoz PA, Munoza M, Casasa JA, Rodrigueza JJ (2015) Trends in the intensification of the Fenton process for wastewater treatment—an overview. Crit Rev Environ Sci Technol. http://dx.doi.org/https://doi.org/10.1080/10643389.2015.1025646

Poulopoulos SG, Yerkinova A, Ulykbanova G, Inglezakis VJ (2019) Photocatalytic treatment of organic pollutants in a synthetic wastewater using UV light and combinations of TiO2, H2O2 and Fe(III). PLoS One 14(5):e0216745. https://doi.org/10.1371/journal.pone. 0216745

Rajoriya S, Carpenter J, Saharan VK, Pandit AB (2016) Hydrodynamic cavitation: an advanced oxidation process for the degradation of bio-refractory pollutants. Rev Chem Eng 32(4):379–411. https://doi.org/10.1515/revce-2015-0075

Rajoriya S, Bargole S, Saharan VK (2017) Degradation of a cationic dye (Rhodamine 6G) using hydrodynamic cavitation coupled with other oxidative agents: reaction mechanism and pathway. Ultras Sonochem 34:183–194. https://doi.org/10.1016/j.ultsonch.2016.05.028

Raut-Jadhav S, Saharan VK, Pinjari DV, Saini DR, Sonawane SH, Pandit AB (2013) Intensification of degradation of imidacloprid in aqueous solutions by combination of hydrodynamic cavitation with various advanced oxidation processes (AOPs). J Environ Chem Eng 1:850–857

Rice RG (1995) Applications of ozone for industrial wastewater treatment—a review. Ozone: Sci Eng 18(6):477–515. https://doi.org/10.1080/01919512.1997.10382859

Ried A, Mielcke J, Wieland A (2009) The potential use of ozone in municipal wastewater. Ozone Sci Eng 31(6):415–421. https://doi.org/10.1080/01919510903199111

Ronquillo MG, Hernandez JCA (2017) Antibiotic and synthetic growth promoters in animal diets: review of impact and analytical methods. Food Control 72:255–267. https://doi.org/10.1016/j.foodcont.2016.03.001

Saharan VK, Pandit ABS, Kumar PS, Anandan S (2012) Hydrodynamic cavitation as an advanced oxidation technique for the degradation of Acid Red 88 dye. Ind Eng Chem Res 51(4):1981–1989. https://doi.org/10.1021/ie200249k

Saharan VK, Badve MP, Pandit AB (2011) Degradation of reactive red 120 dye using hydrodynamic cavitation. Chem Eng J 178:100–107. https://doi.org/10.1016/j.cej.2011.10.018

Sangave PC, Pandit AB (2004) Ultrasound pre-treatment for enhanced biodegradability of the distillery wastewater. Ultras Sonochem 11(3–4):197–203. https://doi.org/10.1016/j.ultsonch.2004.01.026

Scott JP, Ollis DF (1995) Integration of chemical and biological oxidation processes for water treatment: review and recommendations. Environ Prog 14(2):88–103. https://doi.org/10.1002/ep.670140212

Shah NS, Khan JA, Nawaz S Khan HM (2014) Role of aqueous electron and hydroxyl radical in the removal of endosulfan from aqueous solution using gamma irradiation. J Hazard Mater 278:40–48. https://doi.org/10.1016/j.jhazmat.2014.05.073

Sharma VK, Feng M (2017) Water depollution using metal-organic frameworks-catalyzed advanced oxidation processes: a review. J Hazard Mater 372:3–16. https://doi.org/10.1016/j.jhazmat.2017.09.043

Stasinakis A (2008) Use of selected advanced oxidation processes (AOPs) for wastewater treatment—a mini review. Global NEST J 10(3):376–385

Stuber F, Font J, Fortuny A, Bengoa C, Eftaxias A, Fabregat A (2005) Carbon materials and catalytic wet air oxidation of organic pollutants in wastewater. Top Catal 33(1–4):3–50. https://doi.org/10.1007/s11244-005-2497-1

Thanekar P, Murugesan P, Gogate RP (2018) Improvement in biological oxidation process for the removal of dichlorvos from aqueous solutions using pretreatment based on hydrodynamic cavitation. J Water Proc Eng 23:20–26. https://doi.org/10.1016/j.jwpe.2018.03.004

Wang A, Guo W, Hao F, Yue X, Leng Y (2014) Degradation of Acid Orange 7 in aqueous solution by zero-valent aluminum under ultrasonic irradiation. Ultrason Sonochem 21(2):572–575. https://doi.org/10.1016/j.ultsonch.2013.10.015

Wang C Yediler A, Lienert D, Wang Z, Kettrup A (2003) Ozonation of an azo dye C.I. Remazol Black 5 and toxicological assessment of its oxidation products. Chemosphere 52:1225–1232. https://doi.org/10.1016/S0045-6535(03)00331-X

Wang J, Chu L (2016) Irradiation treatment of pharmaceutical and personal care products (PPCPs) in water and wastewater: an overview. Radiat Phys Chem 125:56–64. https://doi.org/10.1016/j.radphyschem.2016.03.012

Wang JL, Xu LJ (2011) Advanced oxidation processes for wastewater treatment: formation of hydroxyl radical and application. Crit Rev Environ Sci Technol 42(3):251–325. https://doi.org/10.1080/10643389.2010.507698

Wang S, Wang J (2018) Degradation of carbamazepine by radiation-induced activation of peroxymonosulfate. Chem Eng J 336:595–601. https://doi.org/10.1016/j.cej.2017.12.068

Wang X, Jia J, Wang Y (2017) Combination of photocatalysis with hydrodynamic cavitation for degradation of tetracycline. Chem Eng J 315:274–282. https://doi.org/10.1016/j.cej.2017.01.011

Xu LJ, Chu W, Graham N (2014) Atrazine degradation using chemical-free process of USUV: analysis of the micro-heterogeneous environments and the degradation mechanisms. J Hazard Mater 275:166–174. https://doi.org/10.1016/j.jhazmat.2014.05.007

Yazdanbakhsh A, Eslami A, Moussavi G, Rafiee M, Sheikhmohammadi A (2018) Photo-assisted degradation of 2, 4, 6-trichlorophenol by an advanced reduction process based on sulfite anion radical: degradation, dechlorination and mineralization. Chemosphere 191:156–165. https://doi.org/10.1016/j.chemosphere.2017.10.023.

Zazo JA, Pliego G, Blasco S, Casas JA, Rodriguez JJ (2010) Intensification of the Fenton process by increasing the temperature. Ind Eng Chem Res 50(2):866–870. https://doi.org/10.1021/ie101963k

Zhang H, Zhang D, Zhou J (2006) Removal of COD from landfill leachate by electro-Fenton method. J Hazard Mater 135(1–3):106–111. https://doi.org/10.1016/j.jhazmat.2005.11.025

Zhang M, Dong H, Zhao L, Wang D, Meng D (2019) A review on Fenton process for organic wastewater treatment based on optimization perspective. Sci Total Environ 670:110–121. https://doi.org/10.1016/j.scitotenv.2019.03.180

Chapter 4
Microbial Aspect in Wastewater Management: Biofilm

Rajani Sharma and Shubha Rani Sharma

Abstract Water, a universal solvent, is a significant component of biotic and abiotic factors. The increasing population is creating the advancement of industries and agriculture. But this advancement is also affecting the environment. Each sector needs to dispose of its waste product, and most of these are polluting surface and groundwater. The discharge includes hazardous pollutants, including heavy metals, radioactive elements, organic as well inorganic waste. This pollutant also causes eutrophication, inviting the growth of algae and bacteria in surface water, increasing the oxygen demand, and disturbing its physical and chemical properties. Water is also a natural habitat of many organisms and even supports life in the lithosphere—water pollution affects both animal and plant kingdom. Extra care is needed while purification of water by various chemical and physical methods.

Further, the chemicals may affect the microenvironment. Biodegradation can overcome the problem of pollution by chemicals. It is convenient as well as safer to use. Bioaugmentation is popularly used for decades to enhance the degradation procedure. The microbes metabolize the pollutant into a less toxic product. This chapter deals with these microbes and their mechanism involved in treating wastewater.

Keywords Wastewater treatment · Biofilm · Biodegradation · Bioremediation · Bioaugmentation · Activated sludge · Exopolysaccharides

Abbreviations

BOD	Biochemical Oxygen Demand
CFIC	Continuous Flow Intermittent Cleaning

R. Sharma (✉)
Department of Biotechnology, Amity University Jharkhand, Ranchi 834002, India
e-mail: rsharma@rnc.amity.edu

S. R. Sharma
Department of Bio-Engineering, Birla Institute of Technology, Mesra, Ranchi 835215, India
e-mail: srsharma@bitmesra.ac.in

S. Roy et al. (eds.), *Advanced Industrial Wastewater Treatment and Reclamation of Water*, Environmental Science and Engineering,
https://doi.org/10.1007/978-3-030-83811-9_4

COD	Chemical Oxygen Demand
CPHEEO	Central Public Health and Environmental Engineering Organization
DO	Dissolved Oxygen
EPS	Exopolysaccharides
FBR	Fluidized-Bed Reactors
HRTs	Hydraulic Loading Rates
HyVAB	Hybrid Vertical Anaerobic Biofilm Reactor
MBBR	Moving Bed Biofilm Reactors
MBR	Membrane Biofilm Reactors
MFCs	Microbial Fuel Cells
PAHs	Polycyclic Aromatic Hydrocarbons
SS	Suspended Solids
TC	Total Coliform
TF	Trickling Filter

4.1 Introduction

Water, an indispensable compound for life, is composed of hydrogen and oxygen. It exists in all three forms- solid, liquid, and gaseous. Water constitutes 71% of the Earth's part and 65–79% of the human body. It plays a vital role in the evolution as well as the survival of life. Due to its importance, it is a global concern for the conservation of water. The major challenge for water conservation is increasing population, rapid urbanization, and industrialization (Hassan Rashid et al. 2018). The said factors are continuously improving the demand along with hampering water quality. A living body is affected mainly by water's physiological and chemical properties (Riveros-Perez and Riveros 2018). Industrialization, municipal discharge, chemicals used in agriculture, and household discharge are the significant factors disturbing the physical and chemical properties of water. A report by the UN says that the primary concern is the treatment of only 20% of the sewage before disposal (WWAP 2017). The industrial discharge is rich in nitrogen and phosphorous, which can be reused for irrigation purposes.

But along with this, nutrition wastewaters may also contain heavy metals or other toxins (Almuktar et al. 2018). As per the Central Public Health and Environmental Engineering Organization (CPHEEO), the permissible organic load ranges from 11.0 to 28.0 kg/ha/day. This is required to maintain the microenvironment and toxic level of nutrition in soils and plants, respectively (Schellenberg et al. 2020). This necessitates the primary treatment and chemical as well as physical analysis even before reuse. Untreated sewage can disrupt soil pH and salinity, resulting in soil clogging and increased permeability of heavy metals. The presence of total suspended solids in the sewage may lead to soil clogging, leading to anaerobic conditions (US EPA. Overview of Watershed Monitoring 2019). Disturbance of pH causes various metals'

Table 4.1 Indian act to protect environment against Sewage disposal

Environmental act/policies	Year	Function
Water prevention and control of pollution act	1974	First regulation for the prevention and control of water pollution and led to the establishment of responsible bodies at the central and state level for implementation
Environment protection act	1986	Targeting protection and improvement of the broader human environment
National urban sanitation policy	2008	Total coverage of sanitation in all Indian cities and towns
Ministry of Jal Shakti	2019	Overall planning, policy formulation, coordination, and guidance for water resources

solubilization, increasing their uptake by plants causing toxicity (WHO 2016). Industrial waste is also rich in many salts, which affect the electrical conductivity of the soil. So, direct disposal of the sewage in agricultural land causes increased absorption of ions, which is toxic to plants (Beltrán 1999).

Britain was the first nation to pay concern about the industrial sewage discharge in water bodies. They focused on pollutant minimization instead of reusing wastewater and formulated the regulation to minimize pollutants' effect. As per their regulation, the sewage should be diluted to reduce the impact of contaminants. The Royal Commission on Sewage Disposal (1898) set the permissible limit for Biochemical Oxygen Demand (BOD) and suspended solids (SS) to 20 and 30 mg/l, respectively. Indian regulation has mainly focused on reusing wastewater in fishery, toilet flushing, floor cleaning, horticulture, forestry, and industrial use (CPHEEO 2013). Table 4.1 represents a few of the acts and policies as an initiative for water pollution in India.

The various regulations mentioned in the Table 4.1 were also meant for awareness. "Water for Life" (2005–2015) by the UN office is the best knowledge hub for understanding related to sustainable water management and sanitation (UNDESA 2015). But along with the individual effort, there is the requirement of large-scale sensitization and treatment process.

The unique property of water that is 'Universal Solvent' contributes to further toxicity of the water. As per WHO, polluted water constitutes approximately 30–40% of disease and death (WHO 2009). Typhoid, cholera, and giardia are common diseases caused by polluted water. Not only humans but plants are also equally influenced by contaminated water. Pollutants from water accumulate in the soil absorbed by the plant roots and increase bacterial and fungal infections (Chibuike and Obiora 2014). Ruminants still occupy a significant position in the agricultural field, badly influenced by the polluted water. Pollution affects the thyroid gland, cardiovascular system, reproductive organ, and immune system (Rhind et al. 2010) and affects the ruminants' milk productivity.

4.2 Sources Causing Pollution in Water

Increasing population is the primary cause of water pollution. The alarming population is causing a burden over land and the general necessities of life. Several chemical and technical industries were established to satisfy the hunger and comfort of the increasing population. A study conducted on the Kelani river in Sri Lanka estimated that the growing population in the watershed area causing increased biochemical oxygen demand (BOD), dissolved oxygen (DO), and total coliform (TC) with approx. correlation coefficient 0.7 (Liyanage and Yamada 2017). Figure 4.1 represents the current scenario of pollution across the world concerning their population. The waste discharge from this industry directly mixes up with water sources. Table 4.2 present the list of pollutants and diseases caused by them in humans. Water is also a niche for various organisms that also affect this pollution factor, causing imbalance to the ecosystem.

Heavy metals are the most challenging pollutant as this is nonbiodegradable. Even a micro amount can be hazardous. Waste from ore mines, tannery, and burning of fossil fuels is the major source of heavy metal pollutants (Tchounwou et al. 2012). Its ionic property makes it more reacting and hence a matter of concern. The worst part is that its solubility increases with the decreasing water pH hence more toxic in

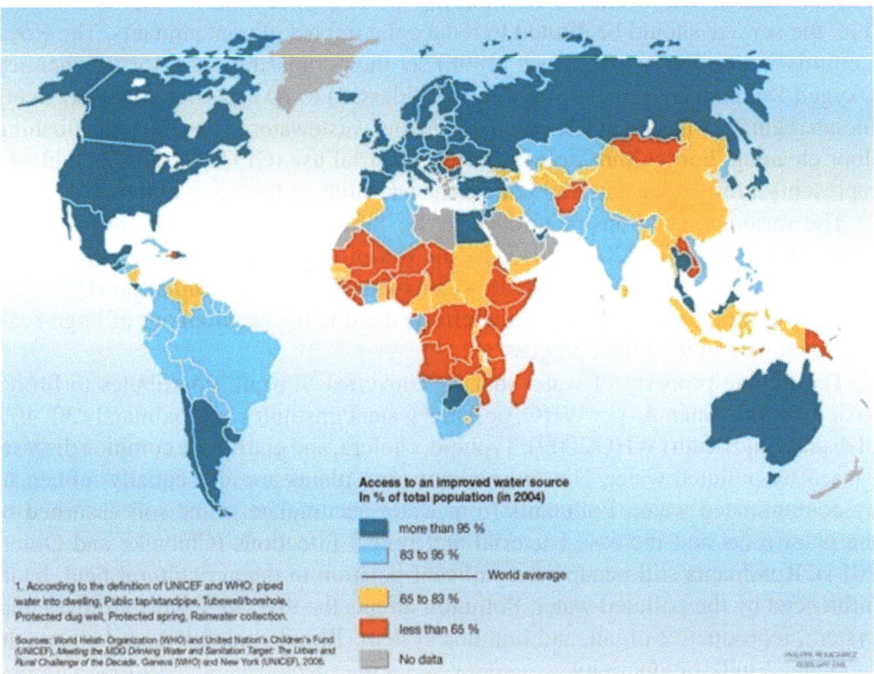

Fig. 4.1 Current scenario of water pollution due to increasing pollution (https://earthhabitat.files.wordpress.com/2010/02/total-population-access-to-an-improved-water-source.jpg)

Table 4.2 Table presenting the pollutant in water and their hazardous effect

Water pollutant	Hazardous effect	Microbes involved	Mechanism	Reference
Chromium (Cr)	Oxidative damage and denaturation of microorganisms as well as weakening the bioremediation capacity of microbes	*Klebsiella pneumoniae* and *Mangrovibactery-ixingensis*	Internalize Cr(VI) and reduce to Cr (III) along with activating enzyme involved in DNA repair mechanism	(Sanjay et al. 2020)
Cadmium (Cd)	Damage nucleic acid and cell membrane	*Pseudomonas aeruginosa*	Microbes act as biosorbent	(Chellaiah 2018)
Copper (Cu)	Production of ROS via Fenton and Haber-Weis reactions, which will act as soluble electron carriers. This can cause severe injury to cytoplasmic molecules, DNA, lipids, and other proteins	*Escherichia coli* and *Salmonella enterica, Salmonella ancillary*	Reduces toxicity by secreting metallochaper-ones	(Pontel et al. 2014; Giachino and Waldron 2020)
Lead (Pb)	Damage cell membranes, and destroy the structure of DNA	*Corynebacterium, Pseudomonas, Cupriavidusmetal-lidurans*	Expresses leadbinding proteins which detoxify the Pb and assist in its metabolism	(Kalita and Joshi 2017)
Nuclear waste	Causes genetic mutation leading to cancer, birth and reproductive defects	*Bacillus sphaericus, Pseudomonas spp., Deinococcus-radioduran*	Bioaccumulation	(Phillips et al. 1995)
Pesticides and herbicides	Leukemia, brain cancer, effect reproductive organ	*Pseudomonas, Rhodococcus* and *Achromobacter, Streptomyces, Arthrobacter*, and *Achromobacter,* fungus *Trame-tesversicolor, Botrytis cinerea*	Exctrete extracellular enzyme which degrade the organic compound into inorganic constituents	(Briceño et al. 2020)

(continued)

Table 4.2 (continued)

Water pollutant	Hazardous effect	Microbes involved	Mechanism	Reference
Oil spills	Liver and lung damage, cancer,	*Pseudomonas spp., Marinobacter, Bacillus spp*	Utilize hydrocarbons as a sole source of energy	(Xu et al. 2018)
Sewage discharge	Hepatitis, cardiovascular degeneration, muscular, respiratory disorder	*Rhodospirilaceae spp.* and *Caldilineaceae spp*	Metabolize as a source of energy and convert to the less toxic compound	(Zhang et al. 2018)
Domestic waste	The flammable, reactive, toxic content causes respiratory disorder, skin and eyes irritation, fever, headache, musculoskeletal damage	*Clostridium* sp., *Aspergillus, Trichoderma, Sclerotium, Cellulomonas, Pseudomonas, Bacillus spp.* and *Thermoaction-mycetes*	Recycle the minerals by metabolizing	(Gutberlet and Uddin 2017)
Septic tank	Gastrointestinal disorder, worm infection, and hepatitis	*Bifidobacterium spp., Actinomyces spp., Clostridium spp., Propionibacterium spp.*, and *Peptostrepto-coccus spp*	Degrade the organic materials by enzymatic action into inorganic content	(Shaw and Dorea 2021)

freshwater. It accumulates in the living organism causes a long-term effect. Mercury toxic in fish well explain the situation (Wuana and Okieimen 2011). The radioactive elements present in the water bodies are even more challenging than heavy metals (Al-Akeel 2018). Nuclear thermal plants and pharmaceutical industries are the primary sources of radioactive pollutants.

The agricultural industry is a boon for us, but they equally contribute to water pollution. Increasing demand has elevated the role of chemical fertilizer, pesticides, and other chemicals. These chemicals rush to the water bodies through running water. Few of these chemicals can even leach up to the underground water. Again, animal discharge dumping engaged in agriculture causes eutrophication in water bodies (Moss 2008). Eutrophication supports many parasites' growth, increasing the oxygen demand and affecting the organisms present in water bodies (Dincer and Abu-Rayash 2020).

Soil erosion is also one of the factors which affect agriculture as well as water quality. The running water, which can be even rainwater or flow out from houses and factories, mainly causes pollution. They directly transport the topsoil to the water bodies. The topsoil may be consisting of a pollutant from medical, industrial, and household waste (Aqeel et al. 2014; Ayilara et al. 2020). Even the flow of pesticides and other chemical-based products used in agriculture goes from farmland to nearby water bodies. These water bodies may be the drinking water source for life around those areas, including both animals and humans.

4.3 Physical and Chemical Methods in Wastewater Treatment

The government has adopted various wastewater treatment methodologies to overcome polluted water's hazardous effect on the ecosystem. Physical and chemical methods are in use for decades but need extra care and precaution. Further, it requires expert handling. Such a technique uses traditional methods, including filtration, flocculation, ion exchange resins, electrochemical treatment, evaporation, and activated charcoal. Such a treatment process includes neutralization of the water pH (Crini and Lichtfouse 2019; Gurreri et al. 2020). Lime (CaO), soda ash (Na_2CO_3), and calcium hydroxide ($Ca(OH)_2$) are commonly used to neutralize the acidic condition, whereas sulfuric acid (H_2SO_4) and carbonic acid (H_2CO_3) proved promising in controlling the basic pH (Kapil and Bhattacharyya 2017). Soluble waste from wastewater is removed through absorption using resins, activated charcoal, hydroxides, and alumina. Charged particles are a prevalent pollutant in industrial waste which can be removed using ion exchange.

Some biotic pollutants are also present in water, including protozoa, bacteria, fungus, etc., which are disinfected using physical and chemical methods. Chlorine, ozone, UV, and bromine treatment are well-established methods. These techniques were effective decades back, but the increasing consumption has burdened such methodologies. New technologies like the use of reverse osmosis, ultrasound, and nanoparticles are used nowadays. The implementation of these techniques largely depends on the quality as well as quantity of the pollutant.

4.3.1 Disadvantages of Physical and Chemical Treatment

The physical and chemical treatment process also suffers from many disadvantages. Along with the high maintenance cost and energy requirement, it is inefficient against metal pollutants. One of the draw backs is that the chemicalsused during the process cannot be reused again. The reactions are affected by the change in pH. Another major challenge is that these processes require pre-treatment. The intermediates formed during the chemical treatment may contain volatile or aromatic amines. Physical methods like filtration and adsorption are pretty non-specific and need high maintenance costs. Also, the regeneration of the material used for filtration and adsorption is expensive. Activated carbon which is used for adsorption, suffers from its loss during regeneration. Above all, such methods are inefficient against the pollutant's significance (Crini and Lichtfouse 2019).

4.4 Microbial Aspect of Wastewater Treatment

The mentioned disadvantages of physical and chemical treatment have intrigued the researcher to look for natural, economic, and efficient processes. There are various organisms, including bacteria, algae, fungi, and plants which can utilize the waste products in their metabolism process. The metabolized product is either less toxic or non-toxic. The metabolic activity of the microbes is further enhanced by microbial engineering. This chapter mainly focused on the microbes forming biofilm and their utility in wastewater management. Biofilm has been even thriving in removing heavy metals, nitrates, sulfates, quinolines, etc. Biofilm is equally valuable and harmful. A side of the coin says that it makes microbes resistant to antibiotics and the host immune system, while the other side can treat wastewater.

Extracellular polymeric substances are the primarybiofilm component composed of proteins, DNA, RNA, lipids, and carbohydrates. Along with adhesion and cohesion, biofilm also provides mechanical strength promoting microbial colonization (Jiao et al. 2010; Di Martino 2018). Depending upon biofilm, biological treatment can be of dispersed growth system or attached growth system. In a dispersed growth system, the bacteria remain attached to the liquid's whole surface but establish less contact with the substrate. The dispersed growth system depends on hydraulic retention time, which should be greater than bacterial doubling time (ElNaker et al. 2018). The attached growth system, which uses biofilms, has much more extensive contact with the substrate, leading to a more efficient organic waste metabolism. In this system, microbes forming biofilm are attached to any surface and can be used for water treatment. This system's significant advantages are that the support system can be liquid, water itself, or solid, including natural (rock) and artificially synthesized. The biofilm-based system has self-cleaning activity and efficiently removes nitrates, ammonia, and heavy metals. It is best suited for treating wastewater with BOD/COD (Chemical Oxygen Demand) ≥ 0.5 (Srinivasa Raghavan et al. 2017). Biofilm has proved effective even in running water as in the river. Another positive aspect is that it supports the coexistence of microniches (Davey and O'toole 2000). Biofilm can accumulate major pollutants from running water, including heavy metals. The accumulated contaminant is transformed into a less toxic substance or absorbed during metabolic activities. Biodegradation, bioaccumulation, biosorption, biomineralization, and bio-immobilization of the toxins by the biofilms can be used for wastewater treatment.

4.4.1 Biofilm Based Bioreactors

Application of biofilm in wastewater treatment is not limited to bacteria; instead, other microbes, including algae, fungi, and yeast, are also preferred in many studies. They are well-chosen for organic wastewater treatment and proved more economical when compared to the chemical and physical methods. Another advantage of biological therapy is that it can be used in both aerobic and anaerobic environments. Anaerobic treatment is preferred when the contaminant is high with elevated COD concentration. But it cannot always be used because it does not support microbial growth and inefficient in flocculation, which produces a low quality of effluent. The aerobic bioreactor has advantages over these drawbacks. The biofilm-based bioreactor has been proved efficient in mass transfer, self-regenerating system, low sludge production, and reduced membrane cost.

There are five types of bioreactor (i) trickling filter (TF); (ii) microbial fuel cells (MFCs); (iii) fluidized-bed reactors (FBR); (iv) moving bed biofilm reactors (MBBR), and (v) membrane biofilm reactors (MBR).

4.4.2 Trickling Filter (TF)

TF consist of a layer of microbial slime over the fixed bed. The bed may be slag, rock, coke, etc. But now a day plastic is more preferred (Wilson et al. 2015). The wastewater is evenly distributed with the help of a rotator fixed at the top of the TF. During the water springing over the bed, the air is sucked from the reactor's bottom by the chimney effect. This process maintains the aerobic condition inside the reactor. The reactor is specialized for organic and ammonia treatment. It requires comparatively less maintenance than suspended culture (Tekerlekopoulou et al. 2020).

Along with bacterial biofilm, fungal and algal growth in the trickling filter play a significant role. The presence of fungi facilitates filtration even at the lower pH required for industrial waste treatment (Scholz 2016; Zhang et al. 2020). Whereas algal growth keeps the upper systems aerated. Unlike the chemical and physical method of treatment, the system itself replaces the biofilm layers. The continuous growth of bacteria over the fixed bed increases biofilm thickness, which prevents the passage of oxygen and organic matter to the innermost layer of biofilm. The algal growth loosens the attachment of biofilm to the fixed bed, due to which the water circulation washes the biofilm layer, and new biofilm is grown. The trickling filter also contains protozoa and rotifers, which feed on bacteria, algae, and fungi, maintaining the culture surface and preventing the accumulation of inactivated biofilm (Cooke 1959). A microscopic study reveals the three microbial layers over the surface of the solid bed. The innermost layer is invaded by bacteria, followed by protozoa and algae (Mack et al. 1975). The filter is efficient in treating domestic wastewater and industrial wastewater through low rate and high-rate trickling filters. The significant bacteria genus isolated from the filter is *Achromobacter, Alcaligens, Flavobacterium, Pseudomonas, Klebsiella, Proteus, Citrobacter Zooglea* (Jia and Zhang 2020). Like all the processes, it also suffers from a major disadvantage which is clogging.

4.4.3 Microbial Fuel Cells (MFCs)

MFCs serve two purposes, along with wastewater management it also generates energy. This type of reactor has two compartments- anode and cathode. Specific microbes which can produce energy during the metabolism of the waste substrate are used here. At the anode, the organic wastes are oxidized by microbial activity, and the electron generated is transferred to the cathode side, where electron acceptors are reduced (O_2). This flow of electrons generates energy. The chamber can also be modified to produce dihydrogen (H_2) instead of electricity. The MFC is also efficient in the recovery of methane (CH_4) and heavy metals. But the major disadvantage is that there need to manage the size of MFCs reactor because increasing size decreases the efficiency of electricity production (Cao et al. 2019). Another disadvantage is that the anode used in this cell should be biocompatible, which prevents highly conducting metals, which inhibits bacteria, such as steel, platinum, and gold (Ezziat et al.) Porous surface is proved more compatible for bacterial growth. Hence, carbon-based material including graphite and carbon paper is used. With the advancement of techniques and the introduction of nanotechnology, the carbon nanotube is more preferred.

On the other hand, platinum-based carbon is used. But the high cost of platinum has also directed the researcher to use a cheaper and efficient alternative. Even cathode biofilm is preferred to catalyze the reduction reaction. In these cells, the microbes efficient in electron transport are preferred. These include *Pseudomonas spp., Micrococcus spp., Shewanella spp., Thiobacillus, Clostridium spp.,* and Ng et al. (2017).

4.4.4 Membrane Biofilm Reactor (MBR)

MBR has proved promising against both municipal and industrial wastewater treatment for decades. It is preferred over the conventional method due to the lower production of net sludge and high effluent quality. This is in use since the nineteenth century and consumes less energy hence proved economic (Ivanovic and Leiknes 2012). But like other methods, it also suffers from certain anomalies. There is a need for maintenance costs due to fouling and clogging of the membrane. In the twentieth century, MBR was tagged with airlift to minimize membrane exchange duration but still do not cut down the maintenance cost.

Further, to minimize these disadvantages, researchers are switching to biofilm-based MBR instead of activated sludge. Initially, a combination of both biofilm and activated sludge is used. This has drastically improved the quality by reducing solid retention time and enhanced the nutrition removal. Very soon, Activated Biofilm-MBR and Pure Biofilm-MBR were preferred. This maintains the quality of the effluent, too (Du et al. 2020). The addition of biofilm to the system has cut down the solid retention time by increasing the surface area. Nitrogen and phosphorous recovery rate have improved on the introduction of biofilm to the conventional activated sludge membrane. *Proteobacteria, Nitrosomonas,* and *Nitrobacter* are commonly used microbes in MBR (Baek and Pagilla 2009).

4.4.5 Fluidized-Bed Reactors (FBR)

Another bioreactor, FBBR, also serves a similar technique but is preferably used for tertiary denitrification of municipal wastes. It is wholly based on biofilm and works under anaerobic conditions. There is a continuous flow of effluent and influent, which creates turbulence and provides maximum contact with the biofilm (Leion et al. 2018; Shim et al. 2020). Either gas or liquid medium can be used to fluidize the reactor. Like MFCs, it also produces hydrogen gas as a byproduct which can be further utilized (Timmerberg et al. 2020). As it is wholly based on biofilm, the EPS secreted by the microbes make it compatible for both higher and lower rate hydraulic loading rates (HRTs). Its major advantages are uniform heat and mass exchange. This creates a better choice over a fixed bed reactor. Like other reactors, it also used nitrogen and phosphorous metabolizing microbes.

4.4.6 Moving Bed Biofilm Reactors (MBBR)

MBBR is in use as an efficient wastewater treatment reactor since the nineteenth century. The reactor creates different layers of oxygen from aerobic to anaerobic conditions. This favors nitrification and denitrification under a single platform (Qiqi et al. 2012). This technique can be easily integrated with other newly developed and conventional methods, like continuous flow intermittent cleaning (CFIC), activated sludge, hybrid vertical anaerobic biofilm reactor (HyVAB), and anaerobic digestor to increase its efficiency (Islam 2020). Being biofilm a fundamental component in this reactor, it produces less sludge and has higher biomass retention and lower hydraulic retention time. Like another biofilm-based reactor, it also has the advantage of a self-regenerating cleaning system and a higher absorption rate.

4.5 Activated Sludge

Activated sludge is a biological aid to treat wastewater. The microbial growth is activated by growing in an aerated tank and then transferred to the sewage system. The microbes feed on organic waste and produce high-quality effluent. The microbial action also led to the flocculation of organic matter and solid particles present in the wastewater, leading to these waste settlements (Li et al. 2020). The process needs proper aeration to keep solid particles suspended.

4.6 Film and Floating Plants

Floating plants or wetlands are designed to treat wastewater under both aerobic and anaerobic conditions. In this technique, the plants are grown on a buoyant raft. The root touches the surface of wastewater, increasing the surface area for absorption of harmful molecules. Their root exudate establishes a symbiotic relationship with the biofilm-producing microbes. Plant roots provide a surface for the attachment to the microbes, while microbes increase the plants' bioavailability. The technique has been proved promising in industrial wastewater, sewage treatment, and eutrophic water (Almuktar et al. 2018). The associated microbes with the roots assimilate the organic and inorganic compounds present in the wastewater reducing the BOD and COD. *Staphylococcus aureus, Pseudomonas aeruginosa, Cyanobacteria, Rhizobium spp., Nitrosomonas spp., Azotospirillum, Klebsiella, Vibrio,* and *Bacillus subtilis* are commonly supported colonies by the floating roots.

Such a system has also been proved efficient in heavy metals removal. Microbes' action on the heavy metals increases its bioavailability to plants in the association of phytochelatins and metallothionein.

4.7 Heavy Metal Remedy

Heavy metals are boon to industries but equally affect the environment. Even traces of heavy metals can disrupt the cell or damage their gene. Biofilm has become a preferable choice for heavy metal bioremediation (Ojuederie and Babalola 2017). Though heavy metals are hazardous even for microbes, engineered microbes or mutated ones are adapted to metabolize the heavy metals. Arsenic, mercury, and nickel are widespread heavy metals that affect enzyme activity, whereas microbes' growth is affected by chromium, copper, selenium, zinc, etc. Bioremediation of heavy metals is highly influenced by heavy metals' concentration, oxidation state, pH, and temperature. Microbes have adapted themselves to metabolize heavy metals. They secrete EPS to form biofilm and a metal-binding cysteine-rich protein known as metallothioneins to adapt for heavy metals metabolism (Medfu Tarekegn et al. 2020). They also utilize their internal environment, like enzymes' use to make microbes compatible for metal oxidation state, methylation, and permeability, etc. A study says that *E. coli* secrets metallo-regulatory proteins can accumulate cadmium toxicity (Reyes-Caballero et al. 2011).

4.8 Biofilm Controls Oil Spill Pollution

Oil spillage is the primary concern for marine as well as human beings. It led to the accumulation of polycyclic aromatic hydrocarbons (PAHs) in the food chain (Honda and Suzuki 2020). Oils also contain alkanes and naphthenes, which are also a matter of concern. Physical and chemical methods, including chemical oxidation, photooxidation, and volatilization, are not efficient in removing PAHs (EFSA 2012). Failing of physical and chemical methods always direct to use the natural or biological process. A marine ecosystem consists of multiple microbes which metabolize hydrocarbons. But the major drawback is the bioavailability of organic waste to the microbes. Chemotactic microbes overcome the limitation with biofilm-forming tendency. *Pseudomonas spp., Marinobacter, Bacillus spp.* are commonly isolated microbes involved in metabolizing hydrocarbon as a sole source of carbon. They release carbon dioxide (CO_2) and water (H_2O) as a byproduct that is, non-toxic (Xu et al. 2018).

Not only marine life rather soil is also polluted with petroleum. At this level, evena biological process is preferred. Multiple microbes are involved in the degradation process, including native bacteria, engineered microbes, and MFCs. Petroleum contamination to the soil causes an alteration in soil pH and carbon: nitrogen ratio (Khalil et al. 2005). This influences fertility as well as an abiotic factor present in the soil. The slower process of hydrocarbon breakdown led to the long-term effect.

4.9 Future Aspect

Waste water management is a major challenge for both developed as well as developing countries. With increasing demand there is the requirement for resources recycling. There are various physical and chemical methods for waste water treatment but both has its own anomalies. The chapter highlighted the biological aspect of waste water treatment which is economic and have less hazardous effect. The microbes need less resources for the survival and uses the waste material itself as the source of nutrition. This serves the most preferred future aspect to use microbes for waste water treatment in economic manner. Further its byproduct are less toxic and can be used in agricultural application. Microbes metabolise the waste at physiological pH which do not effect the other symbiotic organisms. The microbial technique need less training and manpower which is suited for its easy set up.

4.10 Conclusion

The increasing population has burdened the ecosystem which is now a global concern. To satisfy the elevating demand, industrialization and urbanization has polluted our atmosphere. Water pollution is also one of them. To fight against this challenge various methodologies has been implemented but biodegradation is always preferred. Microbes are ubiquitinous in nature. They have the natural ability to develop resistance. Excreation of exopolysaccharide which is a major component of biofilm can be exploited for water treatment. The major advantages of using microbes is that they need less care and expertisation. Also the plantation cost is comparatively less and have self-regeneration ability.

References

Al-Akeel K (2018) The pollution of water by trace elements research trends. In Advances in bioremediation and phytoremediation. InTech. doi: https://doi.org/10.5772/intechopen.72776

Almuktar SAAAN, Abed SN, Scholz M (2018) Wetlands for wastewater treatment and subsequent recycling of treated effluent: a review. Environ Sci Pollut Res 25(24):23595–23623. https://doi.org/10.1007/s11356-018-2629-3

Aqeel M, Jamil M, Yusoff I (2014) Soil contamination, risk assessment and remediation. In Environmental risk assessment of soil contamination. InTech. doi: https://doi.org/10.5772/57287

Ayilara M et al (2020) Waste management through composting: challenges and potentials. Sustainability 12(11):4456. https://doi.org/10.3390/su12114456

Baek SH, Pagilla K (2009) Microbial community structures in conventional activated sludge system and membrane bioreactor (MBR). Biotechnol Bioprocess Eng 14(6):848–853. https://doi.org/10.1007/s12257-008-0303-1

Beltrán JM (1999) Irrigation with saline water: benefits and environmental impact. Agric Water Manag 40(2–3):183–194. https://doi.org/10.1016/S0378-3774(98)00120-6

Briceño G et al. (2020) Pesticide-tolerant bacteria isolated from a biopurification system to remove commonly used pesticides to protect water resources. *PLOS One*, G. Moreno-Hagelsieb (ed) 15(6):e0234865. doi: https://doi.org/10.1371/journal.pone.0234865

Cao Y et al (2019) Electricigens in the anode of microbial fuel cells: pure cultures versus mixed communities. Microb Cell Fact 18(1):39. https://doi.org/10.1186/s12934-019-1087-z

Chellaiah ER (2018) Cadmium (heavy metals) bioremediation by Pseudomonas aeruginosa: a minireview. Appl Water Sci 8(6):154. https://doi.org/10.1007/s13201-018-0796-5

Chibuike GU, Obiora SC (2014) Heavy metal polluted soils: effect on plants and bioremediation methods. Appl Environ Soil Sci 2014:1–12. https://doi.org/10.1155/2014/752708

Cooke WB (1959) Trickling filter ecology. Ecology 40(2):273–291. https://doi.org/10.2307/1930039

CPHEEO (2013) Chapter 7: recycling and reuse of sewage. In Manual on Sewerage and Sewage treatment systems (New Delhi: Ministry of Urban Development, Government of India)

Crini G, Lichtfouse E (2019) Advantages and disadvantages of techniques used for wastewater treatment. Environ Chem Lett 17(1):145–155. https://doi.org/10.1007/s10311-018-0785-9

Davey ME, O'toole GA (2000) Microbial biofilms: from ecology to molecular genetics. Microbiol Molec Biol Rev 64(4):847–867. doi: https://doi.org/10.1128/MMBR.64.4.847-867.2000

Di Martino P (2018) Extracellular polymeric substances, a key element in understanding biofilm phenotype. AIMS Microbiology 4(2):274–288. https://doi.org/10.3934/microbiol.2018.2.274

Dincer I, Abu-Rayash A (2020) Sustainability modeling. In Dincer I, Abu-Rayash A (eds) Energy sustainability. Elsevier, pp 119–164. doi: https://doi.org/10.1016/B978-0-12-819556-7.00006-1

Du X et al (2020) A review on the mechanism, impacts and control methods of membrane fouling in MBR system. Membranes 10(2):24. https://doi.org/10.3390/membranes10020024

EFSA (2012) Scientific opinion on mineral oil hydrocarbons in food. EFSA J 10(6):185. https://doi.org/10.2903/j.efsa.2012.2704

ElNaker NA, Yousef AF, Hasan SW (2018) Effect of hydraulic retention time on microbial community structure in wastewater treatment electro-bioreactors. MicrobiologyOpen 7(4):e00590. doi: https://doi.org/10.1002/mbo3.590

Giachino A, Waldron KJ (2020) Copper tolerance in bacteria requires the activation of multiple accessory pathways. Mol Microbiol 114(3):377–390. https://doi.org/10.1111/mmi.14522

Gurreri L et al (2020) Electrodialysis applications in wastewater treatment for environmental protection and resources recovery: a systematic review on progress and perspectives. Membranes 10(7):146. https://doi.org/10.3390/membranes10070146

Gutberlet J, Uddin SMN (2017) Household waste and health risks affecting waste pickers and the environment in low- and middle-income countries. Int J Occup Environ Health 23(4):299–310. https://doi.org/10.1080/10773525.2018.1484996

Hassan Rashid MA ul, Manzoor MM, Mukhtar S (2018) Urbanization and its effects on water resources: an exploratory analysis. Asian J Water Environ Poll 15(1):67–74. doi: https://doi.org/10.3233/AJW-180007

Honda M, Suzuki N (2020) Toxicities of polycyclic aromatic hydrocarbons for aquatic animals. Int J Environ Res Public Health 17(4):1363. https://doi.org/10.3390/ijerph17041363

Islam R (2020) Study of moving bed biofilm reactor (Mbbr) for secondary treatment of municipal waste water. Int J Adv Sci Technol 29(7):3650–3654. Available at: http://sersc.org/journals/index.php/IJAST/article/view/22100 (Accessed 6 March 2021)

Ivanovic I, Leiknes TO (2012) The biofilm membrane bioreactor (BF-MBR)—a review. Desalin Water Treat 37(1–3):288–295. https://doi.org/10.1080/19443994.2012.661283

Jia S, Zhang X (2020) Biological HRPs in wastewater. In Ren H, Zhang X (eds) High-risk pollutants in wastewater. Elsevier, pp. 41–78. doi: https://doi.org/10.1016/B978-0-12-816448-8.00003-4.

Jiao Y et al (2010) Characterization of extracellular polymeric substances from acidophilic microbial biofilms. Appl Environ Microbiol 76(9):2916–2922. https://doi.org/10.1128/AEM.02289-09

Kalita D, Joshi SR (2017) Study on bioremediation of Lead by exopolysaccharide producing metallophilic bacterium isolated from extreme habitat. Biotechnol Reports 16:48–57. https://doi.org/10.1016/j.btre.2017.11.003

Kapil N, Bhattacharyya KG (2017) A comparison of neutralization efficiency of chemicals with respect to acidic Kopili river water. Appl Water Sci 7(5):2209–2214. https://doi.org/10.1007/s13201-016-0391-6

Khalil MI, Hossain MB, Schmidhalter U (2005) Carbon and nitrogen mineralization in different upland soils of the subtropics treated with organic materials. Soil Biol Biochem 37(8):1507–1518. https://doi.org/10.1016/j.soilbio.2005.01.014

Leion H, Frick V, Hildor F (2018) Experimental method and setup for laboratory fluidized bed reactor testing. Energies 11(10):2505. https://doi.org/10.3390/en11102505

Li H et al (2020) Preparation, performances, and mechanisms of microbial flocculants for wastewater treatment. Int J Environ Res Public Health 17(4):1360. https://doi.org/10.3390/ijerph17041360

Liyanage CP, Yamada K (2017) Impact of population growth on the water quality of natural water bodies. Sustainability 9(8):1405. https://doi.org/10.3390/su9081405

Mack WN, Mack JP, Ackerson AO (1975) Microbial film development in a trickling filter. Microb Ecol 2(3):215–226. https://doi.org/10.1007/BF02010441

Medfu Tarekegn M, Zewdu Salilih F, Ishetu AI (2020) Microbes used as a tool for bioremediation of heavy metal from the environment. Cogent Food & Agriculture. Edited by F. Yildiz 6(1):1783174. doi: https://doi.org/10.1080/23311932.2020.1783174

Moss B (2008) Water pollution by agriculture. Philos Trans Royal Soc b: Biol Sci 363(1491):659–666. https://doi.org/10.1098/rstb.2007.2176

Ng IS, Hsueh CC, Chen BY (2017) Electron transport phenomena of electroactive bacteria in microbial fuel cells: a review of Proteus hauseri. Bioresourc Bioproc 4(1):53. https://doi.org/10.1186/s40643-017-0183-3

Ojuederie O, Babalola O (2017) Microbial and plant-assisted bioremediation of heavy metal polluted environments: a review. Int J Environ Res Public Health 14(12):1504. https://doi.org/10.3390/ijerph14121504

Phillips EJP, Landa ER, Lovley DR (1995) Remediation of uranium contaminated soils with bicarbonate extraction and microbial U(VI) reduction. J Ind Microbiol 14(3–4):203–207. https://doi.org/10.1007/BF01569928

Pontel LB et al (2014) Identification of a Salmonella ancillary copper detoxification mechanism by a comparative analysis of the genome-wide transcriptional response to copper and zinc excess. Microbiology 160(8):1659–1669. https://doi.org/10.1099/mic.0.080473-0

Qiqi Y, Qiang H, T. Ibrahim H (2012) Review on moving bed biofilm processes. Pakistan J Nut 11(9):804–811. doi: https://doi.org/10.3923/pjn.2012.804.811

Reyes-Caballero H, Campanello GC, Giedroc DP (2011) Metalloregulatory proteins: metal selectivity and allosteric switching. Biophys Chem 156(2–3):103–114. https://doi.org/10.1016/j.bpc.2011.03.010

Rhind SM et al (2010) Effects of environmental pollutants on the reproduction and welfare of ruminants. Animal 4(7):1227–1239. https://doi.org/10.1017/S1751731110000595

Riveros-Perez E, Riveros R (2018) Water in the human body: an anesthesiologist's perspective on the connection between physicochemical properties of water and physiologic relevance. Annals Med Surg 26:1–8. https://doi.org/10.1016/j.amsu.2017.12.007

Sanjay MS et al (2020) Isolation and identification of chromium reducing bacteria from tannery effluent. J King Saud Univ Sci 32(1):265–271. https://doi.org/10.1016/j.jksus.2018.05.001

Schellenberg T et al (2020) Wastewater discharge standards in the evolving context of urban sustainability-the case of India. Front Environ Sci 8:30. https://doi.org/10.3389/fenvs.2020.00030

Scholz M (2016) Biological filtration. In Scholz M (ed) Wetlands for water pollution control. Elsevier, pp 81–85. doi: https://doi.org/10.1016/B978-0-444-63607-2.00013-7

Shaw K, Dorea CC (2021) Biodegradation mechanisms and functional microbiology in conventional septic tanks: a systematic review and meta-analysis. Environ Sci: Water Res Technol 7(1):144–155. https://doi.org/10.1039/D0EW00795A

Shim S et al (2020) Design and optimization of fluidized bed reactor operating conditions for struvite recovery process from swine wastewater. Processes 8(4):422. https://doi.org/10.3390/pr8040422

Srinivasa Raghavan DS et al. (2017) Anaerobic treatment of low-strength wastewater. In Lee D-J et al. (eds) Current developments in biotechnology and bioengineering. Elsevier, pp 293–320. doi: https://doi.org/10.1016/B978-0-444-63665-2.00012-6

Tchounwou PB et al. (2012) Heavy metal toxicity and the environment, pp 133–164. doi: https://doi.org/10.1007/978-3-7643-8340-4_6

Tekerlekopoulou AG et al. (2020) Wastewater treatment and water reuse in the food industry. In Galanakis C (ed) The interaction of food industry and environment. Elsevier, pp 245–280. doi: https://doi.org/10.1016/B978-0-12-816449-5.00008-4

Timmerberg S, Kaltschmitt M, Finkbeiner M (2020) Hydrogen and hydrogen-derived fuels through methane decomposition of natural gas—GHG emissions and costs. Energy Conver Manag X(7):100043. doi: https://doi.org/10.1016/j.ecmx.2020.100043

UNDESA (2015) A 10 year story the water for life decade 2005–2015 and beyond. Available at https://www.un.org/waterforlifedecade/pdf/WaterforLifeENG.pdf

US EPA. Overview of Watershed Monitoring (2019) EPA Watershed Academy Training Module. Available at https://cfpub.epa.gov/watertrain/pdf/modules/monitoring.pdf

WHO (2009) *Global health risks: mortality and burden of disease attributable to selected major risks*. Geneva.

WHO (2016) Guidelines for the Safe Use of Wastewater, Excreta and Greywater

Wilson SP et al (2015) Adopting primary plastic trickling filters as a solution for enhanced nitrification. Water Environ Res 87(1):80–87. https://doi.org/10.2175/106143014X13975035525988

Wuana RA, Okieimen FE (2011) Heavy metals in contaminated soils: a review of sources, chemistry, risks and best available strategies for remediation. ISRN Ecology, 1–20. doi: https://doi.org/10.5402/2011/402647

WWAP (2017) United nations world water assessment programme. The United Nations world water development report 2017. Wastewater: the untapped resource. Paris. Available at https://reliefweb.int/sites/reliefweb.int/files/resources/247153e.pdf

Xu X et al (2018) Petroleum hydrocarbon-degrading bacteria for the remediation of oil pollution under aerobic conditions: a perspective analysis. Front Microbiol 9:2885. https://doi.org/10.3389/fmicb.2018.02885

Zhang B, Xu X, Zhu L (2018) Activated sludge bacterial communities of typical wastewater treatment plants: distinct genera identification and metabolic potential differential analysis. AMB Express 8(1):184. https://doi.org/10.1186/s13568-018-0714-0

Zhang Y, Liu J, Li J (2020) Comparison of four methods to solve clogging issues in a fungi-based bio-trickling filter. Biochem Eng J 153:107401. doi: https://doi.org/10.1016/j.bej.2019.107401

Chapter 5
Heavy Metals Removal Techniques from Industrial Waste Water

Nimmy Srivastava, Jayeeta Chattopadhyay, Arushi Yashi, and Trisha Rathore

Abstract Regarded as pollutants heavy metals like Arsenic (As), Hg (Mercury), Zn (Zinc), Cd (Cadmium), Thallium (Tl) and Pb (Lead) not only pose a serious threat on water ecosystem but also on human health owing to their highly toxic, non-biodegradable, and carcinogenic behavior. With industrial advancement it has become one of the foremost sources of heavy metals instigating it into different component of ecosystem air, water and soil. Various techniques are currently available which are used for the treatment of discharge of various industries often laden with noxious metal particles. This chapter discusses in brief current wastewater treatment procedures for removal of heavy metals.

Keywords Heavy metals · Human health · Carcinogenic · Waste water · Treatment

Abbreviations

CNTs	Carbon Nanotubes
EDTA	Ethylenediamine Tetra acetic acid
IP	Interfacial polymerization
NTA	Nitrilotriacetic acid
NF	Nanofiltration
MF	Microfiltre
PAN	Polyacrylonitrile
PES	Polyethersulfone
PP	Polypropylene
PVDF	Polyvinylidene fluoride

N. Srivastava (✉) · A. Yashi · T. Rathore
Amity Institute of Biotechnology, Amity University, Ranchi, Jharkhand, India
e-mail: nsrivastava@rnc.amity.edu

J. Chattopadhyay
Department of Chemistry, Amity School of Engineering and Technology, Amity University, Jharkhand, Ranchi, India

S. Roy et al. (eds.), *Advanced Industrial Wastewater Treatment and Reclamation of Water*, Environmental Science and Engineering,
https://doi.org/10.1007/978-3-030-83811-9_5

RO	Reverse Osmosis
UF	Ultra-filtration

5.1 Introduction

Since time immortal availability of fresh and potable water sources has remain one of the vital needs of human and some great civilizations have flourished near the bank of water reservoirs only. However, with the leap and bounces by which world is progressing for achieving a comfortable life the accessibility of clean water sources is decreasing even at a more rapid pace with some contributing factors like increasing population over the world, industrialization and globalization which has led to an excessive exploitation and degradation in quality of elementary water sources owing to toxic adulterants discharged from fabrication units along with the used water (Malat et al. 2009).

Wastewaters discharged from anthropogenic industries like mining or agricultural plants lacking eco-friendly production methods and natural forces viz. volcanoes, earthquakes, or storms are fount of noxious heavy metals in water bodies. Heavy metals are considered highly toxic due to their carcinogenic nature having inimical effect on health of human and effacing of metals particles from the drinking water sources is highly essential before their accretion start in the food chain. Shortage of pure drinking water is achievable by methods of rain water harvesting but for a short period however the premier way is treatment of wastewater loaded with pollutants and its reuse (Chong et al. 2010).

5.1.1 Noxious Heavy Metals

They are those elements which belong to the fourth period in the periodic table and their density outstrips 5 g per cubic centimeter. Mostly some heavy metals are naturally found in the environment but in this decade a substantial shoot up in the number of heavy metals more than the stipulated limits are recorded at most of the region across the world (Gupta et al. 2013). These noxious ions infiltrate the food chain and swing in into the human body (Merzouk et al. 2011). Once upon accumulation inside any human organ more than standard limits (Table 5.1), they result in severe health problems (Kurniawan et al. 2006). As for example, presence of excessive Zn can genesis various health problems like skin rashes and itching, nausea, and stomachache (Fu and Wang 2011), excessive Ni led to lung cancer and kidney ailments subjection to mercury (Hg) and lead (Pb) leads to the evolution of autoimmune disease and causes rheumatoid arthritis. Due to its non-biodegradable nature heavy metals keep on inside the body of human for a prolonged duration making its impact more severe (Hahladakis et al. 2013).

Table 5.1 The MCL standards for most toxic heavy metals (*Source* Babel and Kurniawan 2003)

Heavy metal MCL (mg/L)	
Zinc	0.80
Nickel	0.20
Mercury	0.00003
Lead	0.006
Copper	0.25
Chromium	0.05
Cadmium	0.01
Arsenic	0.050

Wastewater treatment methods for elimination of heavy metals includes processes like physicochemical, electrochemical, and oxidation processes Physicochemical methods includes techniques like filtration through membrane, precipitation by chemical means, ion change, and adsorption. Techniques like Electrocoagulation, Eectroflotation, and electrodeposition are grouped under electrochemical techniques. Photocatalysis is preeminent recent state of the art oxidation processes. Among these available techniques the most reliable by means of cost effectiveness, eco-friendly nature, and high efficiency rate is selected (Dionisi 2014).

5.2 Wastewater Treatment Method

Various treatment procedure for complete erasure of metals particles from industrial waste water are discussed below.

5.2.1 Physico-Chemical Methods

Industrial effluents are often laden with noxious metal particles and over the years various techniques has been developed for improving the quality of wastewater after treatment. These techniques have their own advantages and certain limitations which have been discussed below.

5.2.1.1 Physical Method

Old techniques such as physical severance are used by researchers for separation of primarily suspended minute metals particles or metal-loaded particles (Dermont 2008). Using techniques essentially like mechanical shielding, magnetic dissolution, electrostatic separation, hydrodynamic classification, gravity concentration, flotation, and attrition scrubbing. Proficiency of these techniques depends on some characteristics features of soil like particle size, particle shape, moisture load, clay content, humus content, magnetic attribute, and hydrophobic properties of particles present on surface (Smith 1995).

5.2.1.2 Chemical Precipitation

Other most commodiously used metal particles removal techniques used for treatment of inorganic effluent is chemical precipitation. In this technique use of lime and limestones are done as the precipitating agents owing to their easy accessibility and potency for treatment of inorganic effluents with metal concentration range of about 900–1000 mg/L (Mirbagherp and Hosseini 2004; Aziz et al. 2008). Schematic representation of various stride involved in this technique has been unveiled in Fig. 5.1. As this method is highly efficient and economical this method is encyclopedically used over the world but use of superfluous chemical for reduction in the level of

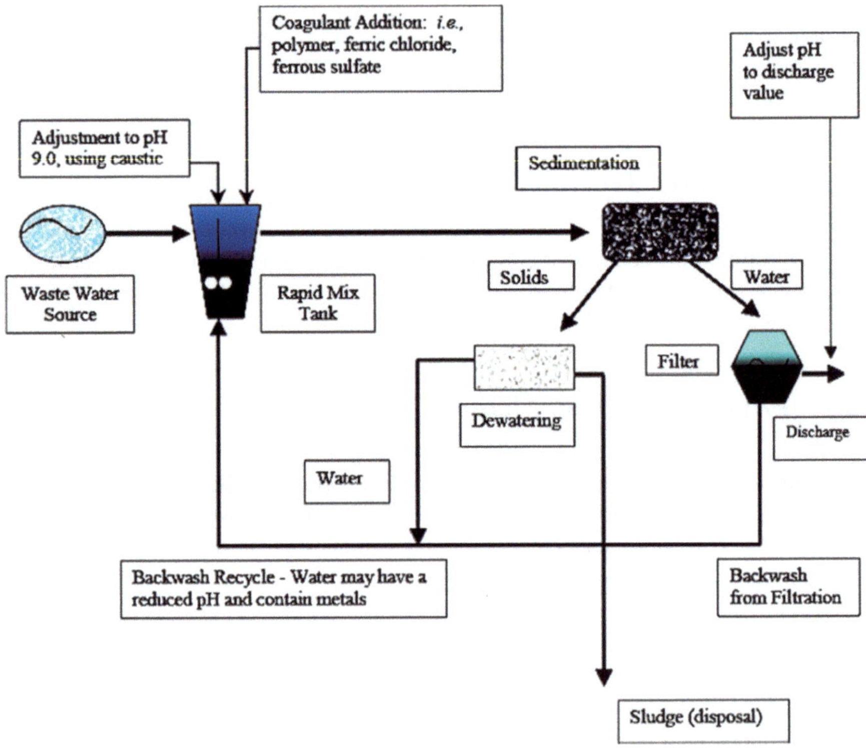

Fig. 5.1 Schematic representation of steps involved in chemical precipitation method (*Source* Barakat 2011)

metal concentration (Galla et al. 2000) and excessive production of after treatment sludge compelled scientist to search for some newer techniques.

Types of Chemical Precipitation Utilized in the Heavy Metals Erasure from Wastewater

A. Hydroxide Precipitation
B. Sulfide Precipitation

A. **Hydroxide Precipitation**

This process aims for settling down of heavy metals by utilizing various coagulants for instance salts of iron, salt of aluminum, and polymers which enhance the rate of metal particles severance from industrial wastewater. The metal particles being soluble precipitate as hydroxides which are removed by either refinement or sedimentation process. Herein operation sometimes various alkaline agents are also exploited to upraise the pH of H_2O (Wang et al. 2004). Alkylating agents have the propensity of reducing solubility tendencies of metal ions which later fling out. Existence of organic radicals affects the process of hydroxide precipitation.

The hydroxide precipitation reaction is given below where A represents the heavy metal

$$A^{2+} + 2OH^{-} \rightarrow A(OH)_2$$

$A(OH)_2$ is the insoluble form.

B. **Sulfide Precipitation**

An identical process hydroxide precipitation is sulfide precipitation which is often used to precipitate metal particles in form of metal sulfides and sludge generated afterwards is filtrated out from the dissolvent by help of gravity settling or by refinement method. This process includes pretreatment and post-treatment steps and a meticulous measurement of chemical reagent added is much indispensable owing to the toxic temperament of the sulfide ion and H_2S. Despite this the most proffer method for heavy metal erasure is only the hydroxide precipitation method owing to minimal load of toxic chemicals used in various steps (Tünay 2003).

5.2.2 *Ion Exchange*

Yet other method sometimes employed for erasure of heavy metals particles from inorganic sewerage (Dabrowski et al. 2004 based on the process of exchange of ions betwixt the solid and liquid phase (Kurniawan et al. 2006). In an ion exchanger cations and anions are exchanged from the surrounding materials with the solid core.

At the stoppage of the process hydration reaction takes place at fringe of the solution or on the pores present on adsorbent surface (Ferreira et al. 1999). Numerous factors like temperature, time, pH etc. related to the operation system influence the expertise of ion exchange method (Kang et al. 2004). Synthetic but organic ion exchange resins are time and again used matrices for ion exchange. Resins which are insoluble are basically used to dispose metals particles from wastewater and for release of different ions having like charge without creating any physical change of the resin. Currently synthetic polymer resins, like styrene–divinylbenzene are preferred for this process (Alyüz and Veli 2009).

Others option for resins are gel-like resins which are more methodical and inexpensive. Lewatit S 100 which is a gel form cation exchange resins have an ion-exchange competence of 0.35–0.39 mmol g^{-1} of Cr (III) provided with an ideal conditions and pH maintained around 3.4 (Gode and Pehlivan 2003). Dowex HCR S/S resins are used in sodium form and is highly economical and utilitarian because of their high regeneration capacities principally for Ni and Zn which are eliminated around 96–98% at optimum conditions and pH of 4 (Alyüz and Veli 2009). A basic resin Ambersep 132 with reported adsorption capacity of about 90–92.10 mg g^{-1} mainly employed in the removal of Cr (VI). IRN-77 a cation exchange resin has reportedly shown high absorbing capacities for Cr (III) with concentration of more than 90 mg L^{-1} at optimum pH level of 3.5 (Rengaraj et al. 2001). Use of Macropore resins which have spongy structure providing an optimal physical stability and noted stress relief (Gode and Pehlivan 2003) have also been reported by various research groups.

Ion exchange strategy is most preferred one as juxtaposed to other chemical methods as it has reportedly high efficiency, low maintenance cost, production of minimal mass of sludge, recuperation of metal's value, and significant selectivity. However, this technique cannot effectively handle concentrated solutions of heavy metals and susceptible to various pH range of the suspension.

5.2.3 *Adsorption*

Another physicochemical process utilized effectively for erasure of metals particles from fabrication unit effluent is adsorption based on bulk transfer betwixt two different phase one liquid phase and the supplemental being solid phase known as adsorbent (Babel and Kurniawan 2003). The adsorbents used presents a surface area large enough and high adsorption potential and its efficiency based on different adsorbents used are presented in Table 5.2.

Steps involved for adsorption of metals involve there basic (Barakat 2011):

(i) Piercing of the pollutant present in heavy solution onto the facet of adsorbent.
(ii) Adsorption by adsorbent surface
(iii) Piercing inside the adsorbent structure.

Table 5.2 Adsorbents used for heavy metal removal (*Source* Hu et al. 2009)

Metal	Concentration of metal ions [mg/L]	Initial pH maintained	Time of contact [h]	Adsorbent used	Dose adsorbent [g L]	Removal efficiency (%)
Copper		7	24	Sawdust	5	74.8
	5	7	24	Sawdust	40	85.2
	50	7	24	Sawdust	5	94.8
Mercury	200	5	7	SS-C (Without treatment)	10	58.6
	200	5	7	SS-C	14	83.4
	200	5	7	SS-S (Treated with H_2SO_4)	8	100
	200	5	7	SS-p (Treated with H_3PO_4-)	6	100
	200	5	7	SS-Z (Treated with $ZnCl_2$-)	4	100
Nickel	10	5	1	AC-K9 Prepared at around 900 °C)	0.7	100
Chromium	1	2.85	~30	Oxidized MWCNT	1	~100
	3	2.85	~280	Oxidized MWCNT	1	~100
	5	2.85	~280	Oxidized MWCNT	1	~50

One benefit of this approach is that it can be operated inconvertible mode and the adsorbents used could be reinvigorating through adsorption. Mainly the adsorbents employed in this process can be procured from agricultural scrap or industrial used and trash, or any other organic materials out of which most commonly used are discussed.

5.2.3.1 Activated Carbon

Activated carbon are utilized for erasure of toxic metals (Abdel-Halim et al. 2003) and are drawn up from agriculture derivate using K_2CO_3 with maximum surface area ranging from 1260–3258 m^2g^{-1} (Tsai et al. 2001). Various researches conducted have reported that carbon prepared at a temperature more than 800 °C is more reportedly more beneficial for the erasure of Ni(II) from an aqueous medium while the concentration of adsorbent is 0.25 g. Notable repercussion of the pH on the adsorption was observed and more 97% Ni(II) erasure was attained in pH stretch between 2–5 which is generally considered the most optimum pH values for adsorption of cation.

5.2.3.2 Wood Sawdust

Derived from timber industry wood sawdust has been popularly used as an inexpensive adsorbent for metals particles owing to its lignocellulosic configuration. Wood sawdust is mainly including cellulose (40–49%) and lignin (25–30%), both having the ability for binding with metal cations because of presence of hydroxyl, carboxylic, and phenolic groups in their structure (Yu et al. 2000). Saw dust have reported highest efficiency for erasure of copper- 75%, Zinc- 37%, and Cadmium-32%.

5.2.3.3 Carbon Nanotubes (CNTs)

As confirmed by various researchers CNTs have an immense potential to obliterate heavy metals (Kabbashi et al. 2009) CNTs are debilitated by use of calcium alginate to curb the potential damage caused by discharge of carbon nanotubes in excess amount in water bodies. Scientists throughout the globe have performed several experiments on copper absorption by CNTs and CNTs immobilized with calcium alginate and reported that at pH around 5.0, 75% removal efficiency by the alone use of CNTs and 83.3% removal efficiency by use of CNTs with calcium alginate under experiment conditions (Li et al. 2010).

5.2.3.4 Modified Natural Materials

In the neoteric time unrefined zeolites has gained significant popularity in discipline of heavy metal removal owing to their potency to work as an ion exchanger. Clinoptilolites the most common and researched natural zeolites which has demonstrated a high discernment for heavy metal pollutants such as Cadmium, Lead, Zinc, and Copper. The cation-exchange competence of clinoptilolite rely on the method used for pre-treatment (Babel and Kurniawan 2003).

Currently investigation on utilization of various synthetic zeolite for heavy metals elimination in ongoing and as per reports pH plays major role in considerable adsorption of different metals particles. Basaldella (2007) along with his research group reported that NaA zeolite can be used for erasure of Cr(III) at a neutral pH, while in year 2008 Barakat synthesized 4A zeolite by the dihydroxylation of meagre grade kaolin and reported that Cu(II) and Zn(II) were successfully adsorbed at neutral and

Table 5.3 Modified natural materials for heavy metals removal (*Source* Azimi et al. 2017)

Adsorbent	Adsorption capacity of natural material (mg/g)					
	Pb2$^+$	Cd2$^+$	Zn^{2+}	Cu^{2+}	Cr^{6+}	Ni^{2+}
Zeolite, clinoptilolite	1.6	2.4	0.5	1.64		0.4
Modified zeolite, MMZ	123					8
HCl-treated clay	63.2	83.3				
Clay/poly (methoxyethyl) Acrylamide	81.02		20.6	29.8		80.9
Calcined phosphate	155.0					
Activated phosphate	4					
Zirconium phosphate	398					

alkaline pH, whereas Cr(VI) was engrossed at an acidic pH and adsorption of Mn(IV) was delineated at an alkaline pH. Synthetic zeolite prepared after modified with iron oxide (MMZ) it was delineated that it shows a soaring adsorption capacity for lead (Nah et al. 2006).

Natural clay minerals after modification with a polymeric material also serves the purpose of removal of heavy metals present in solutions. These natural adsorbents are labelled as clay–polymer composites (Vengris et al. 2001). New form of adsorbents like Phosphates with nitric acid and zirconium phosphate are now being employed for erasure of heavy metals particles (Moufliha et al. 2005; Pan et al. 2007). Table 5.3 depicts the adsorption scope of various modified natural material-based absorbent.

5.2.4 *Electrochemical Treatments*

Till date electrochemical treatment methods of wastewater have not been extensively manoeuvre due to huge capital investment which is required and high rate of electricity and its production cost. Some of the methods used are discussed below.

Electro coagulation (EC)

It is an effortless yet prolific procedure and generally used in meager wastewater treatment plant although it is not a very well-grounded method due to less electrode reliability. A vivid range of environmental pollutants are eliminated by electrolysis.

Electrodes fabricated out of stuffs like Iron and aluminum are used in this method. At the initial step pollutants like heavy metal particles present in the industrial effluent remains bonded together by the electrical charges present on the surface or hydrogen bond and soon after by passing electric current, the electrostatic charges grasping together the suspended contaminants particles are neutralized and afterwards they start coagulating. Heavy metal particles in an aggregate mass behave like small magnets known as sludge or floc. The final end products, flocs produced through this method are large with negligible bond with water molecules and are more unwavering which can be eliminated by further processing like filtration.

In an experiment conducted by Mansoorian and his group it was observed that coagulation done by using hydroxide iron and steel flocs in form of absorbents at maintained current densities of about 6 and 8 mA cm–3, the metal obliteration capacity was found for lead removal was more than 90 and 95%, respectively and for zinc removal was also more than 90% at both current densities.

This technique is an ecofriendly since no chemicals are added so chances of genesis of further pollutants and flocs generation is reduced.

Electro-Floatation (EF)

For effluents containing heavy metals in much diluted form i.e., below 50 mgm^{-3} technique of electro floatation are highly employed by floating heavy metals particles on the facade of the liquid solution· The reason for popularity of this technique is its simple design, ecofriendly operation, low cost of maintenance. Generally, this technique is employed for erasure of oil from an oil–water emulsions, purification of groundwater, fruit juice specially apple juice and food processing end products, dangling particles, sewage water.

Heavy metals are basically adsorbed by the oxygen and hydrogen molecules and resultant emulsified particles are destabilized and post treatment flocs are formed and afterwards the collected pollutants in precipitate form are removed and discarded by filtration methods.

Electrodeposition

Amidst the electrochemical methods electrodeposition is an empirical approach for heavy metals removal. In this technique no chemicals are required nor there is formation of any active sludge. Dissolved metal ions present in the wastewater are transfigured into solid phase in form of minuscule particles by depositing them on anode and cathode to protect them from corrosion. On cathode tube reduced heavy metals ions are electroplated.

The whole exercise efficiency usually depends on few determinants as concentration of pollutants in the waste solution, some physical parameters for reactor operation such as temperature and pH prolongation accompanying the habitation of some complexing and chelating agents like EDTA, NTA and citrate.

5.2.5 *Recent Advance Techniques*

During early 1980s development of impressive techniques for heavy metal erasure from wastewater was on its way. During that time membrane filtration processes was developed and soon it became one of the most preferred method owing to their higher metal erasure efficiency, consumption of lower energy than any other techniques used. These added advantages helped membrane filtration processes to be used in various industries including food processing and beverage industries, filtration of water and in industrial wastewater treatment reactor.

5.2.5.1 Membrane Characteristics

A selectively permeable layer having either porous or non-porous structure used for erasure of pollutants of various sizes is known as membrane. Mostly three kinds of membranes are effectively been utilized in this process viz. liquid, pressure-driven, and hybrid membranes (Singh 2015). Varied parameters affect the functionality of a membrane like size of pores present on membrane surface and composition of the material of membrane (Koyuncu et al. 2015). Materials which are employed in membrane fabrication are either metallic, ceramic, and any composite material is used. And most updated membrane materials been used currently are nano-based materials along with some biological materials which are purposefully modified are also been used. One of the most commonly used composite materials for pollutant filtration membrane preparation are polymers which are employed successfully in different filtration processes like microfiltration (MF) and reverse osmosis (RO). Polymers utilized for membrane preparation are mostly polypropylene (PP), polyvinylidene fluoride (PVDF), cellulose acetate, poly acrylonitrile (PAN), and polyether sulfone (PES). Many a times in certain cases some ceramic materials are considered impressive to polymeric materials owing to their cramped pore size and reportedly higher stability at thermal and chemical parameters (Koyuncu et al. 2015). Various materials used specifically to preparation of ceramic membranes are zirconia, titania, silica, alumina, oxide mixtures, and sintered metals.

5.2.5.2 Various Membrane Processes

The advance technique of membrane filtration includes various processes for effluent treatment in which only pore size and its dispensation over the membrane surface, permeability percentage of membrane and operating milieu of the reactors plant differs.

Micro and Ultrafiltration (MF and UF)
In twain techniques porous membranes are used and are operated at a very insignificant pressure. In MF technique pore sizes of membranes is of 0.02 to 10 mm in and pressure is about 0.1 to 2.1 bar but in UF membranes size of pore is smaller and ranges from 1 nm to around 100 nm with a working pressure of about 1 to 5 bar is maintained.

Nanofiltration (NF)
With the aid of this method separation of relatively large pollutant molecules is possible even through small size pores. To make this technique more efficient use of various technologies like interfacial polymerization (IP), incorporation ultraviolet (UV) and treatment by nanoparticles are combined. NF technique till date has been employed for the disinfection of underground water, surface water, and industrial effluent treatment strategies. Since the last 20 years NF use for erasure of heavy metals has progressively upsurge due to its effortless operation.

Mainly this technique revolves around three major steps for erasure of heavy metals. The beforehand step is a pre-treatment step in which effluent is feed before entering the system to decrease any chances of fouling. Then the next step is of separation where membrane is utilized and the termination step is of post treatment.

AFC- 80 membranes have been utilized in nanofiltration for toxic heavy metals erasure when they are treated by NF solutions accommodating lead in an excessive concentration range and also cadmium.

5.2.6 Photocatalysis Technique

A new era of technology marked its way with Advanced oxidation processes (AOPs) for purification of air and water (Choi et al. 2006). Photocatalysis is a well-known AOP process well investigated by various researchers. This method make use of non-toxic semiconductors that efficiently harness light energy within a suitable wavelengths instead of making use of any chemical substances and thus preferred over chemical methods. This technique has become famous as a well advanced and effective method owing to its effortless design, less operation and maintenance cost, and immense removal efficiency owing to which is used in many broad areas like disintegration of water to form hydrogen gas, disinfection of water, organic compounds degradation, fussy oxidation, producing fuels in liquid form. The best utilization of this technique is erasure of heavy metal from sewerage.

A multi-step process is operated in a photocatalytic machine. Firstly, the pollutants particles present in aqueous solution propel to the adsorbent surface where it is absorbed through the semiconductor surface. In the next step photocatalytic reactions take place and eventually, the products are depraved and waive out from the interface area. This is regarded as the main step from where the photocatalysis process starts. In the entire procedure light is used as an activator which brings about the excitation of photon of the semiconductor. Meanwhile visible light having energy commensurate to or more than the band gap energy of the semiconductor is passed, electrons which are in the valence band (VB) gets promoted to conduction band (CB) thus forming an electron-hollow pair leaving a hole inside the Valence Band and a photon-electron (e–) in the Conduction Band, illustrated by the subsequent equations

$$SC + hv \rightarrow e^{-} + h^{-}$$

Reduction and oxidation of the pollutant occur upon formation of electron–hole pairs and exchange of photo-holes and photoelectrons takes place. Based on the redox ability of different contaminants, electron-holes proceed towards the acceptor molecules and photo-holes get transferred to the donor. Ions are formed which revert to form the intermediate product and final product.

Metal's erasure can take place on being deposited at the facade of photocatalyst by the following response:

$$Hn^{+} + e^{-} \rightarrow H$$

Deposition takes place in form of small crystallites and transforming the ionic species into meager toxic forms. The following pattern of reactivity was observed under specific operation condition Ag > Pd > Au > Pt > > Rh > > Ir > > Cu = Ni = Fe = 0.

Generally, the semiconductors that are utilized in photocatalysis are metals oxides like TiO_2, ZnO, CeO_2, WO_2, and metal sulfides like CdS, ZnS, WS_2, etc. belonging to the chalcogenide group among which $TiO_2$is the most preferred photocatalyst owing to its ecofriendly character and they are easy to handle, non-toxic and stable.

In spite of being highly efficient photocatalysis techniques has some major drawbacks like combining of electron or hole which led to generation of unwanted byproducts, and the defective absorption of visible light.

5.3 Conclusion and Future Aspects

In nature widely distributed toxic heavy metals are generally evaluated as a harmful agent for human's health and also for the ecosystem wherever they exceed their limits. So, to safeguard humans from these intruding toxic pollutants, it's the obligation of hour to eliminate it from the natural reservoirs in which they are present in form of pollutants one such natural source is our water bodies. Various techniques are present and used for treatment of wastewater before their discharge in water bodies broadly classified under electrochemical and physicochemical methods each one having their own efficiencies and their operational parameters for heavy metals removal. Variable results of efficiency removal percentage have been observed after utilizing different methods for each method which vary from 12 to 100%. Utilization of methods, such as chemical precipitation, NF, adsorption, electro flotation, and electro coagulation, can near about remove (>99%) of heavy metals however, electro deposition methods have a fluctuating erasure efficiency for heavy metals varying from 11 to 93%. With progression in technologies new methods needs to be investigated which is greener approach and cost effective. The current tried and tested methods has various advantages along with certain major drawbacks as most of the methods are chemical based. In future a potential research area for treatment of heavy metal erasure is bio-based techniques like use of biological membrane. More biological materials need to be searched out which can be efficiently utilized in field of waste water treatment.

References

Abdel-Halim SH, Shehata AM, El-Shahat MF (2003) Removal of lead ions from industrial waste water by different types of natural materials. Water Res 37(7):1678–1683

Alyüz B, Veli S (2009) Kinetics and equilibrium studies for the removal of nickel and zinc from aqueous solutions by ion exchange resins. J Hazard Mater 167(1–3):482–488
Azimi A, Azari A, Rezakazemi M, Ansarpour M (2017) Removal of Heavy Metals from Industrial Wastewaters: A Review. Chem Bio Eng Rev 1:37–59
Aziz HA, Adlan MN, Ariffin KS (2008) Heavy metals (Cd, Pb, Zn, Ni, Cu and Cr(III)) removal from water in Malaysia: posttreatment by high quality limestone. Bioresour Technol 99:1578–1602
Babel S, Kurniawan TA (2003) Low-cost adsorbents for heavy metals uptake from contaminated water: a review. J Hazard Mater B97:219–243
Barakat MA (2011) New trends in removing heavy metals from industrial wastewater. Arab J Chem 4(4):361–377
Basaldella EI, Patrica GV, Lucolano F, Caputo D (2007) Chromium removal from water using LTA zeolites: effect of pH. J Colloid Interf Sci 313:574–578
Chong MN, Jīn B, Chow CW, Saint C (2010) Recent developments in photocatalytic water treatment technology: a review. Water Res 44(10):2997–3027
Dabrowski A, Hubicki Z, Podkościelny P, Robens E (2004) Selective removal of the heavy metal ions from waters and industrial wastewaters by ion-exchange method. Chemosphere 56(2):91–106
Dermont G, Bergeron M, Mercier G, Richer-Laflèche M (2008) Metal-contaminated soils: remediation practices and treatment technologies. Pract Period Hazard Toxic Radioact Waste Manag 12(3):188–209
Dionisi D (2014) Potential and limits of biodegradation processes for the removal of organic xenobiotics from wastewaters. Chem Bio Eng Rev 1(2):67–82
Ferreira SL, de Brito CF, Dantas AF, Lopo de Araújo NM, Costa AC (1999) Nickel determination in saline matrices by ICP-AES after sorption on Amberlite XAD-2 loaded with PAN. Talanta 48(5):1173–1177
Fu F, Wang Q (2011) Removal of heavy metal ions from wastewaters: a review. J Environ Manage 9(3):407–418
Galla U, Jüttner K, Schmieder H (2000) Electrochemical approaches to environmental problems in the process industry. Electrochemical Acta 45:2575–2594
Gode F, Pehlivan E (2003) A comparative study of two chelating ion-exchange resins for the removal of chromium(III) from aqueous solution. J Hazard Mater 100(1–3):231–243
Gupta VK, Ali I (2013) in Environmental water, Ali VKG (ed). Elsevier, Amsterdam. 1, pp. 12-28
Hahladakis J, Smaragdaki E, Vasilaki G, Gidarakos E (2013) Use of Sediment Quality Guidelines and pollution indicators for the assessment of heavy metal and PAH contamination in Greek surficial sea and lake sediments. Environ Monit Assess 185(3):2843–2853
Kabbashi NA, Atieh MA, Al-Mamun A, Mirghami ME, Alam MD, Yahya N (2009) Kinetic adsorption of application of carbon nanotubes for Pb(II) removal from aqueous solution. J Environ Sci (china) 21(4):539–544
Kang SY, Lee JU, Moon SH, Kim KW (2004) Competitive adsorption characteristics of Co2+, Ni2+, and Cr3+ by IRN-77 cation exchange resin in synthesized wastewater. Chemosphere 56(2):141–147
Koyuncu I, Sengur R, Turken T, Guclu S, Pasaoglu ME (2015) in Advances in membrane technologies for water treatment, Basile A, Rastogi, ACK (eds). Woodhead Publishing, Cambridge, pp 83–128
Kurniawan TA, Chan GYS, Lo WH, Babel S (2006) Physico-chemical treatment technique for wastewater laden with heavy metals. Chem Eng J 118:83–98
Li Y, Liu F, Xia B, Du Q, Zhang P, Wang D, Wang Z, Xia Y (2010) Removal of copper from aqueous solution by carbon nanotube/calcium alginate composites. J Hazard Mater 177(1–3):876–880
Malato S, Fernández-Ibáñez P, Maldonado MI, Blanco J, Gernjak W (2009) Decontamination and disinfection of water by solar photocatalysis: recent overview and trends. Catal Today 147:1–59
Mansoorian HJ, Mahvi AH, Jafari AJ, Amin MM, Rajabizadeh A, Khanjani N (2014) Bioelectricity generation using two chamber microbial fuel cell treating wastewater from food processing. Enzyme Microb Technol 52(6–7):352–357

Merzouk B, Yakoubi M, Kone M, Leclerc JP (2011) Effect of modification of textile wastewater composition on electrocoagulation efficiency. Desalination 275(1):181–186

Mirbagherp SA, Hosseini SN (2004) Pilot plant investigation on petrochemical wastewater treatment for the removal of copper and chromium with the objective of reuse. Desalination 171:85–93

Moufliha M, Aklila A, Sebti S (2005) Removal of lead from aqueous solutions by activated phosphate. J Hazard Mater B119:183–188

Nah IW, Hwang KY, Jeon C, Choi HB (2006) Removal of Pb ion from water by magnetically modified zeolite. Miner Eng 19(14):1452–1455

Pan BC, Zhang QR, Zhang WM, Pana BJ, Dua W, Lvb L, Zhanga QJ, Xua ZW, Zhang QX (2007) Highly effective removal of heavy metals by polymer-based zirconium phosphate. J Colloidal Interf Sci 310(1):141–198

Rengaraj S, Yeon KH, Moon SH (2001) Removal of chromium from water and wastewater by ion exchange resins. J Hazard Mater 87(1–3):273–287

Singh R (2015) in Membrane technology and engineering for water purification, 2nd ed, Butterworth-Heinemann, Oxford Ch. 3, pp 179–281

Smith L (1995) Contaminants and remedial options at selected metal-contaminated sites. Technical resource report, Battelle, Columbus, OH (United States)

Tsai WT, Chang CY, Wang SY, Chang CF, Chien SF, Sun HF (2001) Preparation of activated carbons from corn cob catalyzed by potassium salts and subsequent gasification with CO2. Bioresour Technol 78(2):203–208

Tünay O (2003) Developments in the application of chemical technologies to wastewater treatment. Water Sci Technol 48(11–12):43–52

Vengris T, Binkiene R, Sveikauskaite A (2001) Nickel, copper, and zinc removal from wastewater by a modified clay sorbent. Appl Clay Sci 18:183–190

Wang LK, Vaccari DA, Li Y, Shammas NK (2004) Chemical precipitation. In Wang LK, Hung YT, Shammas NK (eds) Physicochemical treatment processes, vol. 3. HumanaWiley P. E, Trent, J. D. (2016). Clarification of algae-laden water using electrochemical processes. Water Sci Technol Water Supply 16(2):314–323

Yu B, Zhang Y, Shukla A, Shukla SS, Dorris KL (2000) The removal of heavy metal from aqueous solutions by sawdust adsorption—removal of copper. J Hazard Mater 80(1–3):33–42

Chapter 6
Application of Advanced Oxidation Processes (AOPs) for the Treatment of Petrochemical Industry Wastewater

Shilpi Verma, Praveen Kumar, Vimal Chandra Srivastava, and Urška Lavrenčič Štangar

Abstract Petrochemicals are derived directly or indirectly from petroleum. During the production of petrochemicals, various refractory and toxic pollutants are generated, which are harmful to the aquatic system. These pollutants are genotoxic, carcinogenic, and teratogenic. Although these pollutants are harmful, only a limited number of studies have been performed on their removal or degradation. Conventional wastewater treatment techniques are not efficient in removing these pollutants from wastewater. Therefore, the scientific community is now putting up efforts to make the Advanced Oxidation Processes (AOPs) more efficient. The main aim is to produce very reactive species (especially hydroxyl radicals, $HO^{\cdot}$) in water by the AOPs. This leads to the degradation or transformation of the contaminants or pollutants, resulting in complete mineralization. The most studied AOPs for the degradation of petrochemical wastewater pollutants are photocatalytic oxidation (PCO), ozonation, catalytic oxidation, catalytic ozonation, and electrochemical advanced oxidation processes. This chapter specifically discusses the various AOPs used to remove specific pollutants and chemical oxygen demand (COD) from petrochemical wastewater treatment to meet discharge standards.

S. Verma (✉)
School of Energy & Environment, Thapar Institute of Engineering & Technology, Patiala, Punjab 147004, India
e-mail: shilpi.verma@thapar.edu

P. Kumar · U. L. Štangar
Faculty of Chemistry and Chemical Technology, University of Ljubljana, 1000 Ljubljana, Slovenia
e-mail: praveen@engineering.ucsb.edu; Kumar.Praveen@fkkt.uni-lj.si

U. L. Štangar
e-mail: Urska.Lavrencic.Stangar@fkkt.uni-lj.si

V. C. Srivastava
Department of Chemical Engineering, Indian Institute of Technology, Roorkee, Roorkee, Uttarakhand 247667, India
e-mail: vimal.srivastava@ch.iitr.ac.in

S. Roy et al. (eds.), *Advanced Industrial Wastewater Treatment and Reclamation of Water*, Environmental Science and Engineering,
https://doi.org/10.1007/978-3-030-83811-9_6

Abbreviations

AOP	Advanced Oxidation Processes
API	American Petroleum Institute
ASP	Activated sludge processes
BOD	Biological oxygen demand
BPT	Best practicable control technology
BTEX	Benzene, toluene, ethylbenzene, and xylene
CCL	Candidate contamination list
COD	Chemical oxygen demand
DAF	Dissolved air flotation
DAP	Diammonium phosphate
DCE	1,2-Dichloroethane
DCM	Dichloromethane
DNA	Deoxyribonucleic acid
DOM	Dissolved organic matter
EAOPs	Electrochemical advanced oxidation processes
EF	Electro-Fenton
EPS	Extracellular polymeric substances
GAC	Granular activated carbon
LB-EPS	Lightly bound EPS
MBR	Membrane bio-reactor
MEB	Methyl ethylbenzene
MTBE	Methyl tert-butyl ether
PAC	Powered activated carbon
PC	Peroxi-coagulation
PCO	Photocatalytic oxidation
PEF	Photoelectro Fenton
PTA	Purified terephthalic acid
RBC	Rotating biological contactor
RO	Reverse Osmosis
SBR	Sequencing batch reactor
SPEF	Solar photoelectro-Fenton
TB-EPS	Tightly bound EPS
tBME	T-butyl methyl ether
TDS	Total dissolved solids
TOC	Total organic carbon
TPHI	Total petroleum hydrocarbon index
TPI	Tilted plate interceptor
TSS	Total suspended solids
UF	Ultra Filtration
US	Ultrasound
US EPA	U.S. Environmental Protection Agency

UV	Ultraviolet
VOCs	Volatile organic compounds

6.1 Introduction

With rapid urbanization and industrialization, new products have been introduced in the market. The primary source of these chemicals is petroleum and natural gas. The organic chemical industry depends on seven chemical building blocks, namely methane, xylene, benzene, toluene, ethylene, propylene, and C4 olefins, which are derived from the downstream processing of petroleum and natural gas. Petroleum refineries play a vital role in producing these chemicals, as petrochemical production plants are usually located on the premises of these refineries (Wittcoff et al. 2013). The production of these chemicals generates large amounts of wastewater and produces air pollutants and solid wastes (Jafarinejad 2017a). The amount of water required by a particular petrochemical plant varies with the type of petroleum processed, the size of the petrochemical plant, the variety of petrochemicals produced, and the complexity of the operation (El-Naas et al. 2014; Alva-Argáez et al. 2007). On average, a petrochemical production unit generates about 1500 gallons per minute of wastewater (Crittenden and Kujawski 2020).

Various production processes generate the effluent, such as process water, vapour condensation, catalyst recovery section, spent caustic in crackers and aromatics plants. Cooling tower blowdown is another major source of wastewater. Surface effluent from petrochemical production unit also contributes to the wastewater generation as it is contaminated with various types of chemical solvents, spillage, and distillates (Jafarinejad and Jiang 2019; Jafarinejad 2017a; MIGA 2004; IL & FS Ecosmart Limited Hyderabad 2010; US EPA/310-R-95-013 1995; IPIECA Petroleum refining water/wastewater use and management 2010).

The various types of aliphatic and aromatic toxic substances present in the wastewater stream must be removed before discharge to any surface water body. This will help to reduce the pollution load while making the treated effluent wastewater available for reuse and recycling in various other processes in the petrochemical production unit (Jafarinejad 2017a, b, Zhong et al. 2003).

Petrochemical wastewater is usually treated by biological processes as a secondary treatment step, such as continuous stirred tank reactor, two-stage aerobic treatment, activated sludge process, upflow sludge blanket reactor, anaerobic fixed film reactor, downflow tubular fixed film reactor, and expanded granular activated carbon (GAC). These treatment techniques are very efficient in removing contaminants. However, due to the inherent limitations of microbes, recalcitrant and toxic chemicals cannot be degraded during the secondary treatment (Jafarinejad and Jiang 2019).

Petrochemical wastewater has specific characteristics that make AOPs as a tertiary treatment process an effective tool. Petrochemical wastewater contains several biologically recalcitrant chemicals, both organic and inorganic. However, these chemicals are highly animable to AOPs as they are used during AOPs. Petrochemical wastewaters usually have a small amount of suspended solids and almost negligible amounts of oil and gas. This makes applying different types of AOPs a very effective technology (Crittenden and Kujawski 2020).

6.2 Petrochemical Wastewater Characteristics and Discharge Standards

In petrochemical wastewater, various types of organic and inorganic chemicals are present, along with heavy metals. The physicochemical parameters such as ammoniacal nitrogen, total acidity as acetic acid, total suspended solids (TSS), chemical oxygen demand (COD), biological oxygen demand (BOD), organic carbon (TOC), total petroleum hydrocarbon index (TPHI), total nitrogen, total dissolved solids (TDS) and pH are commonly measured. In addition, specific organic and inorganic chemicals derived from certain petrochemical production units are analyzed, such as bromide, benzoic acid, 4-carboxy benzaldehyde, benzene, p-toluic acid, terephthalic acid, phenol, toluene, sulfides, fluorides, methyl acetate, propyl acetate, etc. Metals such as aluminum, calcium, cobalt, iron, chromium, manganese, magnesium, and sodium are also present in petrochemical effluents (Jafarinejad 2017a; Orszulik 2008; European Commission Industrial Emissions Directive 2010/75/EU 2013; European Commission Industrial Emissions, Directive 2010/75/EU 2015; Nacheva and Jha 2011; Ishak et al. 2012). Karthik et al. (2008) reported a high COD = 9222 mg/l and BOD = 4150 mg/l in the effluent wastewater of the purified terephthalic acid (PTA) production unit. High amounts of total solids, benzoic acid, terephthalic acid, p-toluic acid, TOC, TDS, and phenol are also reported in the effluent wastewater generating in the PTA production unit (Verma et al. 2011; Karthik et al. 2008; Cheng et al. 1997; Yousefi et al. 2019; Jothinathan et al. 2021).

Due to the high concentration and toxic nature of petrochemicals, various environmental agencies, such as the World Bank Group and the US EPA, have established standard limits for the disposal of treated petrochemical wastewater into a water body or for reuse (Environmental, Health, and Safety Guidelines Petroleum Refining, World Bank Group 2016).

The US EPA has added several petrochemicals to the Candidate Contamination List (CCL 4) (Regulatory Determination 2021 and Contaminant Candidate List). This action by the US EPA signifies the use of such wastewater technologies that can degrade or remove these harmful petrochemicals from wastewater before they are discharged into a water body. AOPs have emerged as one of the techniques that can convert the toxic chemicals emanating from the petrochemical plants into carbon dioxide and water through transitory oxidant species such as hydroxyl radicals or sulphate radicals.

The U.S. Environmental Protection Agency (US EPA) has given the guidelines for petrochemical wastewater discharge that represent the level of effluent reduction that

Table 6.1 Shows the effluent limitation guidelines for petrochemical operations (Electronic Code of Federal Regulations 2021)

Pollutant property/Pollutant	Limitations of BPT effluent	
	Maximum for one day	Average of daily values for 30 consecutive days (maximum)
Ammonia as N	23.4	10.6
BOD_5	34.6	18.4
COD	210	109
Grease and oil	11.1	5.9
Hexavalent chromium	0.046	0.020
Phenolic compound	0.25	0.120
pH	6–9	6–9
Sulfide	0.22	0.099
Total chromium	0.52	0.30
TSS	23.4	14.8

Unit: Metric units (kg/1000 m^3 of feedstock)

can be achieved by applying the current best practicable control technology (BPT) to be followed by industries, as shown in Table 6.1.

6.3 Ramifications of Petrochemical Industry Wastewaters on Environment

The chemicals/contaminants contained in petrochemical wastewater pose a threat to humans and aquatic life. Xiu et al. (2014) have found that chrysene, benzo[a]pyrene and benzo[b]fluoranthene are capable of causing damage to scallop DNA. These three chemicals have shown genotoxic and carcinogenic effects on humans (Pogribny 2019; Huang and Penning 2014). Repeated exposure to anthracene has been reported to cause changes in skin pigmentation, but little information is available on the effects of anthracene on the reproductive system. Naphthalene is a common polyaromatic carbon with the ability to cause liver and kidney damage. It is also suspected of being carcinogenic. Exposure to high concentrations can cause reproductive defects (Speight 2011). BTEX (benzene, toluene, ethylbenzene, and xylene) is an extensively used petrochemical. However, due to the toxic nature of BTEX, its widespread use has been criticized. BTEX is a known carcinogenic, neurotoxic, and genotoxic substance (Yu and Gu 2006; Carvajal et al. 2018; Jain et al. 2020). Another petrochemical, MTBE (methyl tert-butyl ether), has shown carcinogenic properties in addition to adverse effects on the respiratory, nervous, and cardiac systems (Yu and Gu 2006). Scientists have also reported the harmful effects of petrochemicals on invertebrates and plants. Microorganisms are more sensitive towards petrochemicals in comparison to invertebrates and plants. Therefore, various ecological processes such as nutrient cycles and biodegradation get negatively impacted (Hentati et al. 2013; Roslev et al. 2015). Phenolic compounds are another widely used petrochemical that has been categorized as priority pollutants due to their toxic effect. The toxicity manifests itself ingenotoxicity, immune toxicity, carcinogenicity, teratogenicity, etc. Even at low concentrations, phenol can impair the function of metabolic and enzyme systems of aquatic microorganisms. In addition, phenolic compounds are challenging to biodegrade or take a long time to degrade. Due to this property, they get accumulated in the tissues and lead to the bio-magnification of these phenolic compounds (Mustapha and Lens 2018).

An acute toxic effect of formaldehyde, 2,4-dichlorophenol, and pyridine present in the effect of a petrochemical plant on *Daphnia magna* was reported by Sponza and Oztekin (2010). Volatile organic chemicals (VOCs) are also the major pollutants coming from a petrochemical plant and are carcinogenic in nature. A study conducted by Hajizadeh et al. (2018) in a petrochemical plant in Iran found that when workers are exposed to VOCs such as phenol, ethylbenzene, benzene, styrene, and phenol in ambient air, they are likely to suffer from cancer. A similar study was conducted Wang et al. (2015) in China on PAH exposure of workers employed in a petrochemical plant. It was observed that the workers had a higher risk of developing cancer compared to normal people. Therefore, it is imperative to treat the effluent wastewater coming from petrochemical plants.

6.4 Petrochemical Industry Wastewaters Treatment Technologies

The operating components of petrochemical wastewater treatment facilities are comparable to those of municipal wastewater treatment facilities (Mantzavinos 2007). Petrochemical wastewater treatment plants have facilities that use the physical, chemical, and biological processes, as shown in Table 6.2 (Pophali et al. 2007). However, the primary wastewater treatment includes two stages, where water and oil are separated. A flow equalization tank is also an essential unit since petrochemical production generates different wastewater streams. Flow equalization tank reduces the fluctuations in the wastewater stream and also dampens the fluctuations in the effluent flow. Concentration balancing systems are also used in some petrochemical plants to prevent the shock loading of recalcitrant organic and inorganic contaminants to the biological treatment plants. Depending on the petrochemical manufacturing process, the flow and concentration equalization systems can be placed either upstream of the DAF separator or upstream of the API separator (Izadyar and Fatemi 2013).

In secondary treatment processes, microorganisms are used to degrade the various contaminants present in petrochemical wastewater. The microorganisms are either acclimatized by the particular petrochemical wastewater treatment plant. Sometimes

Table 6.2 Wastewater treatment technologies employed by petrochemical plants worldwide

Industrial unit	Pre, primary and secondary treatments	Tertiary treatments	Usage of treated wastewater	References
Bharath Petroleum Corporation Ltd. (BPCL), Capacity-Expansion-cum-Modernization Project (CEMP) Program	Dissolved Air Flotation (DAF), Tilted Plate Interceptor (TPI), Secondary Clarifier, Aeration Tank, Bio tower, Filters, Chemical dosing and Handling system	VOC Control System	Surface discharge or recycling	Hindustan Dorr Oliver Limited, EPC: Water & Wastewater Management: Water & Wastewater Management Projects (2021)
HPCL Mittal Energy Ltd.	Integrated concurrent technologies with Sequential Batch Reactor (SBR), Sludge Bio-remediation Membrane Bioreactor (MBR)	VOC Control System	Zero Discharge	Hindustan Dorr Oliver Limited, EPC: Water & Wastewater Management: Water & Wastewater Management Projects (2021)
PEMEX refinery, Mexico	Primary clarifier, DAF, secondary clarifier, activated sludge aeration basin	Membrane filtration RO, and UF	Some amount of water is reused in plant	Wong (2012)
GalpEnergia, Porto oil refinery	API separator, coagulation, neutralization, and flotation, conventional activated sludge process	mechanical aeration clarification, chlorination and filtration	Some of the treated wastewater is recycled and rest discharged into the sea	Santos et al. (2015)
Tehran Oil Refinery Co. (TORC)	API separator, DAF, floatation, Equalizing tank, sedimentation tank, filtration and biological oxidation		Majorly use for forest irrigation, some of the treated water is recycled in cooling tower	Rasouli et al. (2010)

(continued)

Table 6.2 (continued)

Industrial unit	Pre, primary and secondary treatments	Tertiary treatments	Usage of treated wastewater	References
Kaduna Refinery and Petrochemical Co. Ltd.				
Lindsey Refinery, U.K.	API separator, aerated lagoon, clarifier, dissolved air flotation or chemical flocculation, trickling filter, equalization tank		Treated water is discharged to surface water body	Ivana Regina Couto de Brito (2009)
Humber Refinery, U.K.	Equalization tank, ASP, water cascade, aerobic reactors	Air flotation unit for final discharge	Treated water is discharged to surface water body	Ivana Regina Couto de Brito (2009)

specific microbes are also purchased to degrade the recalcitrant target contaminants, which are already acclimatized to degrade the target contaminant (Ishak et al. 2012). A mini-review on phthalates degradation in the environment by the bacteria and fungi under aerobic/anoxic and anaerobic conditions was provided by Liang et al. (2008). Activated sludge processes (ASP) are widely used in petrochemical treatment plants due to their high treatment efficiency and stability. Various other types of aerobic biological processes that have been used to treat the petrochemical wastewater are ASP integrated with powdered activated carbon, membrane bioreactor (MBR), sequencing batch reactor (SBR), aerated lagoons, trickling filters, and rotating biological contactor (RBC) (Izadyar and Fatemi 2013; Jemli et al. 2017; Chavan and Mukherji 2008). Due to certain features of an aerobic process such as high energy consumption, generation of a large volume of sludge, poor sludge settling, filterability, and trained personnel, anaerobic treatment was preferred.

Nowadays, the focus is on utilizing the outstanding features of each treatment process in an integrated biological treatment system. This helps in achieving treatment efficiencies (Wang et al. 2017a, b). A Purified Terephthalic Acid (PTA) production unit in Panipat, India, uses both anaerobic and aerobic biological reactors in series to treat PTA wastewater in the second stage of treatment. For proper growth and metabolism of microorganisms, fertilisers such as diammonium phosphate (DAP) and urea, as well as micronutrients such as ferric sulphate and nickel chloride, may need to be provided (Brief process description of PTA-ETP 2016). The biological treatment processes have proven to be successful in achieving high removal of BOD. The range can vary from 50 to 95%, depending on the characteristics of the wastewater and the type of bioreactor used (Jafarinejad and Jiang 2019; Petroleum refining water/wastewater use and management, IPIECA 2010).

Due to the toxic and recalcitrant nature of the contaminants/impurities present in petrochemical wastewater, it is necessary further to treat this effluent wastewater with advanced oxidation processes.

6.4.1 Advanced Oxidation Processes for the Treatment of Petrochemical Wastewater

Tertiary treatment must be considered to treat effluent petrochemical wastewater because of the stringent discharge limits for various contaminants such as COD, suspended and dissolved metals, TSS, and trace organics such as polyaromatic hydrocarbons (PAHs).

AOPs have the ability to degrade the recalcitrant and toxic components of petrochemical wastewater through the generation of radicals. These radicals could either be the hydroxyl radical or the sulphate radical, depending on the AOP technique (Giannakis et al. 2020). Figure 6.1 shows that there has been an increase in the application of AOPs for the treatment of petrochemical wastewater in the last decade.

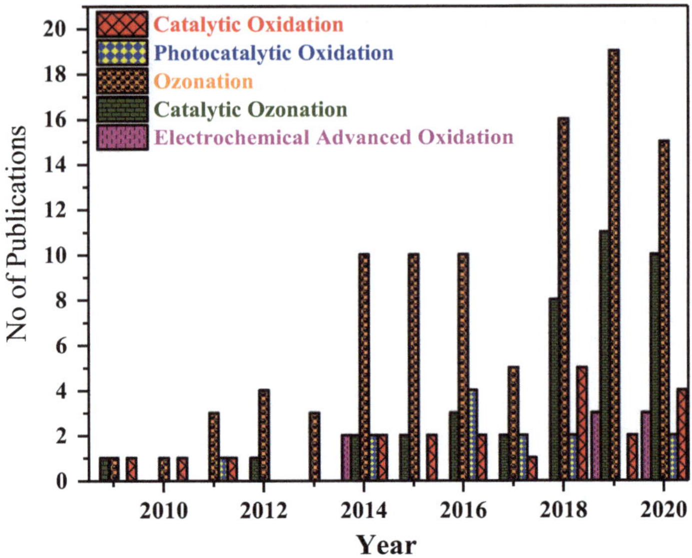

Fig. 6.1 Increase in application of AOPs for the treatment of petrochemical wastewater in last decade

6.4.1.1 Catalytic Oxidation

Recently catalytic oxidation, photocatalytic oxidation, and electrochemical oxidation have been the focus of researchers for the degradation of toxic and recalcitrant contaminants present in petrochemical wastewater. Under the optimum treatment conditions of pH, oxidant, and substrate/pollutant, the catalytic oxidation process can completely mineralize the chemical pollutant into carbon dioxide and water. During the catalytic oxidation of the chemicals with sulfur, nitrogen, and other heteroatoms, nitrate and sulphate anions can also be formed. Noble metal support (Pt, Pd, Ru, and Rh) on TiO_2 as catalysts has shown significant chlorobenzene degradation during the catalytic oxidation process. Ru/TiO_2 showed superior performance by providing the lowest polychlorinated by-products formation and the highest degradation efficiency. The various identified polychlorinated by-products were di-chlorobenzene, trichlorobenzene, tetrachlorobenzene, pentchlorobenzene and hexachlorobenzene (Liu et al. 2019).

Photocatalytic Oxidation (PCO)

There are certain substances or materials that can use light energy to facilitate thermodynamically downward reactions, i.e., the Gibbs free energy is negative. Such materials change the kinetics of the reaction by creating new reaction routes without changing the thermodynamics. These materials are known as photocatalysts (Yang and Wang 2018).

The photocatalytic oxidation process usually uses semiconductor materials that absorb incident radiation. However, there are other materials used as photocatalysts, such as molecular photocatalysts, emerging photocatalysts, plasmonic metal photocatalysts, in addition to traditional photovoltaic-assisted semiconductor-based catalysts (Yang and Wang 2018). With materials other than semiconductors, other principles charge separation processes and govern light absorption during photocatalysis. Since semiconductor-based photocatalysts are mainly used to degrade pollutants/chemicals in petrochemical wastewater, the band theory is suitable for the discussion here. Semiconductor materials have a valence band (full of electrons) and a conduction band (lowest unoccupied band). The energy difference between the two bands (conduction band and the valence band) is called the bandgap energy. When the semiconductor materials are illuminated with light/photons, the electrons (e−) leave the valence band and move to the conduction band, leaving a hole (h+) in the valence band. These electron–hole pairs, h+ e−, can either recombine (either at the surface of the photocatalyst or in bulk) to release the heat, or they can reach the surface of the photocatalyst. The adsorbed pollutant can be oxidized by the hole (h+) and converted into water and carbon dioxide. On the other hand, the electrons will reduce the absorbed oxygen to water (Mills and Soo-Keun Lee 2005).

Photonic efficiency (δ) is an important parameter for assessing the efficiency of semiconductor photocatalytic systems, which can be defined as follows (Mills and Soo-Keun Lee 2005):

$$\delta = \frac{Overall\, rate\, of\ photocatalytic\ process}{Intensity\, of\ incident\ light} \tag{6.1}$$

To reduce the rate of electron–hole recombination, various parameters such as size, chemical composition, electronic properties, and interfacial properties can be fine-tuned during the photocatalyst synthesis (Hayati et al. 2018). Ultraviolet (UV), natural solar, or visible light (Lachheb et al. 2002; Herrmann et al. 2002; Izadyar and Fatemi 2013) can be used as a light source for the excitation of electrons and subsequent production of hydroxyl or sulfate radicals. The photocatalytic oxidation process is effective in the degradation of various organic substances found in petrochemical wastewater such as various chlorophenols (Pera-Titus et al. 2004), n-alkanes (C1–C12), dichloromethane (DCM), tBME—t-butyl methyl ether (tBME), 1,2-dichloroethane (DCE), methyl ethylbenzene (MEB), (Stepnowski et al. 2002), p-nitrophenol, naphthalene, and dibenzothiophene (Vargas and Núñez 2010), etc. In PCO, catalysts are used either as suspended particles or as immobilized beds in the aqueous phase. TiO_2 is a very commonly used photocatalyst. Evonik Degussa P25 is an industrially available TiO_2 catalayst, consisting of anatase (70%) and rutile phases.The electron–hole recombination rate in TiO_2 is very fast, which leads to a short lifetime of the photogenerated electron–holes. Various techniques have been developed to overcome this problem, such as doping of non-metals/metals, introducing heterostructures, using other semiconductors along with TiO_2 or separately and by architectural control, etc. (Hayati et al. 2018).

Various other semiconductor catalysts such as $Bi_4O_5Br_xI_{2-x}$ (Meng et al. 2018), $PdO/Al_2O_3–Nd_2O_3$ (Balgude et al. 2019), PF/TiO_2 (Li et al. 2018), Sn_3O_4microballs, Fe_2O_3 (Vosoughi et al. 2017), La-doped Zn (O, S) nanoparticles (Abdullah et al. 2019), B–GO–TiO_2 (Shokri et al. 2016), Fe_2O_3/RGO nanocomposite (Mohan et al. 2019), MoS_2/rGO, CdS–MoS_2/rGO composite (Peng et al. 2016) have also been shown to be effective in the degradation of various petrochemicals. For the PCO process to be effective, several variables such as light intensity, amount of catalyst, temperature, pH, UV irradiation time, and initial pollutants concentration must be optimized (Shahrezaei et al. 2012). Although through PCO, high degradation of petrochemicals has been achieved, the complete mineralization into water and carbon dioxide has not been achieved. Intermediate chemicals/compounds are formed due to the partial oxidation of contaminants/pollutants. These intermediates may remain in the wastewater as hazardous by-products and reduce the reaction rate and overall removal efficiency (Bahri et al. 2018). Both batch and benchtop reactors have been used to perform PCO. In some cases, a pilot-scale study was also conducted (Pera-Titus et al. 2004; Vargas and Núñez 2010; Shahrezaei et al. 2012). Operation at ambient conditions of pressure and temperature is the major advantage in PCO application. Due to scientific progress in the area of photocatalysts, PCO can be carried out under solar radiation or visible light (Tismanar et al. 2021). Meng et al. (2018) observed that bismuth oxybromide/oxyiodide photocatalyst exhibited significant photocatalytic activities in the decomposition ofo-phenylphenol, resorcinol, and 4-tert-butylphenol under visible light irradiation. $•O_2^-$, •OH, and h^+ were formed during the photocatalytic process. In this process, $•O_2^-$ and h^+ played the main role in the decomposition of phenolic pollutants. The degradation pathway of resorcinol was suggested. Hayati et al. (2018) used GZnTi as a photocatalyst at a dosage of 0.6 g/L and pH = 4 under visible light irradiation to treat the petrochemical wastewater. The TOC and COD removal were found to be 85.9% and 92.1%, respectively, after 440 min. The prolonged time in this petrochemical wastewater treatment can be reduced by adding oxidants such as ozone, peroxymonosulfate (PMS) or hydrogen peroxide. The addition of oxidants reduces the treatment time by accelerating the formation of sulphate and hydroxyl radicals (Bahri et al. 2018). Eslami et al. (2018) observed that UV/MnO_2/PMS completely removed 4-chlorophenol in 30 min. When this integrated process was used to treat real petrochemical wastewater, the COD, TOC, phenol, and BOD_5 removal were 54, 46, 63, and 42%, respectively. A low-cost floatable ZnO/Oak charcoal composite was used as a photocatalyst by Tafreshi et al. (2019) to treat NH_3 industrial wastewater from Kermanshah Petrochemical Company, Iran. An ammonia removal of 86.7% was observed at the optimum treatment condition with a light intensity of 250 W, catalyst dosage of 2 g/L, and pH of 9. The catalyst recycling study revealed that photocatalytic activity decreased after three recycling cycles. This decrease resulted from the poisoning effect of the reactants deposited on the reactive site of the catalyst.

Ozonation/Catalytic Ozonation

Due to the high oxidation potential of 2.07 V and the ability to react with the selection of chemicals, ozone is often used as an oxidizing agent to degrade the recalcitrant contaminants in wastewater streams. According to the valence state theory, ozone derives its strong oxidative strength due to its special molecular formula $O^{+4}(O_2)^{-2}$. According to the resonance structure of ozone proposed by Kalemos and Mavridis (2008), one oxygen atom shows the electrophilic property, while another oxygen atom shows a nucleophilic property. The oxygen atom showing the electrophilic property can react effectively with unsaturated functional groups, amines/sulfides, and other reductive compounds. In contrast, the other oxygen atom can react with carbonyl, strong Brønsted acids, and carbon–nitrogen bond containing substances. This makes ozone a selective oxidizing agent. The ozone can be produced by photochemical excitation, electrochemical reaction, and the reaction of oxygen molecules with oxygen atoms (Wei et al. 2017). Organic contaminants present in petrochemical wastewater are oxidized by hydroxyl radicals and/or ozone generated during the ozonation process. Hydroxyl radicals play an important role in the oxidation of contaminants as they are comparatively less selective and can oxidize the chemicals that are resistant to ozone (Wang et al. 2017a, b). The amount of hydroxyl radicals formed during ozonation (ozone as a single oxidant) is insufficient due to low ozone decomposition. Therefore, to improve the ability of ozone, ozone was integrated with H_2O_2, UV irradiation, and catalyst.

Petrochemical wastewater contains a significant amount of dissolved organic matter (DOM). The removal of DOM is generally measured by total organic carbon (TOC) removal. The presence of particulate matter or DOM in petrochemical wastewater could significantly increase ozone (O_3) consumption during the treatment process. During the secondary/biological treatment of petrochemical wastewater, the extracellular polymeric substances (EPS), both tightly bound EPS (TB-EPS) and loosely bound EPS (LB-EPS), are introduced into the treated wastewater. They can significantly reduce the reaction rate during ozonation or catalytic ozonation treatment of this secondary wastewater (Zhang et al. 2018a). A very low removal of TOC by a single ozonation process, approximately 4.7–15%, has been reported (Fu et al. 2019; Wu et al. 2015). In this process, the hydrophobic components of the petrochemical wastewater were transformed into hydrophilic substances without complete mineralization. This transformation is useful to increase biodegradability. However, when the ozonation process was combined with H_2O_2 (O_3/H_2O_2) and catalysts (O_3/catalyst), the TOC removal was significantly increased to 42.8%. The increased removal of TOC (96%) was also reported by Huang et al. (2019) for the petrochemical wastewater with O_3/catalyst (iron-nickel foam as a catalyst). For this removal, the optimum operating conditions were 10.2 mg/L ozone concentration, catalyst dosage 110 g/L, time duration = 120 min time in a wide pH range. Ahmadi et al. (2017) used PAC@Fe_3O_4 as a catalyst and ozone for the treatment of high salinity petrochemical wastewater. 50.3% TOC and 75.3% COD removal were achieved.

Mohamadi et al. (2020) observed that catalytic ozonation (nanostructured MgO as a catalyst) is more effective for the removal of ethylbenzene, benzene, xylene (BTEX), and toluene from petrochemical wastewater compared to single ozonation. After 30 min of reaction time, the removal of BTEX by catalytic ozonation was 93.8, 83.5, 94.6, and 99.2%, respectively. However, the removal by single ozonation process was 35, 58, 81, and 92%, respectively. Recently, Jothinathan et al. (2021) combined microbubble ozonation with a catalyst. Fe/GAC was used as a catalyst which increased the •OH production during microbubble ozonation of petrochemical wastewater. This resulted in 88% COD removal at optimum operating parameters. This removal is 18% higher than that achieved through microbubble alone and 43% higher than that achieved by the macrobubble ozonation process alone.

Factors Affecting the Catalytic Oxidation

Several key parameters must be optimized for the catalytic oxidation process to degrade the pollutants in petrochemical wastewater effectively. The key parameters are the catalyst loading, the type of support materials, and the type of catalyst suitable for the oxidation of contaminants. Initially, when the catalyst loading is increased, the degradation also increases. However, as the catalyst loading is further increased, the degradation stabilizes or decreases. Overloading the catalyst decreases the surface area where the reaction can occur; consequently, the degradation decreases. Eslami et al. (2018) reported that at both 0.25 g/L and 0.35 g/L MnO_2, the degradation of 4-chlorophenol was complete. Similarly, Rezaei et al. (2013a) used different loadings of Mn (1, 5, 10, and 20%) on alumina to degrade toluene. It was observed that 10% Mn on alumina gave the highest degradation among the different Mn loadings. During characterization, it was found that Mn_2O_3 is the dominant phase at low Mn loading compared to that of MnO_2 and Mn_2O_3. Therefore, at low Mn loading, the transfer of electrons to ozone to initiate ozone decomposition is relatively easy. Consequently, the rate of decomposition of ozone to oxygen atoms increases, leading to enhanced degradation of toluene.

Zhang et al. (2017) studied for the catalytic oxidation of 1,2-dichloropropane in the presence of $LaMnO_3$ perovskites on YSZ, Al_2O_3, CeO_2, and TiO_2. It was observed that $LaMnO_3$ perovskites loaded on CeO_2 exhibited the highest degradation efficiency among the other support materials.

6.4.1.2 Electrochemical Advanced Oxidation Processes (EAOPs)

In the last two decades, electrochemical advanced oxidation processes (EAOPs) have attracted much attention from the scientific community due to their ability to remove persistent organic pollutants from petrochemical wastewater. There are various types of EAOPs currently used for the removal of recalcitrant pollutants, among which Anodic Oxidation (AO) is widely used. In AO, the pollutants are oxidized directly at the surface of the anode by electron transfer and/or indirectly by the hydroxyl radical

(•OH). Sometimes AO is assisted by various other oxidants such as active chlorine species, persulfates, ozone, and hydrogen peroxide, to enhance the degradation of the contaminants. Other EAOPs used to treat various types of wastewater are anodic oxidation with electrogenerated H_2O_2, photoelectro-Fenton (PEF), electrochemical peroxidation, solar photoelectro-Fenton (SPEF), Fered-Fenton, sonoelectro-activated persulfate, peroxi-coagulation (PC), and sonoelectro-Fenton (Moreira et al. 2017).

Yousefi et al. (2019) studied various EAOPs such as persulfate, electro-activated persulfate, sono-activated persulfate, sonoelectro-activated persulfate, and ultrasound processes for the degradation of contaminants in petrochemical wastewater. It was found that the combined sonoelectro-persulfate treatment was the most effective, removing 82.3% COD. The higher removal was the result of the oxidation of the contaminants by both the hydroxyl and sulfate radicals. This was followed by electro-persulfate, sono-persulfate, persulfate, and a single ultrasonic system. A very effective hybrid system consisting of magnetic powdered activated carbon (catalyst), UVirradiation to active H_2O_2, and coupled with ultrasound (US) was applied to treat the petrochemical wastewater by Kakavandi and Ahmadi (2019). About 87% COD removal was achieved with this hybrid treatment with increased biodegradability of the treated wastewater. Hydroxyl radical and $O_2^{\bullet-}$ were involved in the degradation process. A degradation scheme was proposed in which degradation occurred at the surface of activated carbon (PAC) and in solution through the formation of reactive species. The high surface area of PAC played an important role in the adsorption of pollutants and oxidants on the surface of PAC. This effective contact led to the decomposition of H_2O_2, resulting in the formation of •OH, •HO_2 and $O_2^{\bullet-}$. The reaction between H_2O_2 and Fe_3O_4 played a significant role in the formation of the reaction species.

Another route of •OH generation is through the reaction between $O_2^{\bullet-}$ and H_2O; and between $O_2^{\bullet-}$ and H_2O_2. The main purpose of using UV and US here is to accelerate the decomposition of H_2O_2 into reactive oxidizing species. PAC also catalyzes the decomposition of H_2O_2. In the liquid phase, the Fenton mechanism is responsible for the decomposition of H_2O_2 as iron is leached from the catalyst surface and subsequently catalyzes the decomposition of H_2O_2. In the liquid phase, the ultrasonication will decompose the water vapor, followed by the reaction between oxygen and water molecules. In this process, HO_2• and HO• species are formed. These reactive species attack the organic contaminants and decompose them into simpler molecules, water and carbon dioxide. A schematic representation of this process can be seen in Fig. 6.2.

However, these reactive species also recombine to produce water, oxygen, and H_2O_2. In this hybrid process, several chemical reactions such as dihydroxylation, cleavage of the C–C bond, oxidation, and elimination are responsible for degrading organic toxic synthetic compounds in petrochemical wastewater. A very slight decrease in the efficiency of this hybrid process was observed after 5 recycling cycles.

Graphite electrodes were used by Jorfi et al. (2017) for the novel electrokinetic treatment of petrochemical wastewater. Significant removal of COD along with a high

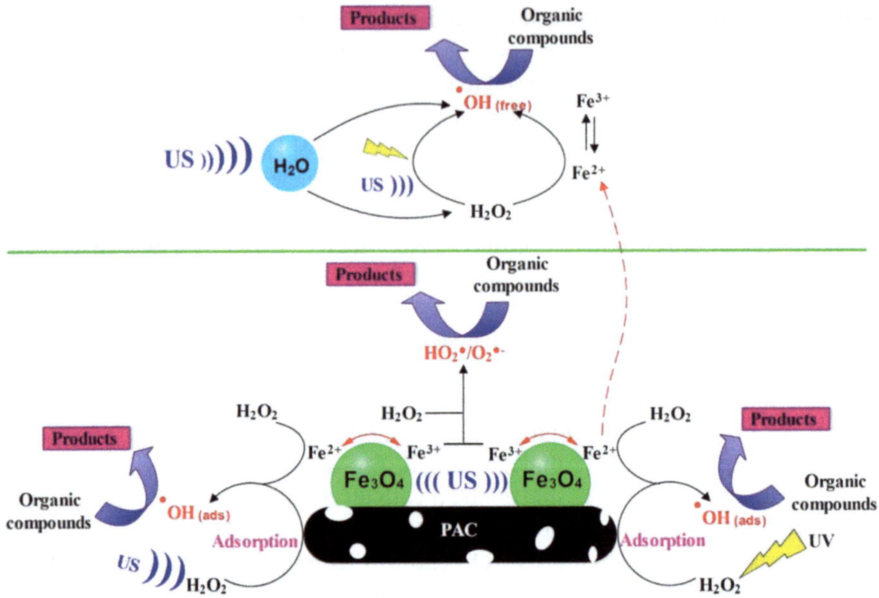

Fig. 6.2 Reactive free radical generation and organic matter degradation during the MPAC/US/UV/H_2O_2 system: a proposed scheme. (Reprinted from Kakavandi and Ahmadi (2019) with permission from Elsevier)

increase in BOD/COD ratio was observed. To evaluate the industrial application of this novel process, the authors scaled up the process. It was found that the energy consumption at the laboratory scale was 32 KWh/m^3, while the energy consumption decreased to 2.1 kWh/m^3 after scaling up 80 times. This shows the feasibility of applying this novel process on an industrial scale. The degradation products such as acetamide, benzaldehyde, dimethylpropane, dodecanoic acid, dodecanol, benzocyclopentane, etc., were identified. The degradation products were formed by the ring cleavage, oxidation, and hydroxylation of the aromatic ring structure of various organic contaminants present in petrochemical wastewater.

An integrated treatment with persulfate and ultrasonication with electro-Fenton (EF) process was applied for the degradation of recalcitrant compounds present in petrochemical wastewater by Ahmadi et al. (2019). It was observed that both the persulfate with EF and ultrasonic treatment with EF removed more than 90% COD. The performance of persulfate with EF was slightly better as it took less time to degrade the contaminants. Improved biodegradability of treated effluent wastewater was also observed. Moattar et al. (2019) applied the photo-electro-Fenton oxidation process for the treatment of saline petrochemical wastewater and achieved 65.9% COD removal.

Influence of Operating Parameters on EAOPs

Various factors affect the efficiency of EAOPs in treating petrochemical wastewater. These operating factors are the initial concentration of various organic and inorganic chemicals, pH, current density, applied current, applied electrode potential, nature of the electrolyte, type of electrodes, temperature, and liquid flow rate. When carbonaceous electrodes are used in any type of EAOP, then the oxygen or air flow rate also affects the degradation process. As with other advanced oxidation processes, the initial concentration of various inorganic or organic materials significantly affects the effectiveness of the degradation. A higher initial concentration results in lower pseudo-first-order kinetic constants, reduced removal of impurities, and inhibition of side reactions. The electrolyte's role is to facilitate the flow of current in the solution. The flow rate of a liquid or the stirring speed also has a significant effect on decomposition. The high flow rate or high stirring will increase the mass transfer of pollutants towards the electrodes and catalyst, or if UV or visible light is used as an irradiation source, then the exposure to the light source will also be increased, leading to effective degradation. The high flow rate or high agitation also reduces the deposition of solids on the electrodes, which increases the degradation of contaminants and extends the life of the electrode.

There are various electrolytes used in EAOPs: sodium chloride, potassium chloride, sodium nitrate, sodium perchlorate, sodium sulfate, and sodium carbonate. The usage of electrolytes adds some ions, and it can either cause the formation of strong oxidants or the scavenging of hydroxyl radicals, or it can produce recalcitrant and toxic by-products or utilize the H_2O_2 or produce some complexes that can alter the amount of photoactive species.

High current density or high applied current leads to a higher generation of oxidizing species and thus higher removal of contaminants from petrochemical wastewater. The temperature and pH of the solution also play an essential role in EAOPs. Temperature does not significantly affect the removal of impurities by EAOPs, but it can be a limiting factor in EAOPs where H_2O_2 electrogeneration occurs. As the amount of dissolved oxygen decreases at high temperatures, the rate of H_2O_2 formation also decreases. In addition, H_2O_2 decomposition also occurs at high temperatures. These two phenomena lead to low degradation of the pollutants. Therefore, in EAOPs where H_2O_2 is electrogenerated, the treatment is generally operated at ambient temperature.

Regarding the effect of pH on EAOP efficiency, there are contradictory observations: (a) the contaminants or pollutants degradation is independent of pH, (b) higher contaminant removal at low pH, i.e., pH 3, and (c) higher contaminant removal at neutral pH. Increased oxygen or airflow rates increase the formation of H_2O_2 and thus the degradation of pollutants. However, very high flow rates are not recommended as this will cause some operational problems (Moreira et al. 2017).

6.5 Energy Consumption in Advanced Oxidation Processes

AOPs are an energy-intensive treatment technology. Therefore, to characterize the performance of an AOP method relative to its alternatives, operational treatment costs are considered. Bolton and Cater (1994) proposed a concept of EE/O to calculate the consumption of electrical energy. The EE/O defines as kWh of electrical energy required to degrade a pollutant or environmental group parameter by order of magnitude in m^3 of contaminated water or wastewater. In the EE/O concept, the degradation of the chemical is assumed to follow first-order kinetics with respect to the chemical concentration in a batch treatment system, using the following formula:

$$\mathrm{EE/O} = \frac{P \times t}{\log\left(\frac{D_0}{D}\right) \times V} = \frac{P \times 38.40}{V \times k_d} \tag{6.2}$$

where, EE/O is the electrical energy (kWh m^{-3}per order), P is the power of the oxidation unit (kW), 38.4 is the conversion factor, V is the volume of wastewater or treated water (L), D and D_0 are the final and initial concentration of the pollutant ($molL^{-1}$), k_d is the pseudo-first-order degradation rate constant for the pollutant (min^{-1}).

When the figure of merit (Olivieri and Escandar 2014) for zero-order decomposition reactions is considered, the electrical energy per mass (EE/M) is considered, which is defined as the kWh of electrical energy required to remove 1 kg of the organic carbon or pollutant.

$$\mathrm{EE/M} = \frac{P \times t}{V \times (C_i - C_t)} \tag{6.3}$$

where, P is the power input (UV lamp in kW), t is the treatment time (h), EE/M is the electrical energy per mass (k Wh kg^{-1} TOC), V is the volume of effluent wastewater or treated water (m^3), C_i, C_t are the collective environmental parameter (initial/final pollutant concentrations), i.e. TOC, COD in kg m^{-3}. Since certain AOPs employs UV light as an artificial light source, calculating TOC electrical energy demand and consumption has been useful. In general, if the EE/O value of an AOP is less than or equal to 10 kWh m^{-3} per order, the AOP can be considered economically feasible (Mills and Soo-Keun Lee 2005).

6.6 Conclusions

The treatment of petrochemical effluents has gained much attention from the scientific community, technologists, and industrialists due to the economic and political facet of the petrochemical industry. Due to the recalcitrant nature of chemicals present in petrochemical effluents, they cannot be treated efficiently by a single process.

The advantage of AOPs over conventional wastewater treatment processes is that they can be used to degrade those recalcitrant chemicals that cannot be degraded by chlorine or ozone. In addition, AOPs require compact operating units because of their fast reaction rates. However, these advantages should be carefully weighed against the operating costs of AOPs and appropriate process control, which may result in less accumulation of oxidation products. One of the bottlenecks of AOPs is that in specific processes, some degradation products are more toxic compared to the starting chemicals due to the extremely fast radical reactions. Highly skilled technical personnel are required to operate AOPs on an industrial scale. In certain AOPs, where a homogeneous catalyst is used, the problem of sludge formation is observed. Recently, SR-AOP has also attracted the research community's attention to treat wastewater containing persistent/recalcitrant pollutants. Sulfate radicals have a similar strong oxidative capacity ($E^0 = 2.60$ V vs. NHE) and short life as the hydroxyl radical. However, the reaction pathway differs from that of hydroxyl radicals, which can be harnessed to treat various recalcitrant chemicals present in petrochemical wastewaters.

6.7 Future Perspectives

Petrochemical wastewater contains several chemicals that are genotoxic, teratogenic, and carcinogenic. Due to the hazardous nature of these chemicals and the stringent discharge limits, it is imperative that the petrochemical wastewater effluents are treated effectively. AOPs have paved the way for the efficient treatment of petrochemical wastewater. The high removal efficiency of various AOPs in removing various petrochemicals from wastewater has been shown in Fig. 6.3

Although recent advances in AOPs were proven effective in degrading these chemicals, focusing on the following aspects of AOPs will help in their widespread application on an industrial scale:

(a) Green synthesis of catalyst: The synthesis of catalysts involves using several organic reagents that may be harmful to the environment due to some inherent chemical properties. Also, in some cases, the amount of catalyst available for use after synthesis is minimal, and large amounts of reagents are wasted. This leads to high costs and the generation of a secondary pollutant stream. Therefore, in the future, the use of new environmentally friendly chemicals will reduce the pollutant load of the catalyst synthesis process.
(b) Development of low-cost and efficient catalysts: Certain metals used in the manufacture of catalysts are rare earth metals, which increases the cost of the catalyst. Therefore, the development of low-cost yet efficient catalysts will help increase the application of photocatalytic AOPs in the industry.
(c) Optimal operating conditions: AOPs are usually tested in laboratories considering the individual components. However, various other chemicals are also present in petrochemical industrial effluents that may affect the operating

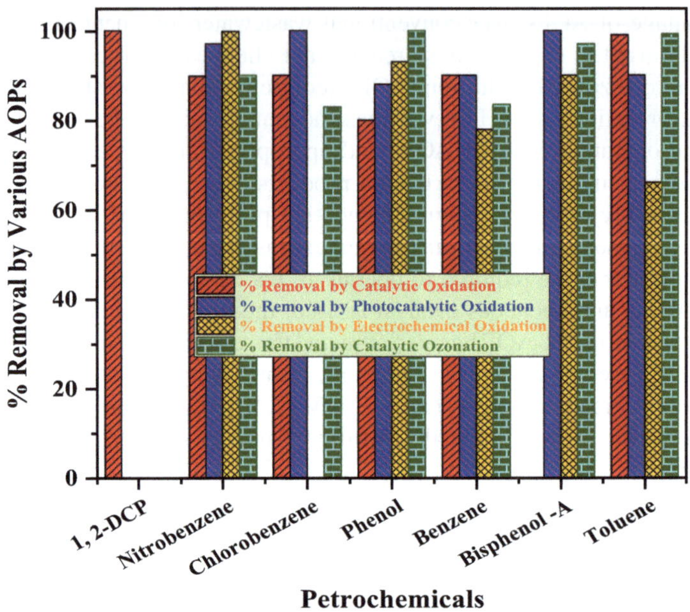

Fig. 6.3 Comparison of various AOPs viz., catalytic oxidation (Hou et al. 2018; Chen et al. 2015; Zhu et al. 2017; Liu et al. 2019; Sable et al. 2018; Zhang et al. 2018b), photocatalytic oxidation (Chen et al. 2020; Nagaraju et al. 2020; Bustillo-Lecompte et al. 2018; Bai et al. 2019; Khaksar et al. 2017; Li et al. 2017), electrochemical oxidation (Dietrich et al. 2017; Cesarino et al. 2013; Zhu et al. 2018; Chen et al. 2014; Gümüs and Akbal 2016) and ozonation/catalytic ozonation (Chen et al. 2020; Zhang et al. 2020a, b; Zhao et al. 2009; Rezaei et al. 2013b) in removing various

conditions optimized for the degradation of a particular component. Therefore, detailed studies with actual industrial effluent wastewater are needed to encourage industries to implement AOPs.

In summary, new developments in the various fields of AOPs will contribute significantly upto the use of this technology by the petrochemical industry to treat its wastewater. Two key factors that may promote industrial application are reduced treatment costs and mitigation of the corrosion problem in AOP reactors.

Acknowledgements The authors would like to thank the Ministry of Education, Science and Sport Republic of Slovenia (Grant No. C3330-19-952015), and the "Slovenian Research Agency", Chemistry Sustainable Development (P1-0134) Program.

References

Abdullah H, Gultom NS, Kuo DH (2019) Synthesis and characterization of La-doped Zn (O, S) photocatalyst for green chemical detoxification of 4-nitrophenol. J Hazard Mater 363:109–118

Ahmadi M, Haghighifard NJ, Soltani RD, Tobeishi M, Jorfi S (2019) Treatment of saline petrochemical wastewater containing recalcitrant organics using Electro-Fenton process: persulfate and ultrasonic intensification. Desalin Water Treat 169:241–250

Ahmadi M, Kakavandi B, Jaafarzadeh N, Babaei AA (2017) Catalytic ozonation of high saline petrochemical wastewater using PAC@FeIIFe$_2$IIIO$_4$: optimization, mechanisms and biodegradability studies. Sep Purif Technol 177:293–303

Alva-Argáez A, Kokossis AC, Smith R (2007) The design of water-using systems in petroleum refining using a water-pinch decomposition. Chem Eng J 128:33–46

Annexure, Brief process description of PTA-ETP (2016). http://www.environmentclearance.nic.in/writereaddata/online/EC/05012018IYUX5675Annexure.pdf. Accessed on 16 January 2021

Bahri M, Mahdavi A, Mirzaei A, Mansouri A, Haghighat F (2018) Integrated oxidation process and biological treatment for highly concentrated petrochemical effluents: a review. Chem Eng Process 125:183–196

Bai X, Yang L, Hagfeldt A, Johansson E, Jin P (2019) D35-TiO_2 nano-crystalline film as a high performance visible-light photocatalyst towards the degradation of bis-phenol A. Chem Eng J 355:999–1010

Balgude S, Sethi Y, Kale B, Amalnerkar D, Adhyapak P (2019) Sn_3O_4 microballs as highly efficient photocatalyst for hydrogen generation and degradation of phenol under solar light irradiation. Mater Chem Phys 221:493–500

Bolton JR, Cater SR (1994) Homogenous photodegradation of pollutants in contaminated water: an introduction. In: Heltz GR, Zepp RG, Crosby DG (eds) Aquatic and surface photochemistry. Lewis Publishers, Florida, pp 467–490

Bustillo-Lecompte C, Kakar D, Mehrvar M (2018) Photochemical treatment of benzene, toluene, ethylbenzene, and xylenes (BTEX) in aqueous solutions using advanced oxidation processes towards a cleaner production in the petroleum refining and petrochemical industries. J Clean Prod 186:609–617

Carvajal A, Akmirza I, Navia D, Pérez R, Muñoz R, Lebrero R (2018) Anoxic denitrification of BTEX: biodegradation kinetics and pollutant interactions. J Environ Manag 214:125–136

Cesarino I, Cesarino V, Moraes FC, Ferreira TC, Lanza MR, Mascaro LH, Machado SA (2013) Electrochemical degradation of benzene in natural water using silver nanoparticle-decorated carbon nanotubes. Mat Chem Phy 141:304–309

Chavan A, Mukherji S (2008) Treatment of hydrocarbon-rich wastewater using oil degrading bacteria and phototrophic microorganisms in rotating biological contactor: effect of N: P ratio. J Hazard Mater 154:63–72

Chen C, Yu J, Yoza B, Li Q, Wang G (2015) A novel "wastes-treat-wastes" technology: role and potential of spent fluid catalytic cracking catalyst assisted ozonation of petrochemical wastewater. J Environ Manag 152:58–65

Chen G, Wang Z, Lin F, Zhang Z, Yu H, Yan B, Wang Z (2020) Comparative investigation on catalytic ozonation of VOCs in different types over supported MnOx catalysts. J Hazard Mater 391:122218

Chen Y, Li H, Liu W, Tu Y, Zhang Y, Han W, Wang L (2014) Electrochemical degradation of nitrobenzene by anodic oxidation on the constructed TiO_2-NTs/SnO_2-Sb/PbO_2 electrode. Chemosphere 113:48–55

Cheng SS, Ho CY, Wu JH (1997) Pilot study of UASB process treating PTA manufacturing wastewater. Water Sci Technol 36:73–82

Contaminant Candidate List (CCL) and Regulatory Determination, https://www.epa.gov/ccl/contaminant-candidate-list-4-ccl-4-0. Accessed on 11 March 2021

Crittenden J, Kujawski D (2020) Oil refining and petrochemical wastewater treatment: creating a better mousetrap. https://www.watertechonline.com/wastewater/article/14176280/oil-refining-and-petrochemical-wastewater-treatment. Accessed on 15 January 2021

Dietrich M, Franke M, Stelter M, Braeutigam P (2017) Degradation of endocrinedisruptor bisphenol A by ultrasound-assisted electrochemical oxidation in water. Ultrason Sonochem 39:741–749

Electronic Code of Federal Regulations (2021) Title 40: Protection of Environment, PART 419—petroleum refining point source category. https://www.ecfr.gov/cgi-bin/text-idx?SID=6b51273d47e8dc451e0aac10f60cdfee&mc=true&node=pt40.31.419&rgn=div5#sp40.31.419.c. Accessed on 25 January 2021

El-Naas MH, Alhaija MA, Al-Zuhair S (2014) Evaluation of a three-step process for the treatment of petroleum refinery wastewater. J Environ Chem Eng 2:56–62

Environmental guidelines for petrochemicals manufacturing, multilateral investment guarantee agency (MIGA) (2004), pp 461–467. https://www.miga.org/documents/Petrochemicals.pdf. Accessed 15 March 2021

Environmental, Heath, and Safety Guidelines Petroleum Refining, World Bank Group (2016), pp 1–35, November 17. https://www.ifc.org/wps/wcm/connect/bde2da1d-3a09-400b-be24-3f6a60353ddc/2016-EHS+Guidelines+for+Petroleum+Refining+FINAL.pdf?MOD=AJPERES&CVID=lxPS7Bu. Accessed 23 February 2021

Eslami A, Hashemi M, Ghanbari F (2018) Degradation of 4-chlorophenol using catalyzed peroxymonosulfate with nano-MnO_2/UV irradiation: toxicity assessment and evaluation for industrial wastewater treatment. J Clean Prod 195:1389–1397

European Commission, Joint Research Center, Best Available Techniques (BAT) Reference Document for the Refining of Mineral Oil and gas, Industrial Emissions Directive 2010/75/EU (Integrated Pollution Prevention and Control), Joint Research Center, Institute for Prospective Technological Studies Sustainable Production and Consumption Unit European IPPC Bureau (2013). https://ec.europa.eu/jrc/en/publication/eur-scientific-and-technical-research-reports/best-available-techniques-bat-reference-document-refining-mineral-oil-and-gas-industrial. Accessed 17 February 2021

European Commission, Joint Research Center, Best Available Techniques (BAT) Reference Document for the Refining of Mineral Oil and Gas, Industrial Emissions, Directive 2010/75/EU (Integrated Pollution Prevention and Control), Pascal Barthe, Michel Chaugny, Serge Roudier, Luis Delgado Sancho, Report EUR 27140 EN (2015) http://eippcb.jrc.ec.europa.eu/reference/BREF/REF_BREF_2015.pdf. Accessed 6 January 2021

Fu L, Wu C, Zhou Y, Zuo J, Song G, Tan Y (2019) Ozonation reactivity characteristics of dissolved organic matter in secondary petrochemical wastewater by single ozone, ozone/H_2O_2, and ozone/catalyst. Chemosphere 233:34–43

Giannakis S, Lin KY, Ghanbari F (2020) A review of the recent advances on the treatment of industrial wastewaters by sulfate radical-based advanced oxidation processes (SR-AOPs). ChemEng J 406:127083.

Gümüs D, Akbal F (2016) Comparison of Fenton and electro-Fenton processes for oxidation of phenol. Procss Saf Environ 103:252–258

Hajizadeh Y, Teiri H, Nazmara S, Parseh I (2018) Environmental and biological monitoring of exposures to VOCs in a petrochemical complex in Iran. Environ Sci Pollut Res 25(7):6656–6667

Hayati F, Isari AA, Fattahi M, Anvaripour B, Jorfi S (2018) Photocatalytic decontamination of phenol and petrochemical wastewater through ZnO/TiO_2 decorated on reduced graphene oxide nanocomposite: influential operating factors, mechanism, and electrical energy consumption. RSC Adv 8:40035–40053

Hentati O, Lachhab R, Ayadi M, Ksibi M (2013) Toxicity assessment for petroleum-contaminated soil using terrestrial invertebrates and plant bioassays. Environ Monit Assess 185:2989–2998

Herrmann JM, Guillard C, Disdier J, Lehaut C, Malato S, Blanco J (2002) New industrial titaniaphotocatalysts for the solar detoxification of water containing various pollutants. Appl Catal B Environ 35:281–294

Hindustan Dorr Oliver Limited, EPC: Water & Wastewater Management: Water & Wastewater Management Projects (2021). http://www.hdo.in/epc/waterWastewaterProjects.html#1. Accessed on 04 February 2021

Hou Z, Feng J, Lin T, Zhang H, Zhou X, Chen Y (2018) The performance of manganese-based catalysts with $Ce_{0.65}Zr_{0.35}O_2$ as support for catalytic oxidation of toluene. Appl Surf Sci 434:82–90

Huang M, Penning TM (2014) Processing contaminants: polycyclic aromatic hydrocarbons (PAHs). In: Motarjemi Y (ed) Encyclopaedia of food safety, vol 2. Academic Press, pp 416–423

Huang Y, Luo M, Xu Z, Zhang D, Li L (2019) Catalytic ozonation of organic contaminants in petrochemical wastewater with iron-nickel foam as catalyst. Sep Purif Technol 211:269–278

IL & FS Ecosmart Limited Hyderabad, Technical EIA Guidance Manual for Petrochemical Complexes, Prepared for the Ministry of Environment and Forests, Government of India, 2010 September. http://environmentclearance.nic.in/writereaddata/Form1A/HomeLinks/TGM_Petrochemical%20Complexes_160910_NK.pdf. Accessed on 5 January 2021

IPIECA, Petroleum refining water/wastewater use and management, IPIECA Operations Best Practice Series, London, United Kingdom (2010). https://www.ipieca.org/resources/good-practice/petroleum-refining-water-wastewater-use-and-management/. Accessed on 02 February 2021

Ishak S, Malakahmad A, Isa MH (2012) Refinery wastewater biological treatment: a short review. J Scient Indust Res 71:251–256

Ivana Regina Couto de Brito (2009) Ecology of nitrification in oil refinery wastewater treatment Systems. PhD Thesis, School of Civil Engineering and Geosciences, University of Newcastle upon Tyne, Newcastle upon Tyne, UK, NE1 7RU. http://theses.ncl.ac.uk/jspui/handle/10443/1161. Accessed on 27 January 2021

Izadyar S, Fatemi S (2013) Fabrication of X zeolite based modified nano TiO_2 photocatalytic paper for removal of VOC pollutants under visible light. Ind Eng Chem Res 52:10961–10968

Jafarinejad S (2017a) Petroleum waste treatment and pollution control, 1st edn. Elsevier Inc., Butterworth-Heinemann, USA

Jafarinejad S (2017b) Recent developments in the application of sequencing batch reactor (SBR) technology for the petroleum industry wastewater treatment. Chem Int 3(3):342–350

Jafarinejad S, Jiang SC (2019) Current technologies and future directions for treating petroleum refineries and petrochemical plants (PRPP) wastewaters. J Environ Chem Eng 7(5):103326

Jain M, Majumder A, Ghosal PS, Gupta AK (2020) A review on treatment of petroleum refinery and petrochemical plant wastewater: a special emphasis on constructed wetlands. J Environ Manag 272:111057

Jemli M, Zaghden H, Rezgi F, Kchaou S, Aloui F, Sayadi S (2017) Biotreatment of petrochemical wastewater: a case study from northern Tunisia. Water Environ Res 89:228–237

Jorfi S, Pourfadakari S, Ahmadi M (2017) Electrokinetic treatment of high saline petrochemical wastewater: evaluation and scale-up. J Environ Manag 204:221–229

Jothinathan L, Cai QQ, Ong SL, Hu JY (2021) Organics removal in high strength petrochemical wastewater with combined microbubble-catalytic ozonation process. Chemosphere 263:127980

Kakavandi B, Ahmadi M (2019) Efficient treatment of saline recalcitrant petrochemical wastewater using heterogeneous UV-assisted sono-Fenton process. Ultrason Sonochem 56:25–36

Kalemos A, Mavridis A (2008) Electronic structure and bonding of ozone. J Chem Phys 129:0543121–0543128

Karthik M, Dafale N, Pathe P, Nandy T (2008) Biodegradability enhancement of purified terephthalic acid wastewater by coagulation: flocculation process as pretreatment. J Hazard Mater 154:721–730

Khaksar A, Nazif S, Taebi A, Shahghasemi E (2017) Treatment of phenol in petrochemical wastewater considering turbidity factor by backlight cascade photocatalytic reactor. J Photochem Photobiol A 348:161–167

Lachheb H, Puzenat E, Houas A, Ksibi M, Elaloui E, Guillard C, Herrmann JM (2002) Photocatalytic degradation of various types of dyes (Alizarin S, Crocein Orange G, Methyl Red, Congo Red, Methylene Blue) in water by UV-irradiated titania. Appl Catal B Environ 39:75–90

Li H, Ji J, Cheng C, Liang K (2018) Preparation of phenol-formaldehyde resin-coupled TiO_2 and study of photocatalytic activity during phenol degradation under sunlight. J Phys Chem Solids 122:25–30

Li J, Xu X, Liu X, Qin W, Wang M, Pan L (2017) Metal-organic frame works derived cake-like anatase rutile mixed phase TiO_2 for highly efficient photo-catalysis. J Alloy Comp 690:640–646

Liang DW, Zhang T, Fang HH, He J (2008) Phthalates biodegradation in the environment. Appl Microbiol Biotechnol 80:183-198
Liu X, Chen L, Zhu T, Ning R (2019) Catalytic oxidation of chlorobenzene over noble metals (Pd, Pt, Ru, Rh) and the distributions of polychlorinated by-products. J Hazard Mater 363:90–98
Mantzavinos D (2007) Basic unit operations in wastewater treatment. In: Oreopoulou V, Russ W (eds) Utilization of by-products and treatment of waste in the food industry. Springer, Boston, MA. https://doi.org/10.1007/978-0-387-35766-9_3. Accessed on 01 February 2021
Meng S, Bi Y, Yan T, Zhang Y, Wu T, Shao Y, Wei D, Du B (2018) Room-temperature fabrication of bismuth oxybromide/oxyiodidephotocatalyst and efficient degradation of phenolic pollutants under visible light. J Hazard Mater 358:20–32
Mills A, Soo-Keun Lee SK (2005) Semiconductor photocatalysis. In: Parsons S (ed) Advanced oxidation processes for water and wastewater treatment, IWA Publishing, Alliance House, 12 Caxton Street, London, pp 137–166
Moattar M, Lotfi M, Jorfi S (2019) Evaluation of photo-assisted electro-Fenton process for the treatment of a saline petrochemical wastewater. Desalin Water Treat 149:138–149
Mohamadi L, Bazrafshan E, Rahdar A, Labuto G, Kamali AR (2020) Nanostructured MgO-enhanced catalytic ozonation of petrochemical wastewater. Bol Soc Esp Ceram Vidrio. https://doi.org/10.1016/j.bsecv.2020.06.002
Mohan BS, Ravi K, Anjaneyulu RB, Sree GS, Basavaiah K (2019) Fe_2O_3/RGO nanocomposite photocatalyst: effective degradation of 4-Nitrophenol. Physica B Condens Matter 553:190–194
Moreira FC, Boaventura RA, Brillas E, Vilar VJ (2017) Electrochemical advanced oxidation processes: a review on their application to synthetic and real wastewaters. Appl Catal B 202:217–261
Mustapha HI, Lens PNL (2018) Constructed wetlands to treat petroleum wastewater. In: Elisabet Aranda E (ed) Prasad R. approaches in bioremediation, Springer, pp 199–237
Nacheva PM, Jha M (eds) (2011) Water management in the petroleum refining industry, water conservation. InTech, ISBN: 978-953-307-960-8 http://www.intechopen.com/books/water-conservation/water-management-in-the-petroleumrefining-industry. Accessed on 4 January 2021
Nagaraju P, Puttaiah SH, Wantala K, Shahmoradi B (2020) Preparation of modified ZnO nanoparticles for photocatalytic degradation of chlorobenzene. Appl Water Sci 10(6):1–15
Olivieri AC, Escandar GM (2014) Analytical figures of merit. Practical Three-Way Calibration, 93–107.
Orszulik ST (2008) Environmental technology in the oil industry. Springer
Peng WC, Chen Y, Li XY (2016) MoS_2/reduced graphene oxide hybrid with CdS nanoparticles as a visible light-driven photocatalyst for the reduction of 4-nitrophenol. J Hazard Mater 309:173–179
Pera-Titus M, García-Molina V, Baños MA, Giménez J, Esplugas S (2004) Degradation of chlorophenols by means of advanced oxidation processes: a general review. ApplCatal B Environ 47:219–256
Petroleum refining water/wastewater use and management, IPIECA Operations Best Practice Series (2010). https://www.ipieca.org/resources/good-practice/petroleum-refining-water-wastewater-use-and-management/. Accessed on 01 February 2021
Pogribny IP (2019) Environmental exposures and epigenetic perturbations. In: Boffetta P, Hainaut P (eds) Encyclopedia of cancer, 3rd edn. Academic Press, pp 574–584
Pophali GR, Khan R, Dhodapkar RS, Nandy T, Devotta S (2007) Anaerobic–aerobic treatment of purified terephthalic acid (PTA) effluent; a techno-economic alternative to two-stage aerobic process. J Environ Manag 85:1024–1033
Rasouli KH, Sarrafzadeh MH, Tavakoli O (2010) An investigation on the nitrogen content of a petroleum refinery wastewater and its removal by biological treatment. Iran J Environ Health Sci Eng 7(5):391–394
Rezaei E, Soltan J, Chen N (2013a) Catalytic oxidation of toluene by ozone over alumina supported manganese oxides: effect of catalyst loading. Appl Catal B Environ 136:239–247
Rezaei E, Soltan J, Chen N, Lin J (2013b) Effect of noble metals on activity of MnOx/γ-alumina catalyst in catalytic ozonation of toluene. Chem Eng J 214:219–228

Roslev P, Lentz T, Hesselsoe M (2015) Microbial toxicity of methyl tert-butyl ether (MTBE) determined with fluorescent and luminescent bioassays. Chemosphere 120:284–291

Sable S, Shah K, Chiang P, Lo S (2018) Catalytic oxidative degradation of phenolusing iron oxide promoted sulfonated-ZrO_2 by advanced oxidation processes (AOPs). J Taiwan Inst Chem Eng 91:434–440

Santos CE, Fonseca A, Kumar E, Bhatnagar A, Vilar VJ, Botelho CM, Boaventura RA (2015) Performance evaluation of the main units of a refinery wastewater treatment plant–A case study. J Environ Chem Eng 3(3):2095–2103

Shahrezaei F, Mansouri Y, Zinatizadeh AA, Akhbari A (2012) Process modeling and kinetic evaluation of petroleum refinery wastewater treatment in a photocatalytic reactor using TiO_2 nanoparticles. Powder Technol 221:203–212

Shokri A, Mahanpoor K, Soodbar D (2016) Evaluation of a modified TiO_2 (GO–B–TiO_2) photo catalyst for degradation of 4-nitrophenol in petrochemical wastewater by response surface methodology based on the central composite design. J Environ Chem Eng 4:585–598

Speight GJ (2011) Pharmaceuticals. In: Handbook of industrial hydrocarbon processes. Gulf Professional Publishing, pp 467–497

Sponza DT, Oztekin R (2010) Removals of PAHs and acute toxicity via sonication in a petrochemical industry wastewater. Chem Eng J 162:142–150

Stepnowski P, Siedlecka EM, Behrend P, Jastorff B (2002) Enhanced photo-degradation of contaminants in petroleum refinery wastewater. Water Res 36:2167–2172

Tafreshi N, Sharifnia S, Dehaghi SM (2019) Photocatalytic treatment of a multicomponent petrochemical wastewater by floatable ZnO/Oak charcoal composite: optimization of operating parameters. J Environ Chem Eng 7(5):103397

Tismanar I, Obreja AC, Buiu O, Duta A (2021) VIS-active TiO_2–graphene oxide composite thin films for photocatalytic applications. Appl Surf Sci 538:147833

United States Environmental Protection Agency (U.S. EPA) (1995) Profile of the petroleum refining industry, EPA Office of Compliance Sector Notebook Project, EPA/310-R-95-013, September Office of Compliance, Office of Enforcement and Compliance Assurance, U.S. Environmental Protection Agency, Washington, DC. https://archive.epa.gov/compliance/resources/publications/assistance/sectors/web/pdf/petrefsn.pdf. Accessed on 21 January 2021

Vargas R, Núñez O (2010) Photocatalytic degradation of oil industry hydrocarbons models at laboratory and at pilot-plant scale. Sol Energy 84:345–351

Verma S, Prasad B, Mishra IM (2011) Thermochemical treatment (thermolysis) of petrochemical wastewater: COD removal mechanism and floc formation. Ind Eng Chem Res 50:5352–5359

Vosoughi M, Fatehifar E, Derafshi S, Rostamizadeh M (2017) High efficient treatment of the petrochemical phenolic effluent using spent catalyst: experimental and optimization. J Environ Chem Eng 5:2024–2031

Wang L, Zhao Y, Liu X, Huang T, Wang Y, Gao H, Ma J (2015) Cancer risk of petrochemical workers exposed to airborne PAHs in industrial Lanzhou City, China. Environ Sci Pollut Res 22:19793 19803

Wang HW, Li XY, Hao ZP, Sun YJ, Wang YN, Li WH, Tsang Y (2017a) Transformation of dissolved organic matter in concentrated leachate from nanofiltration during ozone-based oxidation processes (O_3, O_3/H_2O_2 and O_3/UV). J Environ Manag 191:244–251

Wang S, Ghimire N, Xin G, Janka E, Bakke R (2017b) Efficient high strength petrochemical wastewater treatment in a hybrid vertical anaerobic biofilm (HyVAB) reactor: a pilot study. Water Pract Technol 12:501–513

Wei C, Zhang F, Hu Y, Feng C, Wu H (2017) Ozonation in water treatment: the generation, basic properties of ozone and its practical application. Rev Chem Eng 33:49–89

Wittcoff AH, Reuben GB, Plotkin SJ (2013) Industrial organic chemicals, 3rd edn. Wiley

Wong JM (2012) Water reuse for petroleum oil, product processing industries. https://www.watertechonline.com/home/article/14171740/water-reuse-for-petroleum-oil-product-processing-industries. Accessed on 05 February 2021

Wu C, Gao Z, Zhou Y, Liu M, Song J, Yu Y (2015) Treatment of secondary effluent from a petrochemical wastewater treatment plant by ozonation-biological aerated filter. J Chem Tech Biotechnol 90:543–549

Xiu M, Pan L, Jin Q (2014) Bioaccumulation and oxidative damage in juvenile scallop Chlamys-farreri exposed to benzo [a] pyrene, benzo [b] fluoranthene and chrysene. Ecotoxicol Environ Saf 107:103–110

Yang X, Wang D (2018) Photocatalysis: from fundamental principles to materials and applications. ACS Appl Energy Mater 1:6657–6693

Yousefi N, Pourfadakari S, Esmaeili S, Babaei AA (2019) Mineralization of high saline petrochemical wastewater using Sonoelectro-activated persulfate: degradation mechanisms and reaction kinetics. Microchem J 147:1075–1082

Yu XZ, Gu JD (2006) Uptake, metabolism, and toxicity of methyl tert-butyl ether (MTBE) in weeping willows. J Hazard Mater 137:1417–1423

Zhang C, Wang C, Gil S, Boreave A, Retailleau L, Guo Y, Valverde JL, Giroir-Fendler A (2017) Catalytic oxidation of 1,2-dichloropropane over supported LaMnOx oxides catalysts. Appl Catal B Environ 201:552–560

Zhang H, He Y, Lai L, Yao G, Lai B (2020a) Catalytic ozonation of Bisphenol A in aqueous solution by Fe_3O_4–MnO_2 magnetic composites: performance, transformation pathways and mechanism. Sep Purif Technol 245:116449

Zhang J, Zhuang T, Liu S, Zhang GC, Huo K (2020b) Catalytic ozonation of phenol enhanced by mesoporous MnO_2 prepared through nanocasting method with SBA-15 as template. J Environ Chem Eng 8:103967

Zhang S, Wu C, Zhou Y, Wang Y, He X (2018a) Effect of wastewater particles on catalytic ozonation in the advanced treatment of petrochemical secondary effluent. Chem Eng J 345:280–289

Zhang X, Yea J, Yuan J, Yang L, He D, Cai T, Xiao B, Liu Z, Zhao K (2018b) Excellent low-temperature catalytic performance of nanosheet Co-Mn oxides for total benzene oxidation. Appl Catal A 566:104–112

Zhao L, Ma J, Sun Z, Liu H (2009) Mechanism of heterogeneous catalytic ozonation of nitrobenzene in aqueous solution with modified ceramic honeycomb. Appl Catal B Environ 89:326–334

Zhong J, Sun X, Wang C (2003) Treatment of oily wastewater produced from refinery processes using flocculation and ceramic membrane filtration. Sep Purif Technol 32:93–98

Zhu B, Li X, Sun P, Liu J, Ma X, Zhu X, Zhu A (2017) A novel process of ozonecatalytic oxidation for low concentration formaldehyde removal. Chin J Catal 38:1759–1769

Zhu Y, Jin K, Li H, Qian H, Wang H, Zhao L (2018) A novel anode with anticorrosive coating for efficient degradation of toluene. Chem Eng J 334:206–215

Chapter 7
Removal of Endocrine-Disrupting Compounds by Wastewater Treatment

Gargi Sinha Sarkar, Aanchal Rathi, Soumen Basu, R. K. Arya, G. N. Halder, and Sanghamitra Barman

Abstract Endocrine disruptors are toxic substances having adverse effects on the endocrine system even in very fewer amounts which appeared to be a serious concern for human health and water quality. EDCs (Endocrine disrupting compounds) behave like natural hormones in human beings that cause an interruption in endogenous hormonal activities like decreased fertility, changed sexual behavior, inflate abnormalities, and cancers in humans or animals. There are growing concerns about the effects of EDCs on drinking water or human health which are the key environmental problems worldwide. They can be removed from wastewater by many methods such as absorption, adsorption, oxidation, chemical degradation, photocatalytic degradation, membrane separation, biological degradation, transformation, and volatilization. The toxic effect of some EDCs is not fully known and needs further investigations. Detailed treatment methodology for each process is discussed for a better understanding of the scientific community for mitigation and handling these EDCs. The effect of influential operational parameters on the eradication of EDCs from aqueous media via various processes has been highlighted. Finally, the future perspective of the various water treatment techniques along with the key challenges to be faced by the next generation researchers towards expulsion of EDCs is discussed.

G. S. Sarkar
Department of Chemical Engineering, Jadavpur University, Kolkata-32, India

A. Rathi · S. Basu · S. Barman (✉)
School of Chemistry and Biochemistry, Thapar Institute of Engineering and Technology, Patiala, Punjab, India
e-mail: sbarman@thapar.edu

S. Basu
e-mail: soumen.basu@thapar.edu

R. K. Arya
Department of Chemical Engineering, National Institute of Technology, Jalandhar, Punjab, India

G. N. Halder
Department of Chemical Engineering, National Institute of Technology, Durgapur, West Bengal, India

S. Roy et al. (eds.), *Advanced Industrial Wastewater Treatment and Reclamation of Water*, Environmental Science and Engineering,
https://doi.org/10.1007/978-3-030-83811-9_7

Keywords Advanced oxidation process · Adsorption · Chemical coagulation · Electro-coagulation · Zeolites

7.1 Introduction

A category of appearing new noxious pollutants comprised of EDCs (Endocrine disrupting compounds), having a destructive effect on the ecosystem especially on the aquatic animals, water bodies, and humans (Rathi et al. 2019; Usha Vipinachandran et al. 2020). The unrestrained discharge of toxic chemicals into the water bodies due to huge urban industrialization, agricultural activities, and many human actions, causes a serious threat to the earth's ecosystem (Usha Vipinachandran et al. 2020; Rathi et al. 2021). These substances disturb the normal functioning of the endocrine systems by binding with the receptors site and start mimicking their functions which leads to unwanted responses (antagonizing effect/blocking effect) like neurological, reproductive, metabolic, and cardiovascular disorders in the human species (Aanchal et al. 2020; Sharma and Basu 2021). This disruption is one of the increasing concerns of human beings and the health industry. EDCs can be found in water, air, soil, cosmetics, pharmaceutical, personal care products (PPCPs), food sources, and manufactured products which affect the function of the human's endocrine system. From these sources, EDCs have found a way to interfere with everyday human life and increasingly have become one of the rising concerns of health care professionals all across the world, with its severe and evident effects. For example, Bisphenol A (BPA), an EDC majorly affects the cardiovascular system and researchers reported its link with hypertension. By promoting the vascular endothelial growth factor it affects the angiogenesis process which gave rise to uninhibited neovascularization and enhances the thickening of the inter-ventricular septum. This change further gave a harmful effect on the cardiovascular system of human beings (Ohore and Zhang 2019; Steinmetz et al. 1997).

The common EDCs in the environment are BPA, estradiol, triclosan, diclofenac, pharmaceuticals, and personal care products (PPCPs), pesticides, phenolic compounds, etc. shown in Table 7.1 (Rathi et al. 2019; Del Saz-Salazar et al. 2009; Canle et al. 2017). EDCsfrom different sources are exposed to humans in different ways such as breathing, eating food, drinking water and they can even enter our body while exposed to our skin (Wee and Aris 2017, 2019). EDCs in the polluted area are consumed by algae, plants, and microorganisms, and pass through animals during ingestion of plants. The maximum accumulation of these toxic compounds is in the body tissues of human beings. The impairments caused by the EDCs in the endocrine system can cause severe sickness and even death. As the world's scientific and medical community is concerned about the detrimental effects of EDCs on human health, public policies should be established on the latest available scientific evidence. In this chapter, we shall discuss further the sources, impact, and effective processes for the removal of EDCs. The various EDCs along with their chemical structures are introduced in Table 7.1.

Table 7.1 EDCs structures along with their detection limits (Wee et al. 2020) and solubility (Wee and Aris 2017)

EDC category	Structure with name	Detection limit (ng/L)		Water solubility (mg/L)	Usage
		River water	Tap water		
Phenolic compounds	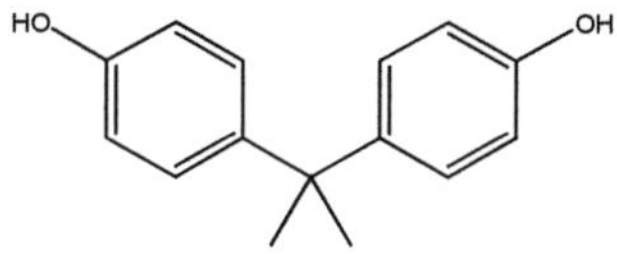 BPA	8.24	66.40	3.00×10^2	– Plasticizers – Surfactants – Lubricants – Fragrances – Antioxidants – Additives
Pesticides	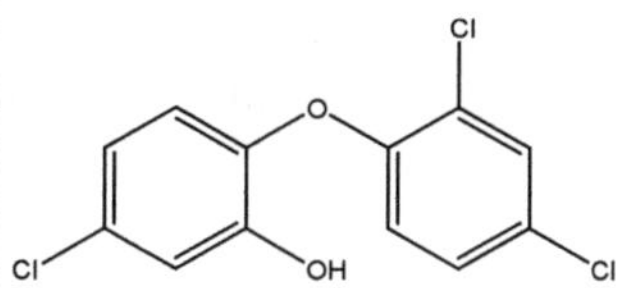 Triclosan	NA	9.74	4.62	– Pest controls – Antimicrobial or antifungal agents
	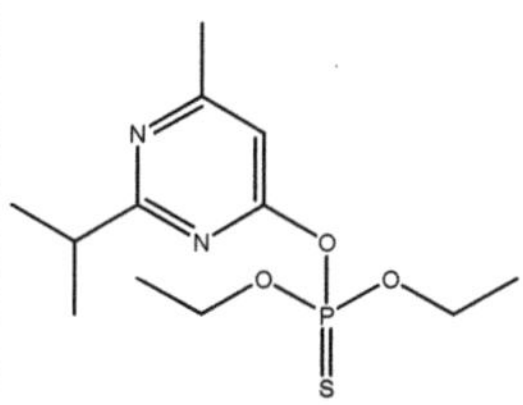 Diazinon	0.01	1.80	40.0	
Hormones	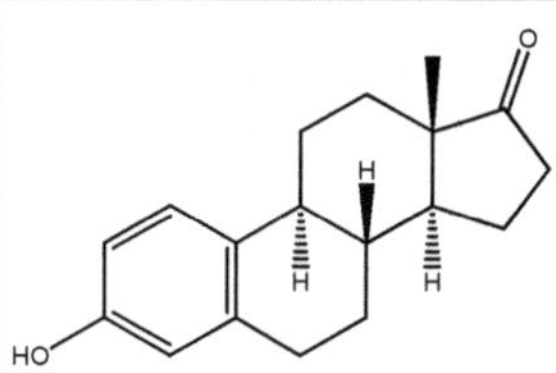 Estrone	0.18	3.42	30.0	– Growth stimulants – Hormonal therapies – Dietary intake
	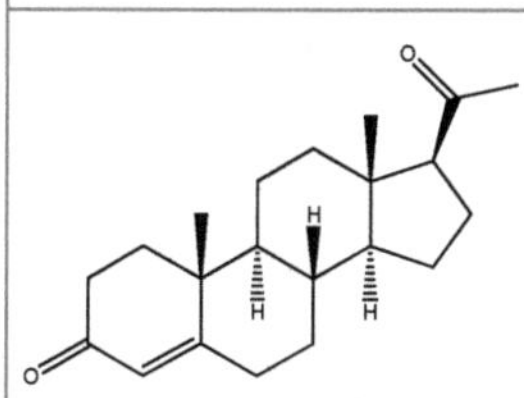 Progesterone	0.03	0.92	8.81	

(continued)

Table 7.1 (continued)

EDC category	Structure with name	Detection limit (ng/L)		Water solubility (mg/L)	Usage
		River water	Tap water		
	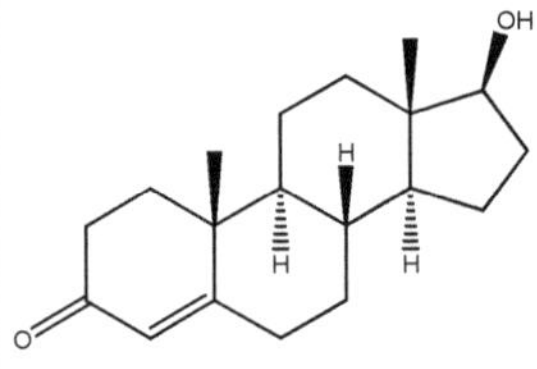 Testosterone	0.04	2.52	23.4	
Drugs and PPCPs	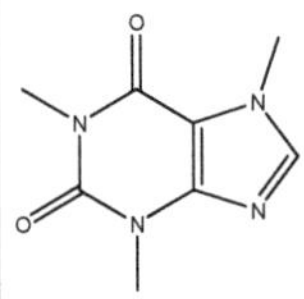 Caffeine	19.33	5.33	2.16×10^4	– Hormonal therapies – Illicit drugs – Veterinary drug uses – Agribusinesses
	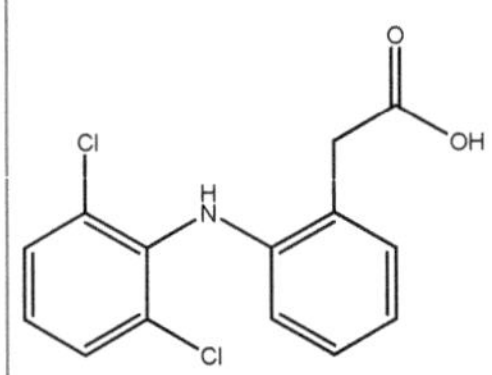 Diclofenac	6.15	21.39	2.37	
	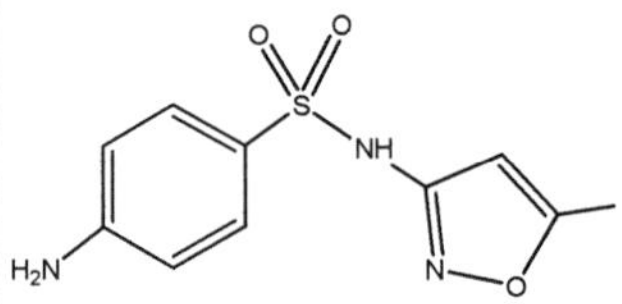 Sulfamethoxazole	0.23	0.90	3.94×10^3	

EDCs: Endocrine disrupting compounds, PPCP: Pharmaceutical and personal care products, BPA: Bisphenol A

7.2 Source of EDCs

Industrialization and agricultural runoff majorly contribute to the contamination of EDCs. EDCs are leached out from groundwater, soil and enters the food chain i.e. animals, fishes, and ultimately into human beings. The excessive usage of daily products such as plastics, detergents, paints, cosmetics, furniture, fragrances, lotions, adhesives, flame retardants, surface cleansers, anti-bacterial soaps, textiles, and

Table 7.2 Different category of EDCs with their sources

Source of EDCs	EDCs category/name
Combustion of by-products	PCBs (polychlorinated biphenyl) and dioxins
Water, food, and soil	Pesticides
Paint, plastic, furniture, food, and electronics	Flame retardant
Nonstick cookware and food packaging	Perfluorinated chemicals
Foods from Soyabean	Phytoestrogens
Plastics, food packaging, food/beverage containers	BPA (bisphenol A)
Cosmetics, cleaning agents, and plastics	Phthalates
Anti-microbial products and personal care products,	Triclosan
Sunscreens and cosmetics	UV filters
Foods/gum as a preservative	BHA (butylated hydroxyanisole)
Deodorants and polyester fabrics	Parabens
Aerospace product, weapon, fireworks, and pharmaceuticals	Perchlorate

electronic appliances, and many more can induce the accumulation of EDCs into the surroundings. The hospital wastes, veterinary therapies also release pharmaceutical contaminants like naproxen, diclofenac, ibuprofen, sulfamethoxazole, etc. (Usha Vipinachandran et al. 2020; Patisaul and Adewale 2009). Phytoestrogens, a group of EDCs produced by plants that mimic estrogen are found in berries, grains, and soy-based food products. EDCs like brominated flame retardants (BFRs) can mimic the estrogen hormone-disrupting endocrine system of humans. One of the well-known EDC is Polychlorinated biphenyls which interacts and changes neurobehavioral effects, especially neurodevelopment and cognitive performance, and also harms reproductive behaviors (Aris et al. 2020). In the coastal environment, EDCs have been revealed to distinctly infect animal populations. The different categories of EDCs along with their sources are described in Table 7.2.

7.3 Hazard Identification Based on the Key Characteristics of Endocrine Disruptors

To regulate any kind of threat, identification and characterization of it are necessary so that arriving at erroneous jurisdiction for hazard evaluation can be avoided or reduced. Hence, it is necessary to identify the chemicals which can be classified as EDCs to follow certain standard procedures to administer them. There is no formal guideline for hazard identification of EDCs; the key characteristics are mainly derived from mechanistic data which is further integrated with epidemiology and animal studies (La Merrill et al. 2020). After a keen evaluation of the data available on EDCs, an estimate of 10 key characteristics was drawn. Under these characteristics, any chemical can be classified to be an EDC. Since a wide range of chemicals were capable of disturbing the endocrine system, and characteristics were lined around the consequences caused rather than their intrinsic properties. These properties might be very diverse and dissimilar for over 1000 known endocrine affecting chemicals. The chemicals can be classified under multiple key characteristics to identify and analyze an EDC hazard as tabulated in Table 7.3.

7.4 Mode of Action of EDCs

The EDCs (natural/synthetic) obstruct the normal functioning of the endocrine systems of vertebrates by producing undesirable effects in the hormonally inhibited systems. These compounds can combine with the receptor site and mimics hormonal synthesis. These effects degrade the plethora of mechanisms by giving ‘fake’ or abnormal responses of the hormones which can alter the regular functioning of the hormonal systems (endocrine). This causes a serious effect on human beings and their mode of action has been depicted in Fig. 7.1 (Schug et al. 2016; Combarnous 2017).

7.5 Impact on Human Health

The endocrine system is responsible for multiple functions in the body starting from growth and development to body temperature regulation. Hormones secreted by the endocrine glands bind the receptors in the target cells to cause cellular changes. The endocrine system is not only responsible for secreting these hormones but also to release the respective hormones to maintain the required hormonal balance in the body. As explained endocrine disruptor scan severely affect the body as their traces can remain long after the exposure. In the former section, how EDCs affect the body is discussed, while in the current section the impacts of EDCs on human beings are reviewed.

EDCs can affect the growth, development, and metabolism of humans. Foetuses, infants, and children are especially vulnerable to such contaminants, which might cause multiple neuro-developmental complexities and malformations leading to various disabilities some resulting in cancer. Continued exposure to these EDCs can cause abnormal hormonal changes in adults which might lead to breast cancer and prostate cancer in females and males respectively, they might even cause infertility, cardiovascular diseases, etc. in both. The Endocrine Society is the leading research

Table 7.3 Key characteristics to identify EDCs supported by examples (La Merrill et al. 2020; Weinhold 2006)

S. no	Main characteristics	Sample of relevant mechanistic evidence	Example
1	Interacting or activating the hormone receptors	Binding with receptor site leading to unnecessary receptor activation causing adverse consequences	– DDT activates estrogen receptors (ERα and Erβ) risking infertility – Hydroxylated-PCBactivates human thyroid receptor
2	Antagonizes hormone receptors	Blocks (prevents) the effect of endogenous hormones by posing as receptor antagonists	DDE binds with androgen receptor (AR) hence impeding androgen disrupting male sexual differentiation in fetuses
3	Changed hormone receptor expression	Interferes with instructions in the DNA of a cell resulting in down regulation or up regulation of that receptor	DEHP downregulates mineralocorticoid receptor (MR)
4	Changed signal transduction in hormone-responsive cells	Transmission of cellular signals is perturbed, an abundance of biochemical transformative enzymes or proteins essential for the synthesis of RNA to DNA such as post-translational modifications, cofactors, and tissue factors	Bisphenol A impedes low glucose-induced Ca^{2+} signals in pancreatic α-cells which secrets glucagon
5	Induces epigenetic alterations in hormone-production or hormone receptive units	Interferes with transcriptions and even permanently affecting non-coding RNA, DNA and modifying histones	Methoxychlor causes epigenetic modifications one of them being DNA methylation with evidence of complex imprinting alterations
6	Changed hormone synthesis	Interferes with intracellular and distant endocrine feedback mechanisms	Perchlorate impedes the absorption of iodine into thyroid cells which in turn represses thyroid hormone synthesis

(continued)

Table 7.3 (continued)

S. no	Main characteristics	Sample of relevant mechanistic evidence	Example
7	Changed hormonal transportation through cell membranes	The active (selective) and passive transport of hormones (in and out of cell membranes/ target tissue) is affected by EDCs	Low doses of DDE reduces the secretion of corticosterone hormone from the adrenal glands of rats
8	Changed hormone distribution/circulating hormonal level	Affects the bioavailability of hormones by disrupting the dissemination or distribution of hormones even relocate the hormones from their binding proteins	Malathion decreases the level of testosterone serum, ovaries, and testis
9	Changed hormonal metabolism	Disrupts hormone elimination and rate of clearance (degradation) consequently changing hormone levels in the body	Coplanar hydroxylated PCB impedes the rate of estrogen clearance from the blood
10	Changing the fate of hormonal production/responsive units	Disrupts the developmental biology of cells (i.e., cell fate) ultimately affecting tissue structure and organization	During pregnancy and lactation in female mice, exposure to oxybenzone rises mammary epithelial cell propagation

DDT: dichloro-diphenyl trichloroethane, DDE: Dichlorodiphenyldichloroethylene, DEHP: di(2-ethylhexyl) phthalate, DNA: deoxyribonucleic acid, PCB: Polychlorinated biphenyl

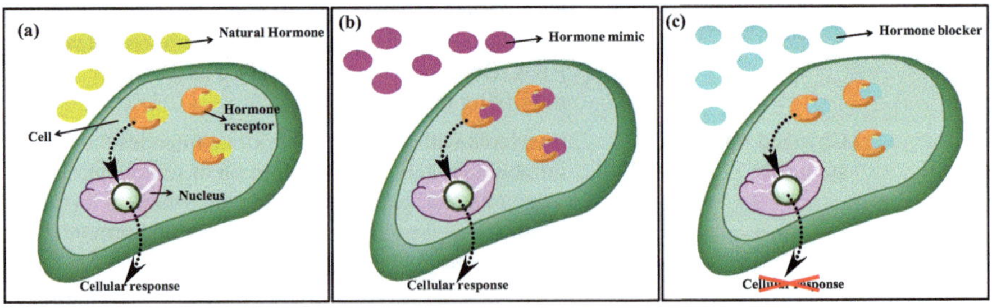

Fig. 7.1 EDC's mode of **a** action and binding to the hormone receptors, **b** hormone mimicking; and **c** hormone blocking with no cellular response

institute in this field; in 2009 based on epidemiological studies, clinical observations of animal models the first scientific statement was released addressing the concerns of public health based on EDCs (Diamanti-Kandarakis et al. 2009). The endocrine disruption steers to toxicity, destructive health effects, and a combination of biological, chemical, physical, psychological, social stressors for life. It was suggested from both basic sciences and regulatory sciences, that there are increased links between physiology, toxicology, biology, and epigenetics which were so significant to raise issues or gives us more future scope for research (Barouki 2017).

7.6 Elimination Techniques of EDCs from Wastewater

As mentioned in the previous sections, with the gravity of the adverse impacts the removal of EDCs is becoming one of the important concerns to the scientific community. The pollutants such as pathogens, nutrients, and organic matters of influent are eliminated via the traditional wastewater treatment plant (WWTP) techniques. The primary treatment includes the removal of oils, greases, and settling of large-sized particles, where as secondary treatment is used to reduce the organic matter in wastewater via microbes. The tertiary actions involve filtration which helps to remove the hazardous bacteria and viruses in the effluent. Additionally, it further reduces the suspended particles and BOD (biological oxygen demand) giving rise to better effluent (Quach-Cu et al. 2018). However, these technologies are inefficient for the complete removal of micro-pollutants like endocrine-disrupting compounds (EDCs) from wastewater as they are still found even in tertiary-treated effluent. Such micro-pollutants are present in wastewater in different concentrations based on the geographic location and population of the area, the removal of pollutants depends on the efficiency of the treatment process employed. Water treatment plants and related industries should modify and execute new technologies to dispense off the EDCs with the proper removal method to maintain an ecological balance. Further some highlighted methods were discussed on which multiple experiments has been conducted for EDCs elimination. These methods can be optimized and scaled up to fit required needs as per the specifications of demography, geography, and industrial need. Various EDCs eliminating techniques are discussed as follows.

7.6.1 Advanced Oxidation Process

Advanced Oxidation Processes (AOPs), are widely known and effective process for wastewater treatment. It has gained a lot of interest for eliminating contaminants from effluent water. Advanced oxidation processes (AOPs) such as Fenton, ozonation, UV photolysis, photocatalysis, wet air oxidation, and sonolysis, are the most suitable techniques for wastewater treatment. The central idea for AOPs is the chemical treatment of the pollutants by oxidation through the use of hydroxyl radicals ($OH^•$).

To perform in-situ (local) formation of highly reactive $OH^{\bullet}$ species strong oxidants are required, where the formation of products is fast such that it has a diffusion-controlled reaction. Oxidation can be achieved by

- Single or multiple primary oxidants (like O_3, H_2O_2, O_2).
- Energy-richUVradiations.
- Catalysts likeTiO_2.

These oxidation methods are used individually or as a combination for various AOPs, a brief description is given below. The underlying mechanism of AOPs is the mineralization of the contaminants present in polluted water to preferably biodegradable by-products with the use of strong oxidizing agents with redox reactions. Efficiencies of the process are therefore dependent on the choice of an oxidizing agent. Overall, when calculated the strength of oxidation can be ordered as: $FeO_4{}^{2-} > O_3 > S_2O_4{}^{2-} > H_2O_2 > Cl_2 > ClO_2$ (Liu et al. 2009). The main points evaluating the research on the removal of EDCs using AOP are—(i) optimization of removal efficiency and operational conditions of AOP (ii) degradation kinetics evaluation on the removal of EDC (iii) pathways for degradation of EDC using AOP.

Fenton's Oxidation Process (Fenton Based AOP)

A solution of hydrogen peroxide (H_2O_2) and ferrous salt (ion) is known as Fenton's reagent, here the ferrous ion acts as a catalyst to obtaina high oxidation potential species ($OH^{\bullet}$). This radical species easily mineralizes the pollutant under observation.

The main chemical reactions involved are as follows (Ifelebuegu and Ezenwa 2011):

$$Fe^{2+} + H_2O_2 \rightarrow Fe^{3+} + OH^{\cdot} + OH^- \tag{7.1}$$

$$OH^{\cdot} + Fe^{2+} \rightarrow Fe^{3+} + OH^- \tag{7.2}$$

$$OH^{\cdot} + \text{Organic Compound} \rightarrow \text{Product}$$

Ferrous ions are produced when the formed ferric ion reacts with H_2O_2. This gave Fenton-based oxidation reactions.

$$Fe^{3+} + H_2O_2 \leftrightarrow FeOOH^{2+} + H^+ \tag{7.3}$$

$$FeOOH^{2+} \rightarrow Fe^{2+} + HO_2^{\cdot} \tag{7.4}$$

$$Fe^{2+} + H_2O_2 \rightarrow Fe^{3+} + OH^{\cdot} + OH^- \tag{7.5}$$

The resultant $OH^{\bullet}$ produced further reacts with organic moieties. The efficiency of the process can be increased through UV irradiation which facilitates the regeneration of ferrous ions from ferric ions, along with the activation of H_2O_2.

Fenton's Oxidation (FO) is a preferred method as shown in Fig. 7.2 for wastewater treatment which is comprised of four stages including pH adjustment, oxidation reaction, neutralization coagulation, and precipitation. The different catalysts and pH adjusting agents are added as per requirement. The by-product or residue can be flocculated and separated easily by using a settling tank. OH radicals are responsible for oxidation, and coagulation is ascribed to the formation of ferric hydroxo complexes. The relative importance of oxidation and coagulation depends primarily on the H_2O_2/Fe^{2+} ratio. Chemical coagulation predominates at a lower H_2O_2/Fe^{2+} ratio, whereas chemical oxidation is dominant at higher H_2O_2/Fe^{2+} ratios (Ghime and Ghosh 2020). Studies and experiments have been carried out for various EDCs such as BPA, E2 (17β-estradiol), and EE2 (17α-ethinylestradiol), These FO reactions gave more than 95% removal and are also thermodynamically stable (Ifelebuegu and Ezenwa 2011). The relatively high cost of H_2O_2 and the huge quantity of ferric sludge produced in the neutralization step of the solution before disposal restricts the application of the Fenton process. To overcome these drawbacks are the major thrust to upgrade the Fenton technology.

Ozone Based AOP

Ozone has a high oxidation potential itself but it exhibits a selective nature, where it prefers to react with the ionized form of the contaminants compared to their neutral form, although it poses no specific selectivity between inorganic and organic compounds. While the free radical causes the oxidation reaction with low selectivity and high redox potential. A peroxone (O_3/H_2O_2) system is used where the O_3 decomposition is enhanced by the HO_2^- species obtained from H_2O_2. The decomposition reaction can be depicted by (Collivignarelli et al. 2018):

$$H_2O_2 + H_2O \rightarrow HO_2^- + H_3O^+ \tag{7.6}$$

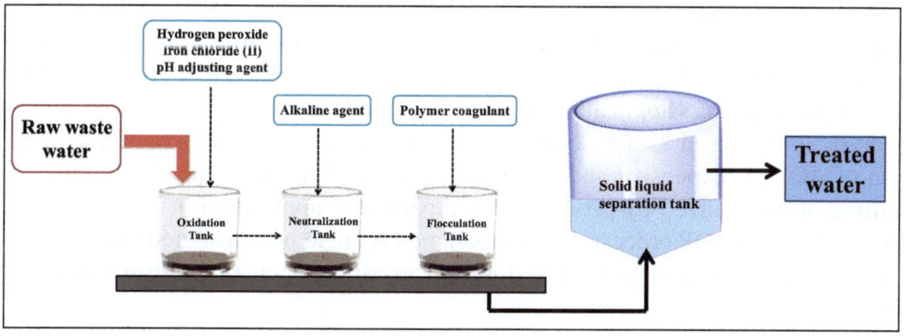

Fig. 7.2 Flow process diagram for reactors used in Fenton's oxidation method

$$H_2O_2 + 2O_3 \rightarrow 2HO^{\bullet} + 2O_2 \quad (7.7)$$

The $OH^{\bullet}$ oxidizes and mineralizes the EDC it attacks. The efficiency of the disinfection process is evaluated by the amount of residual ozone in the effluent water through which the amount of oxidation can be determined. The efficiency can be enhanced further by an increase in peroxide to ozone ratio as it facilitates the formation of free hydroxyl radicals. UVradiation causes the formation of radical reactions when Ozone either in gas or aqueous solution reacts with UV causing a photochemical reaction here, H_2O_2 is generated as an additional oxidant (Deng and Zhao 2015).

$$H_2O + O_3 + hv \rightarrow H_2O_2 + O_2 \quad (7.8)$$

An efficiency of upto100% can be achieved for target EDCs like E1 and Caffeine (Liu et al. 2009). Hence, we can classify the use of ozone in water treatment as (i) Ozonation, (ii) Peroxone (O_3/H_2O_2) and (iii) Photolysis of O_3 by using UV to form H_2O_2

UV Based AOP

UV radiation as seen in the previous sections proves to be very efficient to enhance the formation of radicals. The percentage degradation is significantly increased with the UV-based advance oxidation process. Some examples of photocatalysts are UV, UV/H_2O_2, UV/Fe^{2+}, $UV/H_2O_2/Fe^{2+}$, UV/O_3, $UV/S_2O_8^{2-}$, UV/TiO_2, UV/chlorine, etc.

Photocatalysis is the method where semiconductor compounds such as TiO_2, CeO_2, etc. are used under UV irradiation by hydroxyl radicals in presence of H_2O. The photons activate electrons from the valence band (VB) creating holes (h^+) and the charged electron is excited to the conduction band (CB). Now, the electron aids in hydroxyl species ($OH^{\bullet}$) formation. The $OH^{\bullet}$ species which was generated in the photocatalysis of TiO_2 possess a high reactivity hence they are exhausted fast in the aqueous environment. Experiments have shown a removal efficiency of about 93% of BPA by UV irradiation in presence of peroxide (Xiao et al. 2017). There are other ways of achieving oxidation like sulphate radical AOP, but the ones discussed above are most widely implemented. AOPs although have high efficiencies they have their limitations in process applications where it is not effective in treating a large amount of feed, so it is used as a secondary treatment or pre-treatment. The implementation of AOPs is expensive and not effective in the long run, with the storage and settling of mineralized by-products. For a particular application, it is important to choose the right AOP process. Several parameters are to be considered such as water composition, treatment goals, UV dose, chemical dose, energy consumption, and cost.

Wet Air Oxidation Process

Oxidation of inorganic and organic substances in a suspension utilizing oxygen or air at elevated temperatures and pressures either in the presence or absence of catalysts is called the wet air oxidation process. By this method, in the aqueous phase, the dissolved organic matter is oxidized by the gaseous source of oxygen or air. WAO destroys toxins by breaking down complex molecular structures in industrial wastewater into lighter components such as water and carbon dioxide, without emissions of NOx, SO_2, HCl, dioxins, and furans.

Sonolysis

The ultrasound process is reported to be a very efficient AOP for the degradation of EDCs present in water (Zúñiga-Benítez et al. 2016). Also, it can overcome the restrictions assigned to the use of other AOPscommonly used for water treatment. It is remarkable to highlight that by using sonolysis, mass transfer within the reaction medium is improved which enhances the EC degradation rates.

7.6.2 Photocatalytic Degradation

The Photocatalysis technique is simple, efficient, renewable, and ecologically safe for eliminating contaminants from water bodies. This AOP undergoes the availability of efficient photocatalysts such as metal oxides. The photons were absorbed having energy similar to or greater than that of band-gap energy of VB and CB of photoactive materials. These absorbed photons give rise to the separation of the charges by electron excitation from VB to CB and leave behind h^+ in the VB. The h^+ species react with H_2O molecules to generate $OH^•$ radicals and excited electrons were responsible for reducing adsorbed O_2 to superoxide radicals ($O_2^{•-}$). These free radicals species further react with the harmful contaminants and give rise to harmless intermediates or total mineralization to CO_2 and H_2O molecules (Gusain et al. 2019). The photocatalytic degradation mechanism of pollutants with general photoactive materials was shown in Fig. 7.3 (Nguyen et al. 2019).

7.6.3 Coagulation Processes

In the electrocoagulation (EC) process the contamination level was decreased bypassing the current into the water and through electrolytic oxidation reaction of the anode electrode. The coagulants (e.g. Al, Fe) were produced due to the electrolytic oxidation reaction at the anode electrode. In the EC technique, the amorphous unsolvable metal hydroxide and MO (metal oxides) are produced, which help in the adsorption of contaminants during the precipitation process and makes the separation feasible. The electrode materials Al and Fe were extensively utilized as

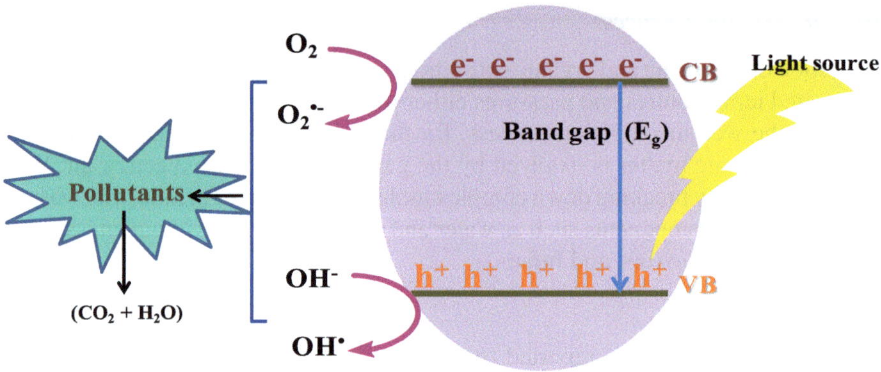

Fig. 7.3 The plausible photocatalytic degradation mechanism of pollutants over photoactive materials

these are cost-friendly and efficient for a variety of contaminants (Cook et al. 2016). While the chemical coagulation processes include the addition of Al salts and polymer substances for the management of wastewater. The flocculant (PACl: Polyaluminium Chloride) was exploited to coagulate fine molecules into aggregated flocs which facilitate the effective succeeding separation phenomenon with sedimentation/filtration processes. Among the conventional flocculants, PACl was prioritized due to its efficacy and cost-friendly nature. Besides, it can efficiently eliminate heavy metals, harmful colors, suspended molecules, colloidal and organic matter from wastewater (Bazrafshan et al. 2012).

7.6.4 Membrane Separation Process

In wastewater treatment technology, membrane technology provides sustainable growth opportunities in the foreseeable future, alongside being adept in solid–liquid separation. The membrane technology is designed based on the selective permeability of various components present in the feed, thus changing the composition of the solution. Separation using membrane technology can be categorized under multiple segments namely—reverse osmosis (RO), nanofiltration (NF), ultrafiltration (UF), Microfiltration (MF) shown in Fig. 7.4. The mentioned separation techniques are pressure-driven i.e.; the driving force is ΔP. In advance wastewater treatment application of Membrane Bioreactors (MBR) is unique which is an integration of all MF/UF/NF with a suspended growth bioreactor. The flow configurations for membranes are typically done in two ways:

(i) Static Filtration (Dead end flow): the thickness of the deposit is proportional and the specific output of filtrate is inversely proportional to time. Backwashing is required to remove the accumulated reject to maintain a steady flow rate.

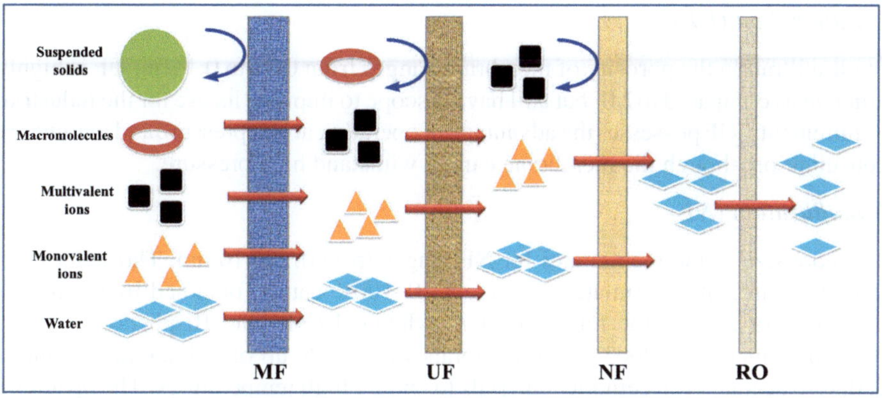

Fig. 7.4 The different types of membrane filtration (Aziman et al. 2019)

(ii) Dynamic Filtration (Cross-Flow Filtration): utilizes a high-pressure feedwater stream across the film. The arrangement is isolated into two sections, a section goes through the layer or channel which is called penetrate, and the leftover outfluxes perpendicular to the pressure feed stream and the film surface without filtration also termed as concentrate. The relation of rates is similar to that of static filtration but they become constant after a certain period.

The processes can be almost automated and operated without any intervention, in case of replacement of frames manual intervention is required due to fouling or clogging. The main criteria for the choice of the process can be classified into three divisions: (1) denial of EDCs due toits physicochemical properties, (2) denial of EDCs due to membrane module impact, and (3) other impact parameters like fouling of the layer or the availability of organic impurities in wastewater (Liu et al. 2009; Del Saz-Salazar et al. 2009).

The equations used in the calculation of flux are Darcy's Law and Fick's Law (Xin 2004). During operation due to various reasons likes caling and clogging caused by deposition of mineral or organic colloids decreases the rate of permeate flux resulting in membrane fouling. Back washing can be used as a method for removal of fouling, i.e., by changing the flow direction to unclog the cavities/pores. Chemical handling was implemented to tackle non-reversible fouling while in extreme cases replacement of the membranes was the only option. The other main processes were also discussed below (Gallenkemper et al. 2003; Zheng et al. 2013).

Microfiltration (MF)

In microfiltration, the pore size of the membranes ranges from 0.1 to 10 μm, though filtration is not adequate and would require further clarification. Sieving and adsorption are the main transfer mechanisms that facilitate separation. Comparatively larger solid particles can be clarified using MF membranes.

Ultrafiltration (UF)

In ultrafiltration the porosity of membrane ranges from 0.01 to 0.1 μm, UF is highly efficient as compared to MF but still having scope to improve its use for the industrial requirements. UF possesses the advantage of operating at low pressure and low energy consumption, though the membrane cannot withstand high pressure.

Nanofiltration (NF)

The pore size of the membranes for NF ranges from 0.1 to 10 nm. Through experimental results, it is estimated that about 70–100% can be obtained by the implementation of NF for the retention of nonylphenol (NP) and Bisphenol A (BPA) (Gallenkemper et al. 2003). The membranes are made up of polyamide polymers, cellulose acetate even ceramic materials to endure high temperatures. The operating pressure requirements for NF are higher so the power requirement is also higher, which consequently increases the cost. Parallel to the former point the initial working, and upkeep cost to set up NF is high.

Reverse Osmosis (RO)

Reverse osmosis has found its wide application in the desalination process used for the purification of water which ultimately provides portable drinking water. The process of reverse osmosis is the application of osmotic pressure passing through the semi-permeable film in the direction opposite to the gradient of natural osmotic pressure resulting in the flowing solvent from a higher to a lower concentration of solute. RO has the potential of achieving the dismissal efficiency of up to 99% dissolved solids. Multiple stages or multiple pass RO membranes are used for better efficiency. The porosity ranges from 0.1 to 1 nm approximately.

Membrane Bioreactors (MBR)

In an MBR system, the membranes are immersed in an aerated biological reactor. Membrane Bioreactors allow for high-quality effluent to be drawn through the membranes and removes typical use of the sedimentation and filtration processes for water treatment. The MBR systems are categorized as vacuum (or gravity-driven) and pressure-driven systems. Vacuum or gravity systems are immersed and normally employ hollow fiber or flat sheet membranes installed in either the bioreactors or a subsequent membrane tank. Pressure-driven systems are in-pipe cartridge systems located externally to the bioreactor. MBR for wastewater management is the biological management method in which the activated sludge is passed through MF, UF, or NF membranes at low pressure. MBR is somehow similar to that of the activated sludge technique. The MBR technique has better elimination efficacy than that of the activated sludge technique (Aziman et al. 2019). It is pore size or contact angle which facilitates membrane separation (filtration) with the target EDCs and removes a diverse number of pollutants.

7.6.5 Adsorption Process

Adsorption is the term that is utilized to portray the grip of a slim layer of atoms to the surfaces of fluids or solids they interact with. Adsorption is a solid–liquid separation process in which when feed containing adsorbable dissolved impurities comes into contact among the rough and permeable porous solid surface. The solute molecules from the feed are retained on the surface of the solid due to solid–liquid intermolecular attraction forces. The main components of any adsorption system are adsorbate and adsorbent. Adsorbate is the concentrated solid retained on the porous solid surface and the porous solid is known as the adsorbent. Hence the adsorption is the surface phenomenon of the aggregation of adsorbate on the surface of the adsorbent. From this technique, the dissolved impurities can be easily eliminated from the bulk liquid phase. Adsorption can be broadly classified into (i) physisorption, and (ii) chemisorption based on the type of bonding between the pollutant (adsorbate) and surface (adsorbent). Physisorption is a type of adsorption caused due to the bonding between the adsorbent and absorbate due to Vander Waal forces of attraction in between them, these bonds are characteristically weak. Chemisorption involves the formation of covalent or ionic bonds between adsorbent and adsorbate through the substantial sharing of electrons.

Adsorption with powdered activated carbon is a widely used method for wastewater treatment for the removal of EDCs and the uses of granulated activated carbon are also reported in the literature. Other materials such as synthetic zeolites (NaA, ZSM-5, or Beta), mesoporous silica, carbon nanotubes, graphene, etc. were also used for adsorption (Aris et al. 2020; Rashed 2013). Different adsorbents are shown in Fig. 7.5. To determine the adsorption capacity and type of adsorption, the equilibrium adsorption data has been validated with the adsorption isotherms.

Adsorption isotherms are utilized to depict the amount of adsorbate on the surface as a component of its pressure at a consistent temperature. The adsorption isotherms (Langmuir and Freundlich) are generally utilized to explore the adsorption information sometimes BET isotherm is also used. The Langmuir isotherm was implied to portray the equilibrium between the pollutant molecules and the surface of the material. The pollutants molecules were getting adsorbed on the homogenous surface with a monolayer and K_L is the Langmuir adsorption coefficient. The adsorption capacity 'Q_m' (mg/g) was calculated from this adsorption isotherm by using the equation given below (in Table 7.4).

Freundlich isotherm reveals the multilayer type of adsorption process on the surface of the material (heterogeneous) with the help of the equation given in Table 7.4, where the K_f constant designates the adsorption capacity (Rathi et al. 2021). And the value of 'n' decides whether the adsorption is favorable ($n > 1$) or not.

The various mentioned techniques along with their parameters/conditions and maximum removal with specific materials are tabulated in the given Table 7.5.

Fig. 7.5 The various types of conventional adsorbents (Aris et al. 2020)

Table 7.4 Equations and parameters for adsorption isotherms

Adsorption isotherms	Equation used	Parameters
Langmuir	$\frac{C_e}{q_e} = \frac{1}{Q_m K_L} + \frac{1}{Q_m} C_e$	C_e: pollutant conc. (mg/L) Q_e: adsorbed quantity (mg/g)
Freundlich	$\log q_e = \log K_f + \frac{\log C_e}{n}$	

7.7 Threat Evaluation and Approaches Towards Preventions

Exposure of harmful EDCs to both human beings and animals is a global problem. Manifestation of adverse effects in humans due to exposure to EDCs is observed in later ages, as they can reside in the adipose tissues for a longer duration of the period. They are generally transmitted through contaminated food, animal skin, and airborne chemicals. And our home, office, farm, water are the potential sources of EDCs. Therefore, a well-designed strategy should be needed to avoid/reduce EDCs exposure. The use of these toxic chemicals, care products, plastics, pesticides, etc. has to be restricted and new remediation technologies in the form of green chemistry

Table 7.5 The various methods for wastewater treatment of EDCs

S. no	Technique used/EDCs name	Material used	Conditions	Removal %	Reference
1	DDT, lindane	$TiO_2/UV/O_2$ system	AOPs	87%, 68%	Aziz et al. (2017)
2	Gemifloxacin	Zn-Co-layered double hydroxide (LDH) nanostructures	Photocatalysis	92.7%	Gholami et al. (2020)
3	Tetracycline (TTC)	Tungsten oxide (WO_3) on carbon nano-tube (CNT)	Sono-photocatalysis process	100%	Isari et al. (2020)
4	Ibuprofen	$BiOBr/Fe_3O_4@SiO_2$	Visible light degradation	–	Khan et al. (2019)
5	Estrone	Polyvinylidene fluoride (PVDF) polyvinylpyrrolidone (PVP)-TiO_2 nano-inorganic modified ultrafiltration membrane	UF	36.1%	Wang et al. (2016)
6		Anoxic and anaerobic tanks/polyvinylidene fluoride microfiltration membrane	MBR	>97%	Kamaz et al. (2019)
7	BPA	Tubular asymmetric ceramic membrane	AS-MF (Activated sludge-MF) AS-NF	~98%	Zielińska et al. (2017)
8	Atrazine	Anoxic and anaerobic tanks/polyvinylidene fluoride microfiltration membrane	MBR	15–30%	Kamaz et al. (2019)
9	Alachlor	Thin-film composite membrane with a polysulfone support layer	NF-UF	>96.2%	Sanches et al. (2012)
10	Lindane	Submerged membrane	MBR	91.6–99.7%	Vieira et al. (2020)
11	Methylparaben	Non-polyamide membrane chemistry using green tannic acid–iron (TA–Fe)	NF	69.2%	Gholami et al. (2020)
12	Benzylparaben	Metal–organic framework-NF	NF	51.1%	Guo et al. (2019)
13	Fipronil	CeO_2-loaded zeolites	Adsorption	93.54%	Rathi et al. (2021)
14	Daidzein coumestrol	Nanozeolite beta modified with tetrasubstituted ammonium cation	Adsorption	(92–98%) (93.5–99%)	Goyal et al. (2018)

(continued)

Table 7.5 (continued)

S. no	Technique used/EDCs name	Material used	Conditions	Removal %	Reference
15	Imidacloprid	WO_3/SiO_2 monoliths	Photodegradation	59%	Sharma and Basu (2020)

must be applied (Wee and Aris 2017; Combarnous 2017). To improve general understanding of EDCs and their potential adverse effect, education programs should be run especially in the area of endocrinology, pediatric, and maternity. Clinical endocrinologists should be well educated on the latest effects of EDCs. On the other way, professional workers who are working with agricultural compounds (insecticides/rodenticides/weedicides), colorful synthetic paints, and harmful chemicals have a greater risk of exposure. Some preventive measures should be implied by the policy creators and government for every individual.

7.8 Conclusion

This chapter introduces different categories of EDCs predominating in the environment. It summarizes the resource and impact of the harmful contaminants available in the water bodies in the ecological environment. The EDCs were commonly used in our daily routine (plasticizers/surfactants/antioxidants/antibiotics) and they were detected in both river water and tap water in ng/L. They cause adverse effects on human health by directly affecting the functioning of the endocrine systems. Various techniques such as the Advanced oxidation process, adsorption, photodegradation, coagulation, filtration, membrane separation were described to remove these EDCs from wastewater.

References

Aanchal, Barman S, Basu S (2020) Complete removal of the endocrine-disrupting compound and toxic dye by visible light active porous g-C_3N_4/H-ZSM-5 nanocomposite. Chemosphere 241:124981

Aris AZ, Zul Adlan MH, Muhammad RR (2020) Metal-organic frameworks (MOFs) for the adsorptive removal of selected endocrine disrupting compounds (EDCs) from aqueous solution: a review. Appl Mater Today 21:100796

Aziman ES, MohdSalehuddin AHJ, Ismail AF (2019) Remediation of thorium (IV) from wastewater: current status and way forward. Sep Purif Rev 1–26

Aziz NAA, Palaniandy P, Aziz HA (2017) Photocatalytic degradation of lindane and DDT using TiO_2/solar light by response surface methodology. Sci Int 29(1):63–63

Barouki R (2017) Endocrine disruptors: revisiting concepts and dogma in toxicology. C R Biol 340(9–10):410–413

Bazrafshan E, Mostafapour FK, Farzadkia M, Ownagh KA, Mahvi AH (2012) Slaughterhouse wastewater treatment by combined chemical coagulation and electrocoagulation process. PLoS One 7(6):e40108

Canle M, Pérez MIF, Santaballa JA (2017) Photocatalyzed degradation/abatement of endocrine disruptors. Curr Opin Green Sustain 6:101–138

Collivignarelli MC, Abbà A, Benigna I, Sorlini S, Torretta V (2018) Overview of the main disinfection processes for wastewater and drinking water treatment plants. Sustainability 10(1):86

Combarnous Y (2017) Endocrine disruptor compounds (EDCs) and agriculture: the case of pesticides. C R Biol 340(9–10):406–409

Cook MM, Symonds EM, Gerber B, Hoare A, Van Vleet ES, Breitbart M (2016) Removal of six estrogenic endocrine-disrupting compounds (EDCs) from municipal wastewater using aluminum electrocoagulation. Water 8(4):128

Dai R, Guo H, Tang CY, Chen M, Li J, Wang Z (2019) Hydrophilic selective nanochannels created by metal-organic frameworks in nanofiltration membranes enhance rejection of hydrophobic endocrine-disrupting compounds. Environ Sci Technol 53(23):13776–13783

Del Saz-Salazar S, Hernández-Sancho F, Sala-Garrido R (2009) The social benefits of restoring water quality in the context of the water framework directive: a comparison of willingness to pay and willingness to accept. Sci Total Environ 407(16):4574–4583

Deng Y, Zhao R (2015) Advanced oxidation processes (AOPs) in wastewater treatment. Curr Pollut Rep 1(3):167–176

Diamanti-Kandarakis E, Bourguignon JP, Giudice LC, Hauser R, Prins GS, Soto AM, Zoeller RT, Gore AC (2009) Endocrine-disrupting chemicals: an endocrine society scientific statement. Endocr Rev 30(4):293–342

Gallenkemper M, Wintgens T, Melin T (2003) Nanofiltration of endocrine disrupting compounds. Water Sci Tech-W Sup 3(5–6):321–327

Ghime D, Ghosh P (2020) Advanced oxidation processes: a powerful treatment option for the removal of recalcitrant organic compounds. Advanced oxidation processes

Gholami P, Khataee A, Soltani RDC, Dinpazhoh L, Bhatnagar A (2020) Photocatalytic degradation of gemifloxacin antibiotic using Zn-Co-LDH@ biochar nanocomposite. J Hazard Mater 382:121070

Goyal N, Bulasara VK, Barman S (2018) Removal of emerging contaminants daidzein and coumestrol from water by nanozeolite beta modified with tetrasubstituted ammonium cation. J Hazard Mater 344:417–430

Guo H, Peng LE, Yao Z, Yang Z, Ma X, Tang CY (2019) Non-polyamide based nanofiltration membranes using green metal–organic coordination complexes: implications for the removal of trace organic contaminants. Environ Sci Technol 53(5):2688–2694

Gusain R, Gupta K, Joshi P, Khatri OP (2019) Adsorptive removal and photocatalytic degradation of organic pollutants using metal oxides and their composites: a comprehensive review. Adv Colloid Interfac 272:102009

Ifelebuegu AO, Ezenwa CP (2011) Removal of endocrine disrupting chemicals in wastewater treatment by Fenton-like oxidation. Water Air Soil Pollut 217(1):213–220

Isari AA, Mehregan M, Mehregan S, Hayati F, Kalantary RR, Kakavandi B (2020) Sono-photocatalytic degradation of tetracycline and pharmaceutical wastewater using WO_3/CNT heterojunction nanocomposite under US and visible light irradiations: a novel hybrid system. J Hazard Mater 390:122050

Kamaz M, Wickramasinghe SR, Eswaranandam S, Zhang W, Jones SM, Watts MJ, Qian X (2019) Investigation into micropollutant removal from wastewaters by a membrane bioreactor. Int J Environ Res Public Health 16(8):1363

Khan M, Fung CS, Kumar A, Lo IM (2019) Magnetically separable BiOBr/Fe_3O_4@ SiO_2 for visible-light-driven photocatalytic degradation of ibuprofen: mechanistic investigation and prototype development. J Hazard Mater 365:733–743

La Merrill MA, Smith VLN, MT, Goodson W. Browne P. Patisaul HB, Guyton KZ, Kortenkamp A, Cogliano VJ, Woodruff TJ, Rieswijk L, (2020) Consensus on the key characteristics of endocrine-disrupting chemicals as a basis for hazard identification. Nat Rev Endocrinol 16(1):45–57
Liu ZH, Kanjo Y, Mizutani S (2009) Removal mechanisms for endocrine disrupting compounds (EDCs) in wastewater treatment—physical means, biodegradation, and chemical advanced oxidation: a review. Sci Total Environ 407(2):731–748
Nguyen TT, Nam SN, Son J, Oh J (2019) Tungsten trioxide (WO_3)-assisted photocatalytic degradation of amoxicillin by simulated solar irradiation. Sci Rep 9(1):1–18
Ohore OE, Zhang S (2019) Endocrine-disrupting effects of bisphenol A exposure and recent advances on its removal by water treatment systems: a review. Sci Afr 5:e00135
Patisaul HB, Adewale HB (2009) Long-term effects of environmental endocrine disruptors on reproductive physiology and behavior. Front Neurosci 3:10
Quach-Cu J, Herrera-Lynch B, Marciniak C, Adams S, Simmerman A, Reinke RA (2018) The effect of primary, secondary, and tertiary wastewater treatment processes on antibiotic resistance gene (ARG) concentrations in solid and dissolved wastewater fractions. Water 10(1):37
Rashed MN (2013) Adsorption technique for the removal of organic pollutants from water and wastewater. Organ Pollut Monitor Risk Treat 7:167–194
Rathi A, Basu S, Barman S (2019) Adsorptive removal of fipronil from its aqueous solution by modified zeolite HZSM-5: equilibrium, kinetic and thermodynamic study. J Mol Liq 283:867–878
Rathi A, Basu S, Barman S (2021) Structural framework effect of various CeO_2-loaded zeolites on the adsorptive removal of fipronil. J Environ Chem Eng 9(2):105167
Sanches S, Penetra A, Rodrigues A, Ferreira E, Cardoso VV, Benoliel MJ, Crespo MTB, Pereira VJ, Crespo JG (2012) Nanofiltration of hormones and pesticides in different real drinking water sources. Sep Purif Technol 94:44–53
Schug TT, Johnson AF, Birnbaum LS, Colborn T, Guillette LJ, Crews DP, Collins T, Soto AM, Vom Saal FS, McLachlan JA, Sonnenschein C (2016) Minireview: endocrine disruptors: past lessons and future directions. Mol Endocrinol 30(8):833–847
Sharma S, Basu S (2021) Fabrication of centimeter-sized Sb_2S_3/SiO_2 monolithic mimosa pudica nanoflowers for remediation of hazardous pollutants from industrial wastewater. J Clean Prod 280:124525
Sharma S, Basu S (2020) Highly reusable visible light active hierarchical porous WO_3/SiO_2 monolith in centimeter length scale for enhanced photocatalytic degradation of toxic pollutants. Separat Purif Technol 231:115916
Steinmetz R, Brown NG, Allen DL, Bigsby RM, Ben-Jonathan N (1997) The environmental estrogen bisphenol A stimulates prolactin release in vitro and in vivo. Endocrinology 138(5):1780–1786
Usha Vipinachandran V, Rajendran S, Badagoppam Haroon KH, Ashokan I, Mondal A, Bhunia SK (2020) Detoxification of endocrine disruptors in water using visible-light-active nanostructures: a review. ACS Appl Nano Mater 3(12):11659–11687
Vieira WT, de Farias MB, Spaolonzi MP, da Silva MGC, Vieira MGA (2020) Removal of endocrine disruptors in waters by adsorption, membrane filtration and biodegradation: a review. Environ Chem Lett 18(4):1113–1143
Wang M, Qu F, Jia R, Sun S, Li G, Liang H (2016) Preliminary study on the removal of steroidal estrogens using TiO_2-doped PVDF ultrafiltration membranes. Water 8(4):134
Wee SY, Aris AZ (2019) Occurrence and public-perceived risk of endocrine disrupting compounds in drinking water. NPJ Clean Water 2(1):1–14
Wee SY, Aris AZ (2017) Endocrine disrupting compounds in drinking water supply system and human health risk implication. Environ Int 106:207–233
Wee SY, Aris AZ, Yusoff FM, Praveena SM (2020) Occurrence of multiclass endocrine disrupting compounds in a drinking water supply system and associated risks. Sci Rep 10(1):1–12
Weinhold B (2006) Epigenetics: the science of change. Environ Health Perspect 114(3):A163
Xiao R, Gao L, Wei Z, Spinney R, Luo S, Wang D, Dionysiou DD, Tang CJ, Yang W (2017) Mechanistic insight into degradation of endocrine disrupting chemical by hydroxyl radical: an experimental and theoretical approach. Environ Pollut 231:1446–1452

XIN T (2004) Reverse osmosis and nanofiltration membrane rejection of endocrine disrupting chemicals. https://scholarbank.nus.edu.sg/handle/10635/14515

Zheng C, Zhao L, Zhou X, Fu Z, Li A (2013) Treatment technologies for organic wastewater. Water Treat 11:250–286

Zielińska M, Cydzik-Kwiatkowska A, Bułkowska K, Bernat K, Wojnowska-Baryła I (2017) Treatment of bisphenol A-containing effluents from aerobic granular sludge reactors with the use of microfiltration and ultrafiltration ceramic membranes. Water Air Soil Pollut 228(8):1–9

Zúñiga-Benítez H, Soltan J, Peñuela GA (2016) Application of ultrasound for degradation of benzophenone-3 in aqueous solutions. Int J Environ Sci Technol 13:77–86

Chapter 8
Ballast Water System Treatment Techniques in Marine Transportation Industry: A Case Study of M/V LOCH MELFORT

Tien Anh Tran

Abstract The maritime industry plays a vital role in the national economies development. The rise of number of ships is bringing more and more the benefit for the national economics. In contrast, the sea environment is gradually influenced by the maritime economic development especial the ocean water environment. In this chapter, the authors propose some modern waste water treatment solutions to restrict the ocean water source from the maritime industry and shipbuilding. The research results are significant to the environmental science and earth in particular the maritime industry.

Keywords Ballast water treatment · Transportation engineering · Sea environment · Ballast tanks · International maritime organization

8.1 Introduction

The sea environment protection plays an important role in the field of the environmental science and engineering. In where, the shipping transportation industry must comply with the strict regulations to ensure both the maritime safety and environmental protection. Currently, there are some international regulations issued to strict the harmful substances emitted into the sea environment. International Maritime Organization (IMO) proposes the international conventionof the sea ocean pollution prevention from ballast water on ships (Fig. 8.1). The ballast water is the storage water during the sailing process in aim with keeping the stable characteristics of vessels when no loading cargo carried. However, this water contains a numerous of chemical toxic that could cause the environmental pollution at some different ocean areas. Therefore, the ballast water onboard is the main cause for invasion in the marine environment (Gollasch 2006).

The ballast water onboard ships stores many microorganisms and diversity organisms, phytoplankton, zooplankton, and other elements so on (Fig. 8.2). Therefore,

T. A. Tran (✉)
Vietnam Maritime University, Haiphong 180000, Vietnam

S. Roy et al. (eds.), *Advanced Industrial Wastewater Treatment and Reclamation of Water*, Environmental Science and Engineering,
https://doi.org/10.1007/978-3-030-83811-9_8

Fig. 8.1 Discharging the ballast water from ship

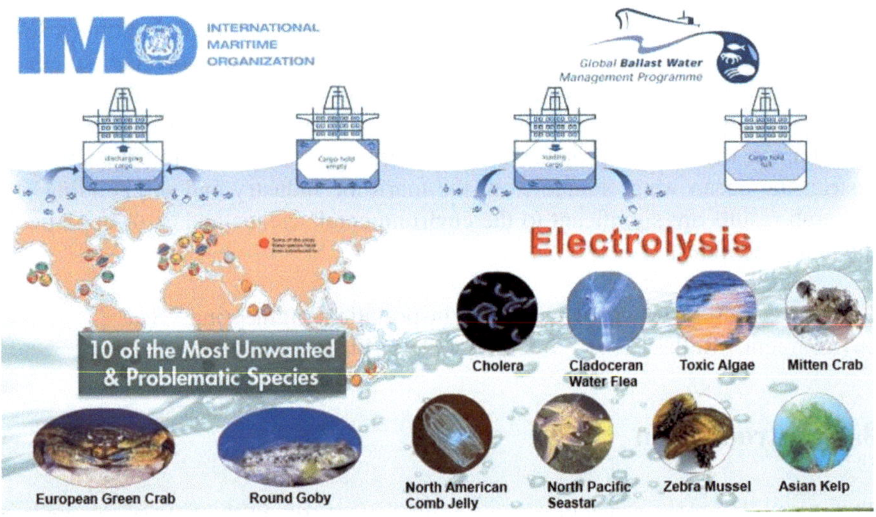

Fig. 8.2 The influences of ballast water to the sea environment

the marine environment will be influenced when these categories discharge into the sea environment through some harmful elements. The waste water management onboard plays an important role to prevent the harmful species into the sea environment. Consequently, that there are some international regulations which have been issued to decrease the harmful factors to the environment. So, there are a lot of countries in the world which enrolled and enforced the strictly regulations to the mandatory regulations in order to participate the international policies that the IMO proposed.

There are two priority strategies of ballast water management, including the open-ballast water management exchange at the different sea ocean areas and the applied ballast water treatment technologies (Matousek et al. 2006). At each specific methods, there are pros and cons between them. The simple method is the ballast-ocean water exchange with less efficiency in the field of removing the organism of the harmful elements into the ballast water. Consequently, the ballast water treatment technologies include the different method, including (1) cyclonic separation, (2) heat treatment method, (3) electric pulse, (4) ultraviolet (UV) method, (5) filtration method, and (6) biocides (Matousek et al. 2006).

The marine policy of ballast water management had been issued by the IMO. The general insight of ballast water management policy had been summarized by Verna and Harris (2016). Additionally, the ballast water management is a cause to make the maritime risk. Therefore, the study of same risk areas has been approached to prevent the invasion risks from the ballast water onboard vessels (Stuer-Lauridsen et al. 2018). The national and international measures had been introduced carefully in the technical documents and guidance codes (Ndlovu 2016).

In this research, the author had investigated the ballast water management technology through referring to the previous results. The detailed study has been conducted to investigate the ballast water management onboard vessels as well as the operational procedures and maintenance on ship. From then, the withdrawn conclusion had been presented to prove the challenges as well as the maritime policies related to the ballast water management.

8.2 Related Works

Currently, there are some different studies which related to the ballast water management and its treatment. Additionally, the development of science and technology is gradually influencing positively to the shipping transportation industries. Herwig et al. had the usage of ozone treatment method for ballast water management for oily tanker, S/T Tonsina (Herwig et al. 2006). An analysis of the enforcement policies of IMO BWM (Ballast Water Management) convention has been conducted by Gerhard et al. (2019). This analysis has been conducted at two ocean areas, including USA and Australia (Gerhard et al. 2019). Effects of the operational factors as well as the environmental aspect have been conducted for the ballast water treatment system (Moreno-Andrès and Peperzak 2019). The input parameters include the type of treatment, the time of holding, the source of carbon, and the substance dosages. The chemical and physical elements consist of the salinity, pH content, sea water temperature, organic matter, etc. which will influence on the water treatment methods. The application of ballast water management onboard had created the decision making system based on the combination between TOPSIS and the Analytic Hierarchy Process (AHP) (Karahalios 2017). This decision-making system has been determined based on a cost–benefit decision making tool to choose the ballast water management for ships in the USA (Karahalios 2017).

There are some benefits which had collected from those research results. The research gap is not stated the operational principles, the diversity application of each water treatment methods. In this research, the author made clear the differences between the waste water treatment technologies from ships. Additionally, the detailed studies have been conducted according to the operational procedures of the ballast water management system on a certain vessel.

8.3 Methodologies

8.3.1 Water Treatment by UV Ray

The water treatment techniques have been conducted by UV ray which provides the physical platform in the technology of ballast water treatment. Firstly, the full flow rate of inlet ballast water will be filtered roughly. After that, it will be continued to treat by the UV ray control box before pumping into the ballast tanks onboard ships which are installed at the double-hull position of vessel. After that, the mechanical filter will remove the raw organisms into the ballast water along with the effective enhancement of this water treatment method.

Consequently, the ballast water will be pumped into the tanks on ships. Afterwards, this water source will be treated by UV ray. The water quality is measured and controlled by the monitoring device before discharging the sea environment. Additionally, the mechanical filter is re-circulated during the de-ballasting process (Fig. 8.3).

The UV treatment method had been used popularly for the small size ships with the simple treatment method. The main energy consumption of UV ray depends on the power consumption (Werschkun et al. 2014). Consequently, the UV method has been employed to comply with the ballast water management convention due to the IMO. The determination of the lowest UV dose has been conducted to estimate the required time for the inactivation when the ballast water had been stored into the ballast tanks on ships after irradiation. Additionally, there are some advanced methods to improve the water treatment quality by UV ray which had been adopted.

Normally, the UV treatment technique will be performed and happened under the low or medium pressure through UV lamps (Oguma et al. 2002).

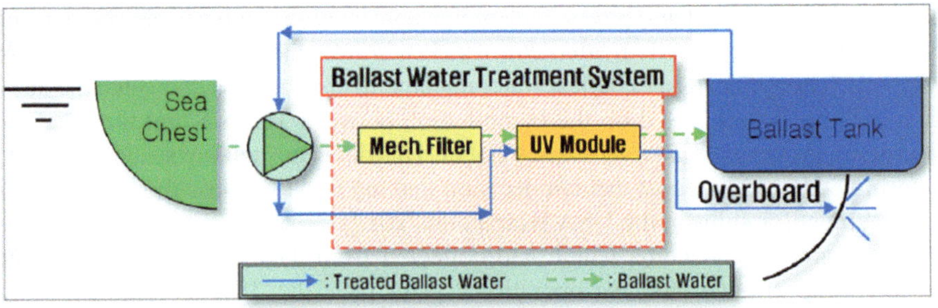

Fig. 8.3 Schematic of ballast water treatment by UV type (Jee and Lee 2017)

8.3.2 Chemical Treatment Method

There are a wide range of chemical treatment which are being applied for the big size ships nowadays. For this method, the secondary treatment technique has been adopted for both the intake water and wastewater which have been pumped into and out the ballast tanks, respectively (Table 8.1).

8.4 Results and Discussion

The application of the ballast water treatment technique has been utilized in this research through the certain vessel which operates at the international ocean sea areas. Then, a study of the ballast water treatment technique has been applied and compared in this research based on the actual vessel as well as related to the environmental aspect and evaluated the environmental indexes. The techniques could be conducted both the onshore and the on board depending on the ballast water management plan of the ship-owners as well as the current regulations of the IMO.

In case of the onshore, there are two separated ways depending on the available facilities at ports. The ballast water management will include to apply the treatment facilities after pumping the ballast water into the ballast water tanks. These facilities will be used to treat when the total volume is stored into the ballast water tanks. Furthermore, the ballast water will be cleaned in case of de-ballasting process (Fig. 8.4).

The onshore ballast water treatment technique is possible to survey 596 ships per year when considering a certain single ballast water treatment (Pereira and Brinati 2012). In a result, the ballast water treatment onshore has been presented economically of full scale. Additionally, the bulk carriers had provided with the different types of treatment facilities.

The onboard vessel is normally carried out through two modes, including the water treatment technologies and the ballast water exchanges (Fig. 8.4). The treatment techniques include a lot of methods however there are two main techniques which many vessels are being applied including the UV ray treatment technique and the chemical treatment technique. These techniques are being utilized popularly for the different sizes of vessel. In where, the UV ray treatment and the chemical treatment have been applied into the small and the big size vessels, respectively.

The evaluation of water treatment technology has been conducted to research and validate the ballast water management system through the proposed methodologies in this research. The study of water treatment has been observed and monitored by the sensors and alarm devices in the emergency situation during the operational system process onboard. In a result, the author had investigated the ballast water management system on a certain vessel. This type of shiphas the ballast water management system according to the IMO. The specification of this ship has been provided in Table 8.2.

Table 8.1 Overview of chemical treatment method

Chemical treatment	Capacity	Tested marine organisms	Removal percentage
Biocides	Not Available	Freshwater	More than 95% (Sano et al. 2005a, b)
Biocides	Lab scale	Bacteria	Effective (Chelossi and Faimali 2006)
Biocides	Lab scale	Marine organism	Effective (Faimali et al. 2006)
Biocides	Lab scale	Microalgae	Effective (Gregg and Hallegraeff 2007)
Biocides (Naphthoquinones)	Lab scale	Dinoflagellates cysts	More than 95% (Wright et al. 2007)
Biocides	Lab scale	Living biomass	More than 99% (Lafontaine et al. 2008)
Sodium hypochlorite	5 mgL^{-1}	Bacteria, Phytoplankton,	Over 85.2% (Zhang et al. 2003) Effective
	2 mgL^{-1}	Zooplankton	Over 15%
Chlorine	Bulk Frotargetina	Bacteria, Phytoplankton, Zooplankton	More than 99% (Vianna et al. 2003)
Chlorine dioxide	Not available	Alexandriumcatenella and Gymnodiniumcatenatum	More than 97% (Ichikawa et al. 1992)
NaOCl	Not available	Cysts	More than 89% (Gray et al. 2006)
Ozone gas	400 L h^{-1}	Bacillus subtilis	Ineffective (Oemcke and Leeuwen 2003, 2004)
	400 L h^{-1}	Dinoflagellate alga Amphidinium sp.	More than 98% (Oemcke and Leeuwen 2005)
Ozone	Not available	Bacteria, Phytoplankton, Zooplankton	99% (Viitasalo et al. 2005)
Ozone	Not available	Species marine organism	More than 95% (Jones et al. 2006)
Electrolysis	1200 m^3 h^{-1}	Phytoplankton	More than 99.99% (Matousek et al. 2006)
		Meso zooplankton	More than 99%
Electronic current	2.5 m^3 h^{-1}	Artemia	More than 95% (Dang et al. 2003)
Magnetic method	100 m^3 day^{-1}	Red phytoplankton	More than 92% (Soho et al. 2002)

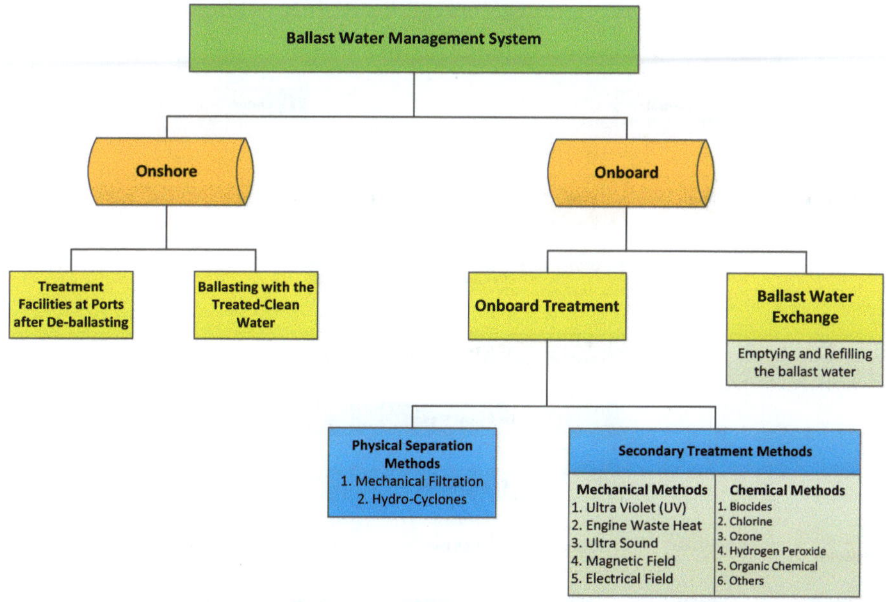

Fig. 8.4 The ballast water treatment technologies at two cases (1) Onshore, (2) Onboard

Table 8.2 The specification of M/V LOCH MELFORT

Number	Category	Parameter
1	Name of ship	LOCH MELFORT
2	IMO Number	9,658,795
3	Type of ship	Bulk Carrier
4	Length of Overall	176.83 m
5	Breadth	28.8 m
6	Depth	14.2 m
7	Draft (Full Load)	9.826 m
8	Dead Weight	33,379 T
9	Type of ship	Bulk Carrier

Consequently, the application of ballast water management treatment has been expressed all ocean-going vessels through connecting the devices and service system in order to treat the harmful substances from the ballast water. Currently, the seafarers and engine officers have to comply strictly the operational procedures of management system. The ballasting process is conducted by the 3rd engine officer on vessel. The de-ballasting process happens when discharging the cargo from ship. Then, the ballast procedure is being conducted when the ship is near the shore or at the ports according to the ballast water management convention of the IMO. The operational process has been monitored through the control panel of the management system like as Fig. 8.5.

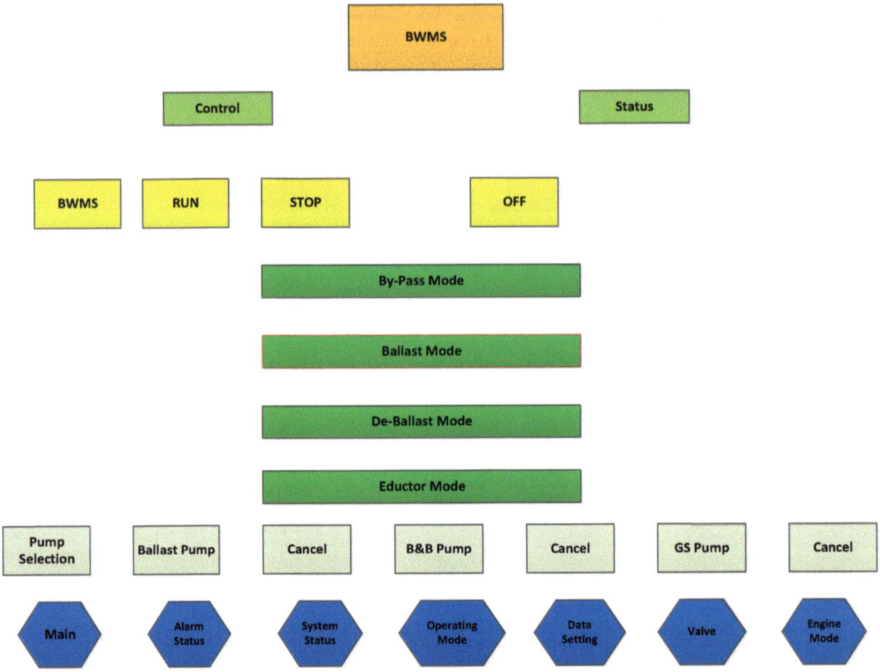

Fig. 8.5 The control panel of ballast water management system of M/V LOCH MELFORT

The control system of ballast water management has been presented in this research. The ballast water treatment system includes some monitoring devices and sensors. In where, there are three controlling modes to operate this system. In where, the ballast valves are controlled by two modes, including the automatic mode and handle mode for controlling the open/close state of these valves.

In this research, the treatment system of a certain vessel has been provided based on the empirical investigation. Normally, this system includes a lot of devices that could conduct some missions to ensure the high waste water treatment onboard vessel. Additionally, the discharged ballast water will be measured by the sensor which is installed at the outlet ballast water line. Afterwards, the ensured ballast water will be discharged when to meet the regulations before discharging the sea environment (Fig. 8.6).

The specific parameters of this unit have been presented in Table 8.3 through investigating the actual devices onboard. The volume rate of this system is high with 750 m^3 h^{-1}. This system is reliable with working pressure under 2 kgF cm^{-2}. Additionally, this system is ensured during the working condition by the seafarers and the 3rd Engineer. Then, this research has been validated through the certain bulk carrier with the UV ray treatment technique of the ballast water management.

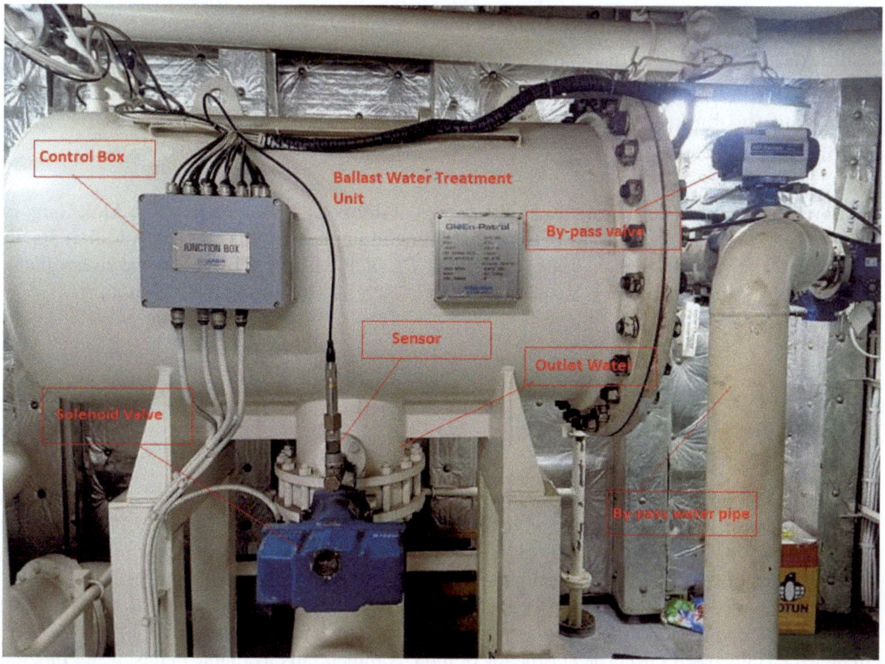

Fig. 8.6 The ballast water treatment unit of M/V LOCH MELFORT

Table 8.3 The specification parameters of ballast water management system

Category	Parameter
Model	GloEn-Patrol PF 750
Capacity *(Q)*	750 $m^3\ h^{-1}$
Maximum working pressure (P_{max})	7 kgF cm^{-2}
Waste water rate	800 litters at P = 2 kgF cm^{-2} and Q = 750 $m^3\ h^{-1}$
Electrical Source	AC 440 V, 60 Hz

8.5 Conclusion Remarks

The ballast water treatment system plays an important role in the marine activities during the ship operation and ship building industries. The study of treatment techniques have been investigated in this research. The indication of different characteristics of the ballast water management has been shown at each specification parameter as well as the actual operation and maintenance of the engine officers. The detailed devices have been presented in this research for a certain object of study. The general study of water treatment techniques has been reviewed in the field of marine transportation industry. The popular treatment has been shown through the empirical investigation onboard vessels. Finally, the ballast water management techniques provide the modern methods to treat the waste water from the sea environment in the field of marine engineering. This study is significant to introduce and investigate the ballast water management system based on the different treatment techniques.

Acknowledgements The author would like to thank Editors Dr. Swapnila Roy, Dr. Alok Garg, Er. Shivani Garg, and Dr. Tien Anh Tran for encouraging to implement this chapter. Additionally, the author appreciate all seafarers on vessel namely M/V LOCH MELFORT. Finally, the author thanks to Vietnam Maritime University, Haiphong City, Vietnam for supporting this research.

References

Chelossi E, Faimali M (2006) Comparative assessment of antimicrobial efficacy of new potential biocides for treatment of cooling and ballast waters. Sci Total Environ 356(1–3):1–10

Dang K et al (2003) Application study of ballast water treatment by electrolyzing seawater. In: IMO Proceedings of the 2nd international ballast water treatment R&D symposium, pp 103–111

Faimali M et al (2006) A new photodegradable molecule as a low impact ballast water biocide: efficacy screening on marine organisms from different trophic levels. Mar Biol 149:7–16

Gregg MD, Hallegraeff GM (2007) Efficacy of three commercially available ballast water biocides against vegetative microalgae dinoflagellate cysts and bacteria. Harmful Algae 6(4):567–584

Gray DK et al (2006) Can sodium hypochlorite reduce the risk of species introductions from diapausing invertebrate eggs in non-ballasted ships? Mar Pollut Bull 52(6):689–695

Gerhard WA et al (2019) Installation and use of ballast water treatment systems-Implications for compliance and enforcement. Ocean Coastal Manag 181:104907

Gollasch S (2006) Overview on introduced aquatic species in European navigational and adjacent waters. Helgol Mar Res 60:84–89

Herwig RP et al (2006) Ozone treatment of ballast water on the oil tanker S/T Tonsina: chemistry, biology and toxicity. Mar Ecol Prog Ser 324:37–55

Ichikawa S et al (1992) Extermination efficacy of hydrogen peroxide against cysts of red tide and toxic dinoflagellates, and its adaptability to ballast water of cargo ships. Nippon Suisan Gakkaishi 58(12):2229–2233 (In Chinese)

Jee J, Lee S (2017) Comparative feasibility study on retrofitting ballast water treatment system for a bulk carrier. Mar Pollut Bull 119(2):17–22

Jones AC et al (2006) Toxicity of ozonated seawater to marine organisms. Environ Toxicol Chem 25(10):2683–2691

Karahalios H (2017) The application of the AHP-TOPSIS for evaluating ballast water treatment systems by ship operators. Transp Res Part D Transp Environ 52A:172–184

Lafontaine Yd et al (2008) Effectiveness and potential toxicological impact of the PERACLEAN® Ocean ballast water treatment technology. Ecotoxicol Environ Saf 71(2):355-369

Matousek RC et al (2006) Electrolytic sodium hypochlorite system for treatment of ballast water. J Ship Product 22(3):160–171

Moreno-Andrès J, Peperzak L (2019) Operational and environmental factors affecting disinfection by products formation in ballast water treatment systems. Chemosphere 232:496-505

Ndlovu FP (2016) The marine environment and ballast water management law. Water Policy 18:964–982

Oguma K et al (2002) Photoreactivation of Escherichia coli after Low or Medium pressure UV disinfection determined by an Endonuclease sensitive site assay. Appl Environ Microbiol 68(12):6029–6035

Oemcke DJ, Leeuwen J (2003) Chemical and physical characterization of ballast water: part 1: effects on ballast water treatment processes. J Marine Environ Eng 7(1):7–64
Oemcke DJ, Leeuwen J (2004) Seawater ozonation of the Bacillus subtilis spores: implications for the use of ozone in ballast water, Ozone. Sci Eng 26(4):389–401
Oemcke DJ, Leeuwen J (2005) Ozonation of the marine dinoflagellate alga Amphidinium sp.-implications for ballast water disinfection. Water Res 39(20):5119–5125
Pereira NN, Brinati HL (2012) Onshore ballast water treatment: a viable option for major ports. Mar Pollut Bull 64(11):2296–2304
Stuer-Lauridsen F et al (2018) Same risk area: An area-based approach for the management of bio-invasion risks from ships' ballast water. Mar Policy 97:147–155
Sano LL et al (2005a) Decay model for biocide treatment of unballasted vessels: application for the Laurentian Great Lakes. Mar Pollut Bull 50(10):1050–1060
Sano LL et al (2005b) Chronic toxicity of glutaraldehyde: differential sensitivity of three freshwater organisms. Aquat Toxicol 71(3):283–296
Soho N et al (2002) Superconducting magnetic separator for ballast water treatment, IMO. In: Proceedings of the 2nd international ballast water treatment R&D symposium, pp 125–137
Verna DE, Harris BP (2016) Review of ballast water management policy and associated implications for Alaska. Mar Policy 70:13–21
Vianna S et al (2003) Use of chlorine for ballast water treatment, IMO. In: Proceedings of the 2nd international ballast water treatment R&D symposium, pp 158–164
Viitasalo S et al (2005) Ozone, ultraviolet light, ultrasound and hydrogen perozide as ballast water treatments: Experiments with mesozooplankton in low-salinen brackish water. J Mar Environ Eng 8(1):35–55
Werschkun B et al (2014) Emerging risks from ballast water treatment: the run-up to the International Ballast Water Management Convention. Chemosphere 112:256–266
Wright DA et al (2007) Naphthoquinones as broad spectrum biocides for treatment of ship's ballast water: toxicity to phytoplankton and bacteria. Water Res 41(6):1294–1302
Zhang S et al (2003) Effects of the chlorination for ballast water treatment, IMO. In: Proceedings of the 2nd international ballast water treatment R&D symposium, pp 148–157

Chapter 9
Science and Technology Roadmap for Adsorption of Metallic Contaminants from Aqueous Effluents Using Biopolymers and Its' Derivatives

Ahmed Elsayid Ali, Zaira Zaman Chowdhury, Rahman F. Rafique, Rabia Ikram, Abu Nasser Mohammad Faisal, ShahJalal Shibly, Arnab Barua, Yasmin Abdul Wahab, and Badrul Mohamed Jan

Abstract Large amount of heavy metal pollutants are discharging by industries which are hazardous for environment. Several metallic pollutants like copper (II), nickel (II), lead (II), cadmium (II), and zinc (II) introduced in water can cause serious threat to aquatic environment. Biopolymers are promising candidate for elimination of heavy metals from waste stream. In this chapter, biopolymers based on polysaccharides can be classified into four sub-categories: homopolymers, blends of polysaccharide, copolymers, and functionalized polysaccharide or their composites. Natural biopolymers like alginate,chitosan, cellulose, chitin, and starch contain numerous surface active groups which can act as binding agent to eliminate heavy metals from water. Introduction of secondary functional groups with biopolymers enhances its' mechanical strength which is essential for recyclability of the adsorbent to ensure economical application of biopolymers for industrial scale waste water treatment plant. This chapter provides the in-depth mechanism of heavy metal interactions with the bio-polymeric surface using π-π, dipole–dipole reaction. Furthermore, it can go for hydrogen network, or weak Van-der-walls attraction with the co-polymers, homopolymer, blending of polymers, as well as biopolymer/composite matrix containing negative ions or electron-enriched groups. The concluding part of the chapter illustrates the adsorptive behavior of cations using kinetics and isotherms analysis of the equilibrium system for waste water treatment.

A. E. Ali · Z. Z. Chowdhury (✉) · A. N. M. Faisal · S. Shibly · A. Barua · Y. A. Wahab
Nanotechnology and Catalysis Research Centre (NANOCAT), University of Malaya, 50603 Kuala Lumpur, Malaysia
e-mail: dr.zaira.chowdhury@um.edu.my

R. F. Rafique
John A. Reif, Jr. Department of Civil & Environmental Engineering, New Jersey Institute of Technology (NJIT), 323 MLK Blvd., New York, NJ 07102, USA

R. Ikram · B. M. Jan
Department of Chemical Engineering, University of Malaya, 50603 Kuala Lumpur, Malaysia

S. Roy et al. (eds.), *Advanced Industrial Wastewater Treatment and Reclamation of Water*, Environmental Science and Engineering,
https://doi.org/10.1007/978-3-030-83811-9_9

Keywords Biopolymers · Homopolymers · Adsorption · Heavy Metals · Kinetics · Isotherm

9.1 Introduction

The exponential growth in the population in our world and the development in the industrialization sectors resulted in critical environmental pollution (Hargreaves et al. 2018; Herath et al. 2016). Several approaches have been suggested to lower the pollution effects. One of the effective adopted approaches is the treatment of waste effluent before it is discharged into the environment (Bhattacharya et al. 2019; Alvarez et al. 2018; Werber et al. 2016; Chowdhury et al. 2017; Lim and Aris 2014; Grant et al. 2012; Loosdrecht and Brdjanovic. 2014). The treatment of effluent wastes is conducted via several techniques such as coagulation, adsorption, flocculation ionexchange, and filtration using membrane (Abussaud et al. 2016). Metal–organic frameworks are among the most adopted materials in the wastewater treatment field (Zhao et al. 2021; Rojas and Horcajada 2020).

In the last decade, several studies focused on developing a green approach for wastewater treatment. Therefore, studies focused on the utilization of agricultural biomass waste to produce new green materials for water treatment. Moreover, biosorption through polymers and biomaterials such as algae, fungi, bacteria, and yeast is proposed as future adsorbents for water contaminants (Ahsan et al. 2019). Polymers can be used in wastewater treatment in natural and synthetic forms. Polysaccharides are types of natural polymers that are abundant in nature. Chitin, chitosan, starch, cellulose, and alginate are the most popular types of polysaccharides that can be used as sorbents (Mogoşanu and Grumezescu 2014; Muzzarelli 2011; Sardar et al. 2018). Typically, these four polysaccharides are connected to synthetic polymers to enhance their hydrophilicity and mechanical stability with metal ions (Mohammadzadeh Pakdel and Peighambardoust 2018; Mohammed et al. 2018; Rojas and Horcajada 2020; Vakili et al. 2014).

Generally, polysaccharides are abundant homopolymers and can be modified easily to other forms such as copolymers and polysaccharides blends. Hence, these materials are inexpensive, easy to handle, and have a great adsorption capability for different types of heavy metals. Polymer blends and copolymers forms are mainly used to eliminate metallic pollutants from contaminated water. Briefly, polymer blends area combination of two polymers or more through physical method to form new polymers with better characteristics. Generally, the new polymer will gain the best features such as higher stability, strong thermal and mechanical strength, as well as better elimination performance of metallic ions from waste effluents (Tabasum et al. 2018; HPS et al. 2016). In comparison to polysaccharide homopolymers, the polymer blend technique can produce highly effective polymers that can be used in various conditions. Polysaccharide blends are used to eradicate metallic cations from contaminated water. However, the blends of polysaccharides are classified into four types, miscible, immiscible, homogenous, and compatible polymers (Cazacu and Popa 2014).

Despite blends polymer, copolymers have revealed a great performance as adsorbents for metal ions.

Basically, copolymers consist of repeated units of backbone strand with more diversity in types, position, and ratio of the sequence. In contrary to homopolymers, copolymers have different repeated units in the structure whereas homopolymer has only one type. Moreover, in the copolymer, the different structural polymers are linked chemically, unlike polymers blends. Thus, the copolymer has higher mechanical and chemical stability which are required to enrich the exclusion of metallic cations from different types of liquid environments (Kanmani et al. 2017; Suhas et al. 2016; Vandenbossche et al. 2014).

Another approach to modifying polysaccharides is the mixing with different forms of inorganic materials to make a composite with improved properties (Stela et al. 2015). Usually, the synthesised composed have better chemical stability, better mechanical stability, and higher adsorption performance. The modification of polysaccharide with organic materials enhances the functionality of nano and microparticles. Thus, heavy metals can be trapped into the functional active groups of the surface of the polymers. Moreover, surface functionality enhances the hydrophilicity of the polymer composite and hence improves the heavy metal removal. For instance, a published study found that the adsorption of mesoporous polystyrene for metal ions can be enhanced using zirconium phosphate (ZrP) in nanoconfinement (Zhang et al. 2020).

This book chapter focus on the efficiency of four types of polysaccharides as adsorbents for heavy metals. The book chapter studies the adsorption mechanism and properties of polysaccharide composites, polysaccharide homopolymers, polysaccharide blends, and copolymers. Lastly, the current and future applications of the adsorbents are covered for further studies.

9.2 Different Types of Biopolymers

9.2.1 Chitin

Chitin ($(C_8H_{13}O_5N)_n$) is a popular biopolymer containing a long chain of N-acetylglucosamine (poly-β-(1,4)-N-acetyl-D-glucosamine) polymer (Fig. 9.1). Chitin is considered the major constituent forming the cell membrane of fungi, yeast, and different types of invertebrates such as crab and shrimp (Schleuter et al. 2013). Additionally, chitin can be found in the structure of 3-D network marine sponges such as Verondiga Sponge (Jung and Song. 2019). Ehrlich et al., have started the experiment on extracting chitin from marine sponge (Ehrlich et al. 2013). In the experiment, they cultivated the sponges using sodium hydroxide (NaOH) and ethanoic acid (CH_3COOH) solution. Similarly, Schleuter et al., adopted ananalogous methodology

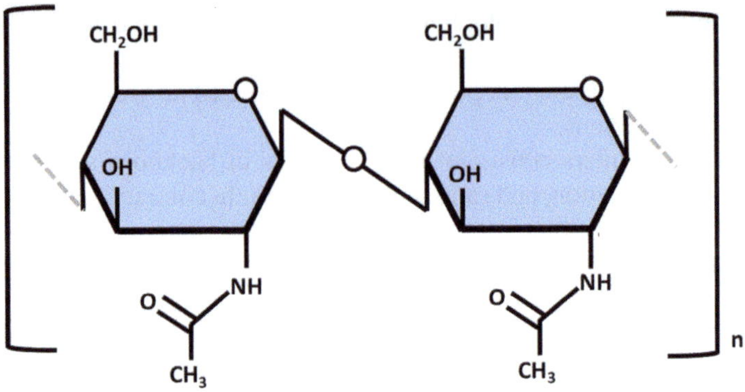

Fig. 9.1 Structure of Chitin

to produce chitin from Aureate sponge (Schleuter et al. 2013). They stated that the interconnection of the network in a hydrate state has great mechanical stability and flexibility. In addition, the extracted chitin was found to be tubular with a large volume of pores present between the weakly connected networks. Thus, the polymer can be modified into different forms. Moreover, to reserve the samples, they freeze-dried the extracted chitin as the 3-D network of the chitin can easily collapse at room temperature. In regards to the adsorption performance, the sponge chitin samples have revealed great results. The samples adsorbed 288 ± 11 mgg^{-1} uranium (U) at pH 6. The adsorption capacity is directly influenced by the structure and porosity of sponge chitin.

An interesting study is published by Liu et al. where they prepared chitin nanofibers, alternatively called chitin nanofibrils (CNFs), from chitin microparticles (CMPs) (Liu et al. 2013). For the preparation process, they wet ground and compressed CMPs following by a freeze-drying process. The adopted approach converts CMPs into nanofibers without changing the crystalline or chemical structures. The study of the adsorption performance is conducted on 6 types of metal ions; chromium (Cr(II)), nickel (Ni(II)), cadmium (Cd(II)), zinc (Zn(II)), copper (Cu(II)), and lead (Pb(II)). The maximum obtained adsorption capacities were 330.15, 134.72, 16.28, 134.30, 303.49, and 141.08 mg/g for Nickel (II), Cadmium (II), Chromium (II), lead (II), Zinc (II), and Copper (II) correspondingly. However, the observation of the data showed that the increase in metal binding is affected by the rigidity of the chitin molecular structure. Hence, the study rebuts the primary fundamental of chitin hydrophobic property. Additionally, the study displays the prominence of the surface active area and porous volume for CNFs adsorption capacity.

Yang et al. have done a research on the functionalization of chitin nanofibers to enhance their biosorption efficiency (Yang et al. 2015). They used a series of base, acid, and mechanical treatments to extract chitin from shrimp shell. Additionally, to obtain functionalized chitin nanofibers they used thiol-modification method. Thiol-modification produced highly efficient thiol-chitin nanofibers. The functionalization through the thiol approach was possible because ofthe amine groups exist on the surface of nanofibers. The produced functionalized chitin nanofibers exhibited a great performance in adsorbing As(III) from contaminated water. The adsorption performance of the nano-fibers are directly influenced by the characteristics and enlarged surface area and porous texture. The As(III) ions removal test was conducted using sodium arsenite (Na_3AsO_3) solution having concentration of 100 ppm and it was diluted to make pH 7 by adding DI water. The adsorption data revealed that non-functionalized chitin nanofibers have a significantly lower adsorption capacity compared to functionalized nanofibers. The maximum sorption capability of nonfunctionalized chitin nanofibers were 58 mgg^{-1}, whereas the maximum sorption capability for thiol-chitin nanofibers were 138 mgg^{-1}. The value of maximum adsorption capacity at pH 7 determined by Langmuir isotherm was 149 mgg^{-1}.

9.2.2 Chitosan

Another kind of biopolymer is Chitosan ($(C_6H_{11}NO_4)_n$). It is polysaccharide having linear chain. The constituent materials for the cellular wall ofcrab shell, fungi, shrimp shell, and shellfish is chitosan. The main structure of chitosan composes of linear (1,4)-2-amino-2-deoxy-β-D-glucan (Fig. 9.2). Typically, chitosan is derived from chitin via a partial deacetylation process (Jung and Song 2019; Ehrlich et al. 2007). The process involved deacetylation of α-chitin in dilute alkali and under temperature between 100 and 160 °C for a short period.

Chitosan can be used effectively to remove metallic ions owing to the existence of $-NH_2$ groups over its external surface. Furthermore, the sorption capability of chitosan can be improved via developing new derivatives with new functional groups. An interesting study is conducted by Negm et al., where they synthesized chitosan mix with glycine and chloro-acetic acid infused chitosan polymers using xylene solvent (Negm et al. 2015). The research group identified that, the sorption aptitude of chitosan derivatives is lower than expected. The poor adsorption performance

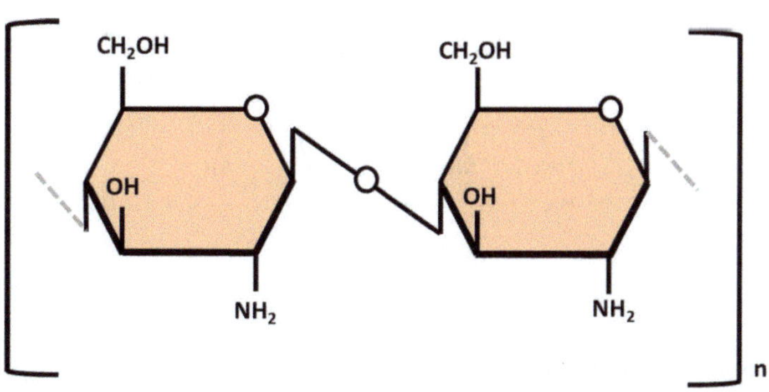

Fig. 9.2 Structure of chitosan

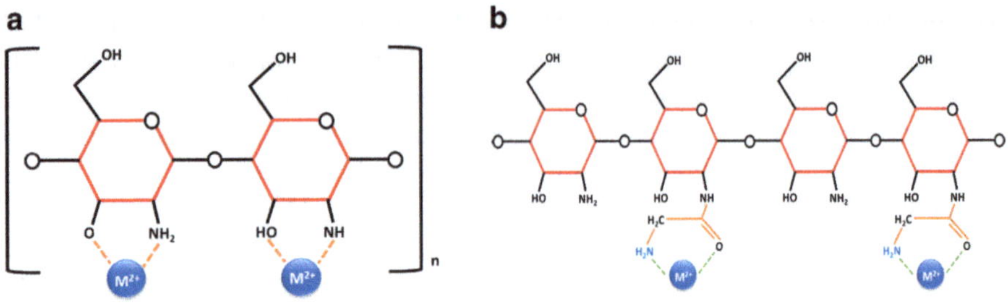

Fig. 9.3 Adsorption behavior of **a** chitosan, and **b** glycine chitosan

of chitosan derivatives was observed (Fig. 9.3a–b).This is due to manifestation of intermolecular networks of hydrogen bond between the monomer units of chitosan. However, the formation of hydrogen bonds is caused by the substituent bonds, which result in a lower amount of free active groups on the chitosan. Therefore, pure chitosan has wider active sites in comparison to glycine and chloroacetic acid derivatives. In general, the sorption performance of chitosan and its functionalized composites are in effective. Therefore, a new approach of chitosan modification via chemicals is proposed by several studies (Varma et al. 2004; Zhao et al. 2013).

The functionalization strategies of chitosan using ethylene glycol-bis (2-aminoethylether) N,N,N',N'-tetraacidic (EGTA) acid was studied (Zhao et al. 2013). The functionalized chitosan is used to eliminate lead (II) and Cadmium (II) cations. The highersorption capacities reportedas0.502 and 0.737 mmolg^{-1} for lead (II), and Cadmium (II) cations correspondingly. The obtained analytical results displayed the form of interaction between the metallic ions and EGTA. Basically, the interaction occurs via the 3 carboxyl groups, 2 ether oxygen, and 1 amino nitrogen of the EGTA. This interaction forms an octahedral structure where the rest of the carboxyl group attaches to chitosan (Fig. 9.4).

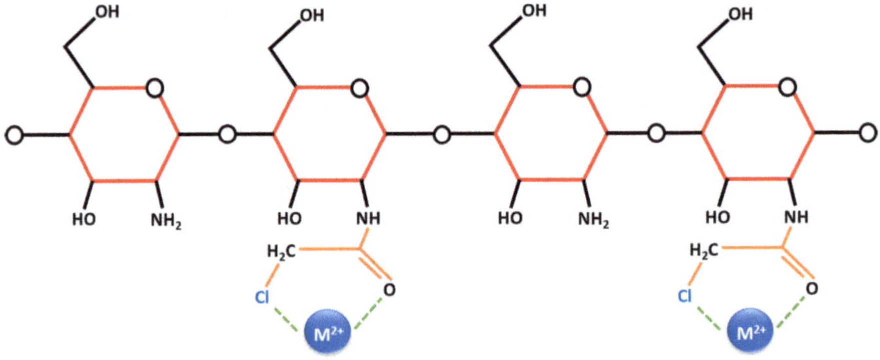

Fig. 9.4 Adsorption behavior of chloroacetic acid-chitosan

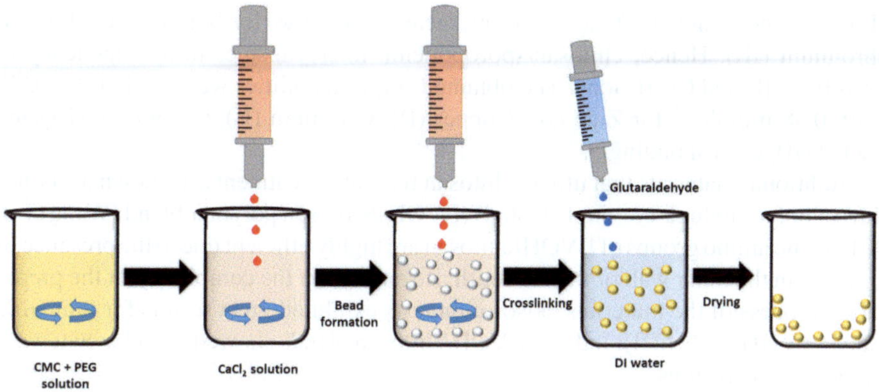

Fig. 9.5 The preparation steps for PCMC beads

An interesting study on the chemical modification of chitosan has been lately published by previous research group (Luo et al. 2018). In their study, they prepared porous carboxymethyl chitosan (PCMC) via an ionic coacervation/chemical cross-linking method (Fig. 9.5). The experiment involved the use of calcium chloride as a physical crosslinker and glutaraldehyde as a chemical crosslinker. Ammine (-NH_2) and hydroxyl (-OH) functional active groups were present over chitosan. Methyl carboxylation reaction was carried out to have –COOH groups. The obtained PCMC is used to eliminate Co(II) ions where the reported maximum sorption was 46.25 mgg^{-1}. The measured surface area of PCMC was 9.35 m^2g^{-1}. The analytical results exhibited that sorption trait obeyed second-order kinetic and Langmuir model.

9.2.2.1 Chitosan and Its Blends

Among the different types of carbohydrates, chitosan has exhibited a great adsorption performance foreradicating metallic cations from polluted water. The adsorption performance of chitosan is influenced by the existed free –NH_2 and –OH groups which provide active sites for the metal ions. Sargin and Arslan have synthesised polysaccharides blends using sporopollenin and chitosan (Sargın and Arslan 2015). Sporopollenin has high mechanical stability and string structure that consist of carotenoid and carotenoid esters. Thus, sporopollenin can resist biological and chemical changes (Ahlers et al. 1999). Sporopollenin polymers can be found in the wall of spores and pollen (Yule et al. 2000). The approach adopted by Sargin and Arslan to synthesis chitosan/sporopollenin microcapsules is polymer cross-linking. The obtained polysaccharide blends is used to adsorb Copper (II), Nickel (II), Chromium (III), Zinc (II), and Cadmium (II) ions from water. The adsorption test is conducted under different conditions including, different temperatures, pH levels, sorption time, and amount of sorbent. Generally, the analytical results showed that microcapsule higher adsorption capacity for Cadmium (II), and Zinc (II) related to pure chitosan.

However, the removal efficiency of microcapsules is low for both copper (II), and chromium (III). Hence, chitosan/sporopollenin microcapsules have a tendency to adsorb Zn(II) and Cd(II) ions. The obtained sorption abilities were 0.71, 1.34, 0.77, 0.99, 0.58 mmolg^{-1} for Zinc (II), Copper (II), Cadmium (II), Chromium (III),and Nickel (II), correspondingly.

Additional materials that utilize chitosan for water treatment are known as monoliths of poly(ethylene-co-vinyl alcohol) (EVOH)/chitosan polymer blend (Wang et al. 2015). The amino groups of EVOH/chitosan are highly efficient due to the presence of chitosan in the inner wall of the monolith. Regardless of the complexity of the preparation process of the materials, adsorption tests exhibited great results for removing Cu(II), Co(II), and Ni(II). Indeed, EVOH/chitosan blends showed a tendency toward adsorbing Cu(II) ions.

A research was carried out on the sorption performance of chitosan nanofibers for Nickel (II) and Copper (II) ions. The study adopted an electrospinning method to fabricate chitosan/poly(N-isopropylacrylamide-co–N-methylolacrylamide) (PNN) nanofibrous mats. The pre-fabrication process involved synthesizing PNN for the blending. The cross-linking of chitosan/PNN nanofibrous mats is caused by N-methylolacrylamid (NMA) on PNN. The study revealed that chtiosan/PNN nanofibers are sensitive materials, and their structure can be changed based on environmental conditions. Hence, the adsorption activities of the nanofibrous mats is prejudiced by several environmental factors. The obtained data show that at 25 °C and pH 2, the size of PNN and chitosan expand. However, at pH 2 and 70 °C, chitosan expands while PNN size dwindle. The crucial change is observed at 25 °C and pH 10 where chitosan starts deprotonation and PNN expand. The deprotonation of chitosan increases the hydrophobicity of the nanofibrous mats and hence the precipitation in an aqueous solution. In regards of the adsorption test, Cu (II) and Ni(II) ions are used with an initial concentration of 100 ppm at pH 4 and 25 °C. The reported adsorption capacities were 29 mgg^{-1} and 20 mgg^{-1} for Cu(II) and Ni (II), respectively (Hande et al. 2017). Indeed, the research stated that the adsorption capacity of nanofibrous mats could be enhanced via washing the samples at pH 7 repeatedly. The washed nanofibrous mats are used to remove 1000 ppm of Cu (II) ions where the obtained sorption capability was 79 $\pm$ 2 mgg^{-1}. The increase in Cu (II) adsorption capacity is related to the increase in the initial concentration and chitosan protonation.

9.2.2.2 Chitosan Based Copolymers

Copolymers are a common approach to modify polysaccharides via chemical bonding. Similar to polymer blends, there are different types of copolymerized chitosan. In the water treatment field, β-chitosan copolymerized with polyamine is extensively used as adsorbents to exclude heavy metals from polluted stream. For instance, β-chitosan Schiff base (BCS) is used as sorbents to eliminatethe Cu(II) and Ag(I) cations from aqueous effluents (Wu et al. 2016). BCS is synthesised using benzaldehyde with BC amine groups. In addition, Epoxide was used to modify the surface of BCS. It was cross-linked with triethylenetetramine or triethylenetetramine to enhance the ratio of chitosan grafting. The reported adsorption capacities were 151.2 and $\leq$ 117.60 mgg^{-1} for Ag(I) and Cu(II) ions. The bonding between BCS and heavy metals occurred via the primary amine and hydroxyl functional groups. In comparison to pure chitosan, BCS derivatives have higher acid-resistivity. Therefore, the modified polymers can have relatively great adsorption capacity at pH 2.

Gopal et al. prepared a copolymerized polymer known asChitosan-g-maleic anhydride-g-ethylene dimethacrylate (CS-g-MA-g-EDMA) to eliminate Cu(II) and Cr(II) ions from wastewater (Gopal Reddi et al. 2017). Ethylene dimethacrylate was first dissolved in ethanol and formic acid to dissolve chitosan-g-maleic anhydride. Moreover, ceric ammonium nitrate is used as an initiator in the synthesis process. Cs-g-MA-g-EDMA interacts with metalliccations through aliphatic, carboxylic, and amid surface active functional groups. As synthesized, Cs-g-MA-g-EDMA is economically efficient and environmentally friendly; therefore it is proposed as a solution for large-scale applications such as treating wastewater from urban areas.

α-aminophosphonate/chitosan is fabricated for removing uranyl (U(VI)) ions via adsorption (Imam et al. 2018). The functionalized chitosan exhibited better physical and chemical properties compared to untreated chitosan. The main active group in the interconnection with U(VI) ion is methyl group. The obtained adsorption data showed the tendency of methyl group to bond with the ions where the adsorption capacity was 245 mgg^{-1} for uranyl ion at pH 5. However, the substitution of the methyl group by phenyl group revealed a dramatic decrease in the adsorption capacity to 114 mgg^{-1}.

Yang et al., has published an interesting study on the formation of the electrospun membrane via chitosan/polyethylene oxide (PEO) (Yang et al. 2019). The synthesised membrane is used as a filter to remove PEO. Additionally, the membrane is modified by pol(glycidyl methacrylate) (PGMA) through the grafting process. The obtained nanofiber membrane can be reused several times due to its high mechanical stability. The adsorption capacities of the nanofiber membrane for Cr(VI) and Co(II) were 138.96 mgg^{-1} and 68.31 mgg^{-1}, respectively.

Recently, Ibrahim et al., have carried out a study on the formation of chitosan grafted with maleic acid through gamma radiation technique (Ibrahim et al. 2019). The grafting efficiency is monitored by using different concentrations of monomers and radiation dosage. The modified chitosan polymers are used as adsorbents to eliminate nickel (II) and Copper (II) ions from wastewater. However, the sorption behavior is conditional and relies mainly on the pH level. Hence, at pH 6, the highest reported sorption capability for Cu (II) was 312.4 mgg^{-1}, whereas the highest sorption capacity for Ni(II) ions was 70.1 mgg^{-1} at pH 8 (Fig. 9.6).

Fan et al. have also reported fabrication of chitosan-based nanocomposites (Fan et al. 2017). In their experiment, they prepared tetraethylenepentamine (TEPA)-modified chitosan/cobalt ferrite ($CoFe_2O_4$) and chitosan/$CoFe_2O_4$ to remove Cu(II) and Pb(II) cations from contaminated solutions. The presence of TEPA has formed more active sites in the composite associated with pure chitosan. Moreover, the composite inherited magnetic properties from $CoFe_2O_4$ which make it easy to remove from the treated solution. The study reported a maximum adsorption capacity of 228.311 mgg^{-1} for Pb(II) ions and 168.067 mgg^{-1} for Cu(II) ions.

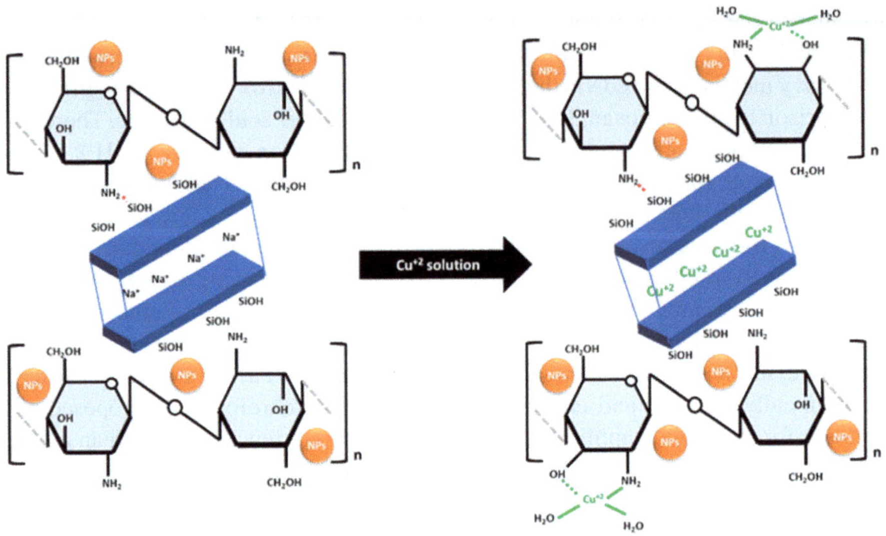

Fig. 9.6 Adsorption behavior of Cu(II) ions via chitosan clay nanocomposites

Further study synthesised chitosan-aluminum oxohydroxide (AlOOH) via mixing aluminum nitrate ($Al(NO_3)_3$) with chitosan (Rajamani and Rajendrakumar 2019). Followingly, the chitosan-$Al(NO_3)_3$mixture is treated with aluminum hydroxide $Al(OH)_3$ to form AlOOH. The bonding between the functional groups of AlOOH and metal ions was in the form of physical forces such as hydrogen bonding and π-π interaction. Indeed, this study has reported oxohydroxide as a novel functional group used for heavy metal adsorption.

Lastly, an interesting study reported the fabrication of unique structured chitosan polymers (Monier et al. 2010). Basically, the study prepared Isatin/chitosan Schiff base via magnetic crosslinkers to eliminate Ni(II), Cu(II), and Co(II) ions from wastewater. The results revealed the composites have a tendency toward Cu(II) ions where the adsorption capacities were 40.15, 103.16, and 53.51 mgg^{-1} for Ni(II), Cu(II), and Co(II), respectively.

9.2.2.3 Chitosan Composites

Polysaccharide composites are considered sufficient adsorbents for heavy metal ions. Nonetheless, polysaccharide composites are costly compared to blends and copolymers (Table 9.1). A study conducted by Kahraman reported the fabrication of chitosan nano-clay (CNC) as a new type of nano-biocomposites (Kahraman 2017). The biocomposites are used to adsorb Cr(VI) from contaminated water. The fabrication process revolved around the use of solvent casting techniques with Cloisite 20A and chitosan. The adsorption performance is influenced by amine and hydroxyl groups. In addition, the existed carbonyl groups increase the adsorption for Cr(VI)

Table 9.1 Types of polysaccharide copolymers and their targeted Heavy Metals

Polysaccharide copolymers	Maximum Adsorption capacity (mgg^{-1})	Adsorption mechanism	Heavy metal	Ref
β-chitosan crosslinked with triethylenetetramine	117.60	Hydroxyl, primary amine	Copper(II)	(Wu et al. 2016)
	151.20		Silver(I)	
Starch/SnO2	333	Hydroxyl, Carbonyl, free oxygen group	Mercury(II)	(Naushad et al. 2016)
Polysaccharide composites Chitosan-nanoclay	128.43	Hydroxyl, Carboxylic, amine group	Chromium(VI)	(Kahraman. 2017)
Alginate-hydrogel	534.25	Sulfonate Carboxylate, groups	Lead(II)	(Mohammad et al. 2017)
	258.6		Cadmium(II)	
	187.0		Nickel(II)	
	224.5		Copper(II)	
CS-g-MA-g-EDMA	–	Amide, Aliphatic, carboxylic functional groups	Chromium(VI)	(Gopal Reddi et al. 2017)
Ion imprinted-interpenetrating network	37.50	Hydroxyl, carboxyl groups	Lead(II)	(Ren et al. 2008)
Polyacrylic acid/starch graft copolymer	588	Carboxylic, swelling	Cadmium(II)	(Abdel-Halim and Al-Deyab. 2014)
CS/AgNP/clay	181.5	Hydroxyl, carboxyl groups	Copper(II)	(Azzam et al. 2016)
AC-CaAlgC	15.7	Carboxylic ion	Lead(II)	(Cataldo et al. 2016)
TEPA-CS/CoFe2O4	168.067	Hydroxyl, carboxyl groups	Copper(II)	(Fan et al. 2017)
	228.311		Lead(II)	
Fe_2O_3 based Chitosan-isatin Schiff's base	40.15	Amine, azomethines, amide groups	Cobult(II)	(Monier et al. 2010)
	53.51		Nickel(II)	
	103.16		Copper(II)	
Chitosan ferrite/Alginate, Titania doped Alginate	–	Carboxylic group	Copper(II)	(Esmat et al. 2017)
			Arsenic (III)	
			Iron (III)	

and form carboxylic groups because of the bonding with the hydroxyl group. The result showed a poor sorption efficiency at pH value above four. The maximum adsorption capacity was 128.43 mgg^{-1} obtained at pH two. The major advantage of CNC as adsorbent is reusability. Basically, CNC can be reused for several cycles after treatment with an acidic solution. Nonetheless, the adsorption capacity of CNC decreases from 93.5% to around 60% after 5 cycles.

Regardless of the several studies conducted on removing heavy metal contaminants from wastewater, there is still a crucial need to enhance the selectivity of utilized adsorbents along with the adsorption capacity. Thus, in the last few years ion-imprinted polymers (IIPs) are analysed to enhance the selectivity of the heavy metal (Ren et al. 2008; Birlik et al. 2006). In the polysaccharides field, IIPs based on chitosan are studied to remove different types of metal ions. However, chitosan-based IIPs have relatively poor selectivitycompared to other types of chitosan-based adsorbents (Ren et al. 2008; Birlik et al. 2006). Hande et al., have reported the fabrication of interpenetrating network of imprinted lead ions (II-IPN) to yield a sorbent with high selectivity and adsorption capacity (Hande et al. 2016). The fabrication process involved the use of lead templates (Pb (II)), methacrylate (MMA) and chitosan as monomers, and ethylene glycol dimethacrylate (EGDMA) as cross-linker. To form the Pb(II) imprinted cavities, II-IPN is washed repeatedly with an acidic solution. The fabrication process is repeated, excluding Pb(II) templates to form a interpenetrating non-imprinted polymeric (NI-IPN) networks. The obtained samples are analysed to study the surface area, pore volume, and adsorption performance. The stated surface areas for II-IPN and NI-IPN were 66.29 m^2g^{-1} and 8.4 m^2g^{-1}, respectively. Moreover, the reported pore volume values for II-IPN were 0.056 cm^3g^{-1}, whereas NI-IPN was 0.015 cm^3g^{-1}. The analytical data show that II-IPN has a developed surface active area and porous volume compared to NI-IPN. Thus, II-IPN has a greater sorption capability. Indeed, the analytical data revealed that the maximum sorption capability of II-IPN is 37.50 mgg^{-1} where NI-IPN is 10.30 mgg^{-1}. Additionally, II-IPN indicated a great selectivity for lead(II) in the occurrence of potassium (K(I)), magnesium (Mg(II), tungsten (W(VI)), zinc (Zn(II)), and copper (Cu(II)) ions. Besides the great selectivity and adsorption performance, II-IPN is reusable where it can be reused for up to five cycles with a few drops in the adsorption performance (11.5% loss).

Recently, researchers studied the modification of chitosan using nanoparticles for water treatment applications. In general, the modification of chitosan as green polymers via nanoparticles can enhance several chemical and physical properties such as the surface area, chemical stability, mechanical stability, and energy surface. Azzam et al. have published a study on the fabrication of gold/chitosan nanoparticles (AuNps) and clay/silver nanoparticles (AgNps) composites to eradicate Cu(II) ions from wastewater (Azzam et al. 2016). The fabrication process revolved around the hybridization of AuNps or AgNps into composites. The nanoparticles could be used to eliminate numerous types of pollutants due to their special surface energy. In comparison to pure Nanocomposites of chitosan, chitosan, have reactive locations for exchange of cations and enlarged surface active area. However, the bonding between chitosan and Cu(II) occurs through hydroxyl and amino groups.

The stated maximum adsorption capacities for chitosan/clay, chitosan/AgNp/clay, and chitosan/AuNp/clay were 134.5, 181.5, and 174.1 mgg^{-1}, respectively. The maximum adsorption capacities are recorded at pH seven and 24.85 °C.

9.2.3 *Cellulose*

Cellulose is considered the most abundant type of biopolymers as it forms the main structure of plants' walls (Fig. 9.7). Cellulose and chitosan are proposed by several studies as future bio-renewable polymers (Jung and Song 2019). In the water treatment field, nanocelluloses are modified via chemicals known as metal-binding groups. These groups include carboxylic, sulfonate, and phosphate groups which used to remove different types of metallic ions from waste stream.

A research group has done a research on the sorption behavior of bioethanol cellulose nanocrystals (CNC_{BE}), sludge cellulose nanocrystals (CNC_{SL}), phosphorylated cellulose nanofibers (phos-CNF_{SL}), and phos-CNC_{SL} form sludge (Liu et al. 2015). The prepared nanocellulose biopolymers are used to adsorb Ag(II), Fe(III), and Cu(II) cations. However, the fabrication of nanocrystals and nanofibers involved different steps. For instance, sludge cellulose is hydrolyzed and sonified to obtain isolated nanocrystals (CNC_{SL}). Similarly, cellulose sludge is grinded mechanically to isolate nanofibers (CNF_{SL}). Nanocellulose gel (CNC_{BE}) was obtained via mixing and homogenizing pure cellulose. The adsorption performance of nanocrystals is influenced by the presence of acid sites. Additionally, different types of functional groups enhanced the bonding with metal ions. For instance, sulfonic groups were observed in CNC_{SL}, whereas carboxylate groups were found in CNC_{BE}. Above all, the modification of nanocellulose biopolymers has significantly increased the surface areas and hence the adsorption capacity.

Recently, Sharma et al., have published a study on the fabrication of carboxycellulose nanofibers (NOCNF) to remove Pb(II) ions from contaminated water using jute fiber (Sharma et al. 2018). Initially, NOCNF has a higher amount of carboxyl

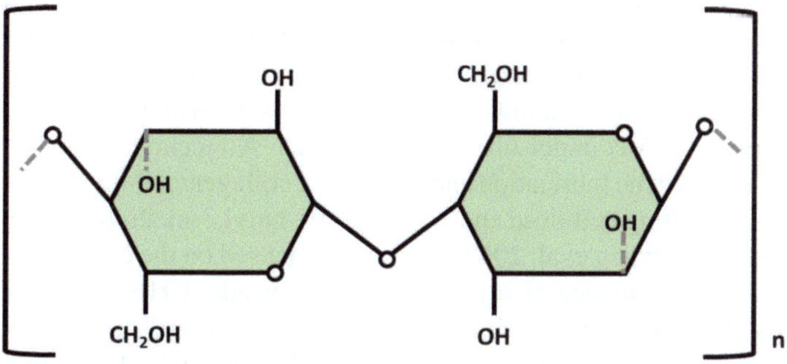

Fig. 9.7 Structure of cellulose

group and larger surface area, which makes it suitable for adsorbing lead (II) cations. The analytical outcomes exhibited that the sorption trait obeys Langmuir isotherm, where the maximum adsorption capacity 2270 mgg^{-1}at pH 7.

An interesting study is conducted on the fabrication and analysis of cellulose microsphere (Luo et al. 2017). This new approach has been introduced to eliminate heavy metals such as Pb(II) from contaminated water. Basically, the study involved using a sol–gel transition method along with solution of cellulose to prepare phosphorylated cellulose microspheres. Moreover, phosphate groups have been introduced during the fabrication to the interface of cellulose microspheres. The adsorption performance is directly related to the bonding formation between phosphate groups and lead ions. The adsorbent surface consisted of various pore sizes. The reported maximum adsorption capacity was 139.38 mgg^{-1}.

Finally, Wang et al. have proposed a new approach of transforming cellulose fibers into carboxymethylated cellulose fibers (CMFs) via controlling carboxymethylation process and the addition of sodium chloroacetate ($C_2H_2ClNaO_2$) solution [66]. The adsorbent is tested on Ni(II) and Cu(II) cations. Indeed, the sorbent exhibited a tendency toward Cu(II) ions, where the maximum sorption capability was 23.48 mgg^{-1}. The sorption behavior of CMFs is influenced by the surface complexity and the electrostatic interaction force. The analytical data revealed that the adsorption follows the Langmuir model. Moreover, the results discovered that the inner charge of the modified cellulose has a significant impact on the adsorption performance. In other words, the primary influential charge inner charge following by the surface charge (Fig. 9.8).

9.2.3.1 Cellulose Based Blends

Cellulose blends are studied as an efficient green solution for water treatment. Riaz et al. have reported the fabrication of compatible polyurethane (PU)/cellulose acetate (CA) blend via a simple blending of CA and PU (Riaz et al. 2016). The blends are prepared with different percentages of CA, 15 to 30%. Additionally, PU/CA blends are used to fabricate membranes by dissolving tetrahydrofuran composition and casting on Teflon plate. However, the adsorption performance of influenced by the preliminary concentration of sorbate and the amount of CA. Indeed, enhancing the content of CA will also enhance the hydrophilicity of the blended synthesized membrane here. In general, a lower amount of CA is associated with a high adsorption capacity of Cr(VI) under ultra-filtration (UF). A recently published study by Wang et al., covered the fabrication and analysis of collagen/cellulose hydrogel beads (CCHBs) via blending cellulose and collagen in 1-butyl,3-methylimidazolium chloride ($C_8H_{15}ClN_2$) (Wang et al. 2013). The study focused on the comparison between the adsorption performance of cellulose hydrogel beads (CHBs) and (CCHBs) for Cu(II) ions. However, the results shown that the structure of CCHBs has high porosity and more binding sites compared to CHBs. The major factor in the adsorption performance of CCHBs is the collagen/cellulose ratio. A higher ratio means

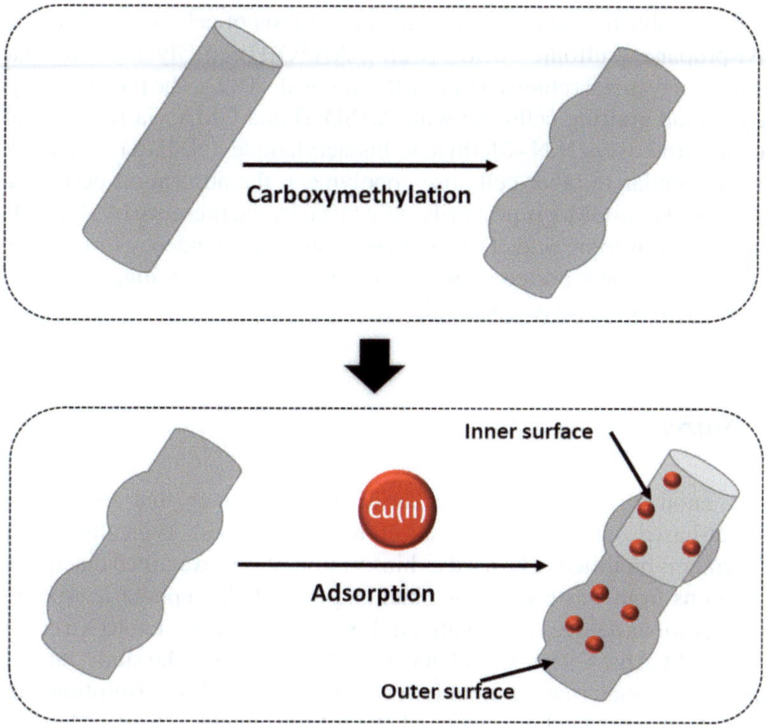

Fig. 9.8 Schematic of adsorption process of CMF for Cu(II) ions

higher adsorption capacity. However, the adsorption capacity is also influenced by the concentration of Cu(II) ions.

9.2.3.2 Cellulose Copolymers

Similar to chitosan copolymers, cellulose copolymers are highly efficient in removing several types of heavy metals from polluted water. Huang et al., have described the synthesis of cellulose/poly(ethylene imine)(PEI) composites via dissolving PEI in lithium hydroxide (LiOH)/urea solution and mixing the solution with epichlorohydrin (ECH) as cross-linker (Ge et al. 2016). The obtained composites are classified into four types according to cellulose (2 to 4 wt%) and PEI (2 to 20 wt%) concentration. The classified composites include sample-1, sample-2, sample-3, and sample-4. Indeed, the findings exhibited that the sorption behavior is directly connected to the concentration of PEI. In addition, the adsorption behavior is also influenced by the presence of carbonyl, phosphate, amino, and sulfo groups. The maximum adsorption capacities of sample 2 were 148.4, 248.2, 112.2, and 30.4 mgg^{-1} for Zn(II), Pb(II), Ni(II), and Cr(II) cations correspondingly.

A recently published study reported the preparation of cellulose-g-2-acrylamido-2-methyl-propane sulfonic acid (Cell-g-$AASO_3H$)-co-Glycidyl methacrylate (GMA) for heavy metal removal (Leyva-Ramos et al. 2012). The preparation process revolved around grafting cellulose with $AASO_3H$ and GMA via free radical polymerization. Moreover, N,N'-Methylene-bis-acrylamide (NMBA) is utilized as a crosslinker. Similar to other cellulose copolymers, the adsorption performance of Cell-g-$AASO_3H$-co-GMA is primarily influenced by the presence of glycidyl group, amid group, and sulfonic acid. The composites showed a tendency toward removing Cu(II) ions where the reported sorption capacities were 69.06 $mgg^{-1}$and 78.24 for Ni(II) and Cu(II) ions, correspondingly.

9.2.4 *Starch*

Starch is another popular polysaccharide homopolymer that exist in several agricultural-based substances such as corn and potatoes. Typically, starch as a homopolymer is modified to be used as biosorption. It was examined that the sorption of lead cations from waste water by obtaining cross-link amphoteric type of starch (CAS) with various degrees of substituted (DS) functionalized starch (Xu et al. 2005). Three types of CASs: CAS1, CAS2, and CAS3 were used in the study, and a 0.3 DS was reported for quaternary ammonium cationic groups. The adsorption capacities defined by Langmuir isotherm were 21.01, 62.11, and 156.25 mgg^{-1} for each of starch, CAS1, CAS2, and CAS3.The obtained results showed thatadsorption capacities are directly related to DS value. Indeed, when the degree of substitution of the carboxy-methyl groups enhances, the sorption capacity increases. This relationship is caused by the reactions between the leadions and anionic carboxy-methyl groups of the modified starch.

Recently, Ma et al., reported the modification of starch with xanthate and carboxylate groups. The composite is used to examine the sorption of Pb(II) cations by preparing carbon disulfide (CD) modified starch citrate (PSC) and starch xanthate (PSX) both of which have porous texture (Ma et al. 2015). Both of the chelation collaboration between lead (II) cations and the xanthate group of the sorption mechanism of PSX and the ion exchange between the carboxylate group and Pb(II)cations of the adsorption mechanism of PSC resulted in an increase in the starch/CD and starch/CA ratios. The high ratios are associated with the intensification in thesorption capability. The analysis of the interactions between particles concluded that eliminating one metal ions is achieved via only one xanthate group. However, eliminating one ion require two carboxylate groups. Hence, more heavy metals are adsorbed viaPSX due to the chelation interaction. The obtained maximum adsorption capacities, at 30 °C, were 109.1 and 57.6 mgg^{-1} for PSX and PSC, respectively. The results shows the dominant performance of PSX adsorption performance over PSC.

Xiang et al. have published an interesting study on the preparation and performance of starch and its composites; functionalized with dithiocarbamate (DTC). The sophisticated study included dithiocarbamate starch (DTCS), dithiocarbamate enzymolysis starch (DTCES), and dithiocarbamate starch (DTCMS) having mesoporous texture (Xiang et al. 2016). It was examined that, the sorption of copper ions using the three types of modified starch obeyed second-order kinetics, having sorption abilities of 1.623, 1.172, and 1.950 $mmolg^{-1}$ for DTCES, DTCS, and DTCMS, correspondingly. Moreover, they obtained the highest possible sorption capability using Langmuir Model. The highest monolayer sorption capability determined was 2.547 $mmolg^{-1}$ for Cu(II) ions sorbed over DTCMS. The chelation between metallic cations and DTC groups was primarily responsible for the adsorption mechanism. Furthermore, the wide and rough surface of DTCMS assisted in the trapping of metalliccations during sorption. The enzyme of pullulanase was used for de-branching of maize starch. It will yield glucan molecules having linear structure. This was dried in freeze to form nanoparticle crystals of starch. To introduce a chlorine functional group, sodium hypochlorite was used for oxidation of this nano crystals of starch. The average size of nanoparticles was 263 nm. Differential Scanning Calorimeter (DSC), exhibited lower magnitude for enthalpy (ΔH) of the oxidized sample, indicating that the nano crystals of starch is now connected with chlorine functional groups (DSC). Charge of the surface of the nano crystals were determined using zeta potential analysis. The positive metallic cations reacted with the negatively charged nano crystals of starch. After adsorption of metallic cations, when the sample was checked with zeta potential, the negative charge of the nano crystals decreased over time. Adsorption kinetics results show that modified starch nanoparticles adsorb less Cu(II) ions around 37.74 g mg^{-1} than Pb(II) ions around 44.64 gmg^{-1} and it obeys the second-order kinetics. Metallic ion sorption follows the Langmuir model.

9.2.4.1 Starch Based Copolymers

Starch also can be used as copolymers for water treatment applications. The graft copolymer of starch/polyacrylic acid was additionally introduced with epichlorohydrin-which was made alkaline first. The resultant mixture yields a cross-linked polymer (PASGc) having superior ability to remove Cd(II) ions (Abdel-Halim and Al-Deyab 2014). Starch extracted from a water slurry was polymerized at certain temperature before being mixed with potassium borate, thiourea dioxide and acrylic acid (AA). As synthesized PASGc was mixed with NaOH to precipitate as hydrogel. The hydrogel contain a great number of surface functional groups, primarily –COOH groups, which can interact with metallic cations to initiate sorption as well as ionic exchange reactions. As the concentration of PASGc was increasing, the content of active functional groups were also increasing, resulting in superior Cd(II) ion sorption having sorption capacity of 588 mgg^{-1}. The first cycle exhibited a 93.8 percent recovery, which dropped to 58.7 percent by the sixth cycle.

9.2.5 Alginate

Alginate is a kind of homo-polymer having biological origin which can be found in nature in an anionic form in bacteria and brown seaweed (Elwakeel et al. 2016). Similar to other polysaccharides, heavy metals tend to link with the carboxyl groups of alginate. Thus, polysaccharide alginate is used assorbent for different types of metal ions in wastewater. Indeed, Lagoa and Rodrigues have conducted a study on the sorption characteristics of gel and dry alginate beads in both wastewater and water contaminated with a single type of metal ion (Lagoa and Rodrigues 2009). Their results revealed that dry alginate has a greater adsorption performance in single metal solution compared to gel alginate. The difference in performance is associated to the percentage of metallic ions interaction sites existing in the dry alginate beads. In fact, the analytical data showed that gel alginate is composed of 97% of water. Thus, gel alginate beads are unstable and have poor adsorption performance, which makes them inefficient for large-scale water treatment.

A further study is conducted by Nussinovitch and Dagan stated the usage of alginate beads for metalliccations elimination (Nussinovitch and Dagan 2015). Nonetheless, in this study hydrocolloid, liquid-core capsules are used to contain non-crosslinked fluidic alginate. The capsules are used to eradicate lead (II), nickel (II), cadmium (II) and copper (II), where the reported adsorption capacities were 267, 65, 197, and 219 mgg^{-1}, respectively. The adsorption performance is directly related to the cross-linking between metal cations and carboxylanions.

9.2.5.1 Alginate Based Composites

The special properties of nanomaterials are combined with alginate polymers to form an efficient, non-toxic, and inexpensive biocompatible metal ions adsorbent. Interesting study have studied the exclusion of Fe(III), As(III) and Cu (II) cations from wastewater using Nanocomposites of alginate having titanate nanotubes (T) (T/Alg) and cobalt ferric (CF) nanoparticles (CF/Alg) (Esmat et al. 2017). The produced nanocomposites have special properties such as the enlarged surface area, high porous volume, as well as ferrimagnetic behavior. The high porosity is formed by the integration of nanoparticles and nanotubes into the polymer structure. Hence, more active sites are formed to bind with metal ions. Moreover, the ferrimagnetic property is inherited from CF nanoparticles, which allow for a regeneration and ease extraction of the composites. The sorption behavior of the composites is influenced by numerous issues such as the temperature, pH value and the existence of carboxylic groups. The results showed an escalation in the sorption efficiency of Fe(III) and Cu(II) with an enhance in pH value from 2 to 6.5. The exclusion efficiencies of T/Alg and CF/Alg for Cu(II) were 98% and 95%, respectively. However, the stated removal efficiencies for Fe(III) were 60% and 82% of CF/Alg and T/Alg, respectively.

Recently published studies have covered the synthesis of active (AC) carbon-calcium salts of alginate composite (AC-CaAlgC) to remove lead (II) ions from wastewater (Cataldo et al. 2016). Activated carbon (AC) is an extensively used sorbent to eradicate several types of contaminants from water. Indeed, AC has a large surface active area, high porous volume, and relatively good chemical and physical stability. Typically, activated carbon are prepared from organic wastes known as biomass. Thus, activated carbon is considered a green solution for water treatment. In general, activated carbon is used to adsorb metal ions due to the presence of oxygen donors on its surface. Hence, -COOH groups in AC-CaAlgC act as active sorption sites for lead (II) ions. Nevertheless, the maximum sorption capability of AC-CaAlgC for Pb(II) was 15.7 mgg^{-1}.

9.3 Ligno-Cellulosic Residues

The crucial need for new water treatment materials has increased with the increase in the water contamination level. As discussed before, several approaches and materials are adopted to remove contaminants such as heavy metals, dye, and biowastes from water. For instance, metal oxide nanoparticles are typically used as adsorbents to remove metal ions from industrial wastes before they are discharged into natural water resources. However, metal oxides are not efficient in removing certain sorts of metallic cations and dyes. Moreover, metal oxides nanoparticles are exceedinglyhazardous to the environment and human wellbeing. Thus, lignocellulosic biomass is introduced as a green alternative to toxic adsorbents used in the water treatment field. Lignocellulosic biomass, also known as photo-mass, is abundant material formed via the photosynthesis process. Lignocellulosic materials are composed mainly of cellulose, hemicellulose, and lignin. Indeed, lignocellulosic biomass materials are also used in the energy field to produce green energy sources such as biofuels and biogases. In general, materials derived from lignocellulosic biomass are inexpensive, efficient, and eco-friendly. In the water treatment field, lignocellulosic biomass substances are used to produce highly efficient adsorbents for contaminated water. However, biomass derivatives such as cellulose and lignocellulose are also used to produce adsorbents. The quality and efficiency of the lignocellulosic-based adsorbents rely on the material composition.

9.3.1 Composition of Ligno-Cellulosic Materials/Residues

Initially, lignocellulosic materials are contains three key compounds; cellulose, hemicellulose, and lignin (Fig. 9.9). The percentage of each compound varies according to lignocellulosic base type, like wood, wheat husk, date seed,etc. Typically, cellulose represents 30 to 50% of the composite, whereas hemicellulose and lignin represent around 20–40% and 15–25%, respectively (Table 9.2) (Gautam et al. 2014). The main properties of lignocellulosic material are inherited from the three main compounds. For instance, cellulose gives different properties such as degradability, charity, and

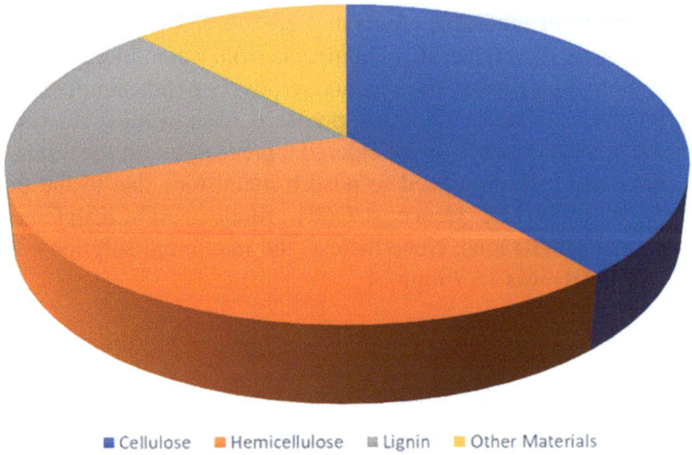

Fig. 9.9 Proportion for the composites of Ligno-cellulosic Materials

Table 9.2 Chemical constituent of some lingo-cellulosic materials

Biomass	Cellulose (%)	Hemicellulose (%)	Lignin (%)
Almond shell	50.7	28.9	20.4
Bamboo	26–43	15–26	21–31
Banana waste	13.2	14.8	14
Barley straw	31–37	24–29	14–15
Corncob	50.5	31	15
Corn stover	28	35	16–21
Cotton seed fibers	80–95	5–20	0
Coastal Bermudagrass	25	35.7	6.4
Hazelnut shell	28.8	30.4	42.9
Jute fiber	45–53	18–21	21–26
Nut shell	25–30	25–30	30–40
Oat straw	31–37	24–29	16–19

hydrophilicity. Additionally, the chemical reactivity of the structure is affected by the existence of cellulose. Hemicellulose is a heteropolymer with an unorganized amorphous phase. Indeed, the structure of hemicellulose includes a group of polysaccharide polymers with an unorganized crystalline structure. On the other hand, lignin is considered the backbone of the lignocellulosic material as it consists of phenyl propane building blocks. Indeed, lignocellulosic materials with a high percentage of lignin are used to produce high-quality activated carbon. Besides cellulose, hemicellulose, and lignin, the structure of lignocellulosic material contains a small percentage of ash, water, organic and inorganic materials, and cyclic hydrocarbons.

9.3.2 Ligno-Cellulosic Materials for Heavy Metals Removal

Lignocellulosic-based materials are considered great sorbents for heavy metals due to their exceptional structure and properties. Typically, heavy metalliccations bind to the lingo-cellulosic adsorbent via the chemical functional/active groups or surface sites on the adsorbent surface. Similar to other types of adsorbents, the adsorption performance of lignocellulosic materials is influenced by several variables such as the type of lingo-cellulosic material, temperature, content of constituent polymers, pH level, particle size, and preparation process. Regardless of the properties of lignocellulosic materials, availability and price are considered the main selection factor of the adsorbent. For instance, sugarcane bagasse is the most studied type of lingo-cellulosic waste as it is widely available and inexpensive (Alomá et al. 2012).

The performance of lingo-cellulosic-based adsorbent is typically improved via chemical modification. Basically, chemical modification enhances the porosity, surface active areas, and functional anchoring groups of the lignocellulosic material. This approach can significantly enhance the exclusion performance of metallic cations. The chemical modification can be used in any type of biomass precursor such as wheatgrass, corncob, rice husk, olive oil, orange peel, barley straw, etc. Usually, the modification process includes chemical agents such as mineral and organic acid, bases, and basic solutions. The most common type of reagent is hydrochloric acid (HCl), sulfuric acid (H_2SO_4), calcium hydroxide ($Ca(OH)_2$ and sodium hydroxide (NaOH), and In general, the chemical modification improves different properties of a lignocellulosic-based adsorbent, such as the hydrophobicity of hydrophilicity, ion exchange capability, thermal resistivity, and elasticity. Table 9.3 displays a summary of diverse categories of biomass-based adsorbents and their targeted metal ion.

9.4 Conclusion

This chapter analyzed the adsorption performance of most known types of polysaccharides, cellulose, chitin, chitosan, starch, and alginate. In general, polysaccharides homopolymers are proposed by several studies as environmentally friendly biosorbents. Moreover, polysaccharides are inexpensive as they can be extracted from an abundant amount of organic wastes such as wheatgrass, wood, Jute, etc. To enhance the adsorption behavior of polysaccharides, methods such as blending, mixing, or grafting can be adopted to incorporate novel active sites or functional groups inside the polymeric matrix. Moreover, the mechanical strength of polysaccharides is enhanced via cross-linking with end functional polymers instead of monomers.

Table 9.3 Different types of biomass-based adsorbents and their Targeted Contaminants

Adsorbent	Agent	Type of the solution	Adsorbate	Maximum Adsorption Capacity (mgg−1)	Treatment Mechanism	Ref
Saccharum officinarum L	Pure/untreated	–	–	2.23	–	(Alomá et al. 2012)
Saccharum officinarum L	Sodium Hydroxide	–	Cadmium	18.32	Complex Formationand ionic interchange	(Velazquez-Jimenez et al. 2013)
Saccharum officinarum L	Hydrochloric Acid	–	Cadmium	13.5	Complex Formation and ionic interchange	(Velazquez-Jimenez et al. 2013)
Saccharum officinarum L	Pure/untreated	–	Cadmium	13.27	Complex Formation and ionic interchange	(Velazquez-Jimenez et al. 2013)
Saccharum officinarum L	Nitric acid	–	Cadmium	12.5	Complex Formation and ionic interchange	(Velazquez-Jimenez et al. 2013)
Saccharum officinarum L	Hydrochloric Acid	–	Lead	54.29	Complex Formationand ionic interchange	(Velazquez-Jimenez et al. 2013)
Saccharum officinarum L	Sodium Hydroxide	–	Lead	50.12	Complex Formation and ionic interchange	(Velazquez-Jimenez et al. 2013)
Saccharum officinarum L	Nitric acid	–	Lead	42.31	Complex Formation and ionic interchange	(Velazquez-Jimenez et al. 2013)
Saccharum officinarum L	Pure/untreated	–	Lead	35.6	Complex Formationand ionic interchange	(Velazquez-Jimenez et al. 2013)

(continued)

Table 9.3 (continued)

Adsorbent	Agent	Type of the solution	Adsorbate	Maximum Adsorption Capacity (mgg−1)	Treatment Mechanism	Ref
Saccharum officinarum L	Ethylene-diamine-Tetraacetic dianhydride	Synthetic	Zinc	105.26	Ionic interchange	(Pereira et al. 2010)
Saccharum officinarum L	Ethylene-diamine-Tetraacetic dianhydride	Real- wastewater	Zinc	45.45	–	(Pereira et al. 2010)
Saccharum officinarum L	Sodium Hydroxide	–	Zinc	20.24	Complex Formationand ionic interchange	(Velazquez-Jimenez et al. 2013)
Saccharum officinarum L	Hydrochloric Acid	–	Zinc	14.43	Complex Formationand ionic interchange	(Velazquez-Jimenez et al. 2013)
Saccharum officinarum L	Nitric Acid	–	Zinc	12.4	Complex Formationand ionic interchange	(Velazquez-Jimenez et al. 2013)
Saccharum officinarum L	Pure/untreated	–	Zinc	7.84	Complex Formation and ionic interchange	(Velazquez-Jimenez et al. 2013)
Hordeum vulgare Straw (Barley)	$C_6H_8O_7$	–	Copper	31.71	Ionic interchange, and chelation	(Pehlivan et al. 2012)
Hordeum vulgare Straw (Barley)	Pure/ untreated	–	Copper	4.64	Ionic interchange, and chelation	(Pehlivan et al. 2012)
Anacardium occidentale Nut shell (Cashew)	Pure/ untreated	–	Nickel	18.86	–	(Kumar et al. 2011)

(continued)

Table 9.3 (continued)

Adsorbent	Agent	Type of the solution	Adsorbate	Maximum Adsorption Capacity (mgg−1)	Treatment Mechanism	Ref
Cocos nucifera (Coconut) shell	Pure/ untreated	Single element	Cadmium	37.78	–	(Sousa et al. 2010)
Cocos nucifera (Coconut) shell	Pure/ untreated	Multi-elements	Cadmium	11.96	–	(Sousa et al. 2010)
Cocos nucifera (Coconut) shell	Pure/ untreated	–	Copper	41.36	–	(Sousa et al. 2010)
Cocos nucifera (Coconut) shell	Pure/ untreated	Single element	Copper	41.36	–	(Sousa et al. 2010)
Cocos nucifera (Coconut) shell	Pure/ untreated	Multi-elements	Copper	20.26	–	(Sousa et al. 2010)
Cocos nucifera (Coconut) shell	Pure/ untreated	Single element	Nickel	16.34	–	(Sousa et al. 2010)
Cocos nucifera (Coconut) shell	Pure/ untreated	Multi-elements	Nickel	3.12	–	(Sousa et al. 2010)
Cocos nucifera (Coconut) shell	–	Single element	Zinc	17.08	–	(Sousa et al. 2010)
Cocos nucifera (Coconut) shell	–	Multi-elements	Zinc	7.32	–	(Sousa et al. 2010)
Cocos nucifera (Coconut) shell	Pure/ untreated	Single element	Lead	54.62	–	(Sousa et al. 2010)
Cocos nucifera (Coconut) shell	Pure/ untreated	Multi-elements	Lead	17.9	–	(Sousa et al. 2010)

(continued)

Table 9.3 (continued)

Adsorbent	Agent	Type of the solution	Adsorbate	Maximum Adsorption Capacity (mgg−1)	Treatment Mechanism	Ref
Corncob	$C_6H_8O_7$	–	Cadmium	49.2	Ionic interchange	(Leyva-Ramos et al. 2012)
Corncob	CH_3OH	–	Lead	43.4	–	(Tan et al. 2010)
Corncob	Pure/Untreated	–	Lead	16.22	–	(Tan et al. 2010)
Corncob	Sodium Hydroxide	–	Lead	7.89	–	(Tan et al. 2010)
Lawn/Garden grass	Pure/ untreated	–	Copper	58.34	–	(Hossain et al. 2012)
Grapefruit skin	Pure/ untreated	–	Cadmium	42.09	Ionic interchange	(Torab-Mostaedi et al. 2013)
Grapefruit skin	Pure/ untreated	–	Nickel	46.13	Ionic interchange	(Torab-Mostaedi et al. 2013)
Olive seeds	Pure/ untreated	–	Nickel	21.3	–	(Fiol et al. 2006)
Orange skin	Pure/ untreated	–	Copper	50.94	Ionic interchange	(Bansal et al. 2009)
Hull of Rice	CH_2O	–	Chromium	10.4	Ionic interchange surface,Complex Formation, and precipitate formation	(Bansal et al. 2009)
Hull of Rice	Pure/ untreated	–	Chromium	8.5	Ionic interchange surface, and Complex Formation, precipitate formation	(Bansal et al. 2009)
Hull of Rice	Alkali (-OH)	–	Lead	58.1	Ionic interchange	(Krishnani et al. 2008)

(continued)

Table 9.3 (continued)

Adsorbent	Agent	Type of the solution	Adsorbate	Maximum Adsorption Capacity (mgg−1)	Treatment Mechanism	Ref
Hull of Rice	Alkali (-OH)	–	Zinc	8.14	Ionic interchange	(Krishnani et al. 2008)
Straw (Rice)	Pure/ untreated	–	Cadmium	13.89	Ionic interchange, chelation	(Ding et al. 2012)
Straw (Rice)	Nitric acid and Sodium Hydroxide	–	Copper	8.13	–	(Rocha et al. 2009)
Straw (Rice)	Pure/ untreated	–	Mercury	22.06	–	(Rocha et al. 2009)
Straw (Rice)	Nitric acid and Sodium Hydroxide	–	Zinc	8.63	–	(Rocha et al. 2009)
Sawdust	Ethylene-diamine-Tetraacetic dianhydride	Synthetic	Zinc	80	Ionic interchange	(Pereira et al. 2010)
Sawdust	Ethylene-diamine-Tetraacetic dianhydride	Real- wastewater	Zinc	47.39	–	(Pereira et al. 2010)
Saccharum officinarum L	Pure/ untreated	–	Mercury	35.71	–	(Khoramzadeh et al. 2013)
Hull of Sunflower	Pure/ untreated	–	Copper	57.14	–	(Witek-Krowiak 2012)
Wheat stalk	Sodium Hydroxide	–	Cadmium	21.84	Complex Formation	(Tan and Xiao 2009)
Wheat stalk	Pure/ untreated	–	Cadmium	11.6	Complex Formation	(Tan and Xiao 2009)
Wheat stalk	CH_3OH	–	Cadmium	0.35	Complex Formation	(Tan and Xiao 2009)

The modification of polymers via blending and mixing enhances the lifetime of the polymers and increases the reusability cycles with few drop-in adsorption capacities.

References

Abdel-Halim E, Al-Deyab SS (2014) Preparation of poly(acrylic acid)/starch hydrogel and its application for cadmium ion removal from aqueous solutions. React Funct Polym 75:1–8. https://doi.org/10.1016/j.reactfunctpolym.2013.12.003

Abussaud B, Asmaly HA, Ihsanullah Saleh TA, Gupta VK, Laoui T, Atieh MA (2016) Sorption of phenol from waters on activated carbon impregnated with iron oxide, aluminum oxide and titanium oxide. J Mol Liq 213. https://doi.org/10.1016/j.molliq.2015.08.044

Ahlers F, Thom I, Lambert J, Kuckuk R, Wiermann R (1999) 1H NMR analysis of sporopollenin from Typha Angustifolia. Phytochemistry 50(6):1095–1098. https://doi.org/10.1016/s0031-9422(98)00225-8

Ahsan T, Tabassum M, Rahman J, Nur M, Siddique EA (2019) Environmental analysis of arsenic in water, soil and food materials from highly contaminated area of Alampur Village, Amjhupi Union, Meherpur. Adv J Chem Sect A 3(2). https://doi.org/10.33945/sami/ajca.2020.2.7

Alomá I, Martín-Lara M, Rodríguez I, Blázquez G, Calero M (2012) Removal of nickel (II) ions from aqueous solutions by biosorption on sugarcane bagasse. J Taiwan Inst Chem Eng 43(2):275–281. https://doi.org/10.1016/j.jtice.2011.10.011

Alvarez PJJ, Chan CK, Elimelech M, Halas NJ, Villagrán D (2018) Emerging opportunities for nanotechnology to enhance water security. Nat Nanotechnol 13(8):634–641. https://doi.org/10.1038/s41565-018-0203-2

Azzam EM, Eshaq G, Rabie A, Bakr A, Abd-Elaal AA, El Metwally A, Tawfik SM (2016) Preparation and characterization of chitosan-clay nanocomposites for the removal of Cu(II) from aqueous solution. Int J Biol Macromol 89:507–517. https://doi.org/10.1016/j.ijbiomac.2016.05.004

Bansal M, Garg U, Singh D, Garg V (2009) Removal of Cr(VI) from aqueous solutions using pre-consumer processing agricultural waste: A case study of rice husk. J Hazard Mater 162(1):312–320. https://doi.org/10.1016/j.jhazmat.2008.05.037

Bhattacharya A, Sarani C, Kumar Roy A, Karthik D, Singh A, Kumari S, Kumar Mishra A (2019) An analysis of arsenic contamination in the groundwater of India, Bangladesh and Nepal with a special focus on the stabilisation of arsenic-laden sludge from arsenic filters. Electron J Geotechn Eng 24

Birlik E, Ersöz A, Denizli A, Say R (2006) Preconcentration of copper using double-imprinted polymer via solid phase extraction. Anal Chim Acta 565(2):145–151. https://doi.org/10.1016/j.aca.2006.02.051

Cataldo S, Gianguzza A, Milea D, Muratore N, Pettignano A (2016) Pb(II) adsorption by a novel activated carbon-alginate composite material: a kinetic and equilibrium study. Int J Biol Macromol 92:769–778. https://doi.org/10.1016/j.ijbiomac.2016.07.099

Cazacu G, Popa V (2014) Blends and composites based on cellulose materials. In: Dumitriu S (ed) Polysaccharides: structural diversity and functional versatility, 2nd ed. New York, Basel, Hong Kong, Marcel Dekker, pp 1141–1177

Ding Y, Jing D, Gong H, Zhou L, Yang X (2012) Biosorption of aquatic cadmium(II) by unmodified rice straw. Biores Technol 114:20–25. https://doi.org/10.1016/j.biortech.2012.01.110

Ehrlich H, Maldonado M, Spindler KD, Eckert C, Hanke T, Born R, Goebel C, Simon P, Heinemann S, Worch H (2007) First evidence of chitin as a component of the skeletal fibers of marine sponges, part I. Verongidae (demospongia: Porifera). J Exp Zool Part B Molecular Develop Evolution 308B(4):347–356. https://doi.org/10.1002/jez.b.21156[32]

Liu D, Zhu Y, Li Z, Tian D, Chen L, Chen P (2013) Chitin nanofibrils for rapid and efficient removal of metal ions from water system. Carbohyd Polym 98(1):483–489. https://doi.org/10.1016/j.carbpol.2013.06.015

Elwakeel KZ, Daher AM, Abd El-Fatah AIL, Abd El Monem H, Khalil MMH (2016) Biosorption of lanthanum from aqueous solutions using magnetic alginate beads. J Dispersion Sci Technol 38(1):145–151. https://doi.org/10.1080/01932691.2016.1146617

Esmat M, Farghali AA, Khedr MH, El-Sherbiny IM (2017) Alginate-based nanocomposites for efficient removal of heavy metal ions. Int J Biol Macromol 102:272–283. https://doi.org/10.1016/j.ijbiomac.2017.04.021

Fan C, Li K, Li J, Ying D, Wang Y, Jia J (2017) Comparative and competitive adsorption of Pb(II) and Cu(II) using Tetraethylenepentamine modified chitosan/CoFe2O4 particles. J Hazard Mater 326:211–220. https://doi.org/10.1016/j.jhazmat.2016.12.036

Fiol N, Villaescusa I, Martínez M, Miralles N, Poch J, Serarols J (2006) Sorption of Pb(II), Ni(II), Cu(II) and Cd(II) from aqueous solution by olive stone waste. Sep Purif Technol 50(1):132–140. https://doi.org/10.1016/j.seppur.2005.11.016

Gautam RK, Sharma SK, Mahiya S, Chattopadhyaya MC (2014) Contamination of heavy metals in aquatic media: transport, toxicity and technologies for remediation. Heavy Metals Water. https://doi.org/10.1039/9781782620174-00001

Ge H, Huang H, Xu M, Chen Q (2016) Cellulose/poly(ethylene imine) composites as efficient and reusable adsorbents for heavy metal ions. Cellulose 23(4):2527–2537. https://doi.org/10.1007/s10570-016-0973-3

Gopal Reddi M, Gomathi T, Saranya M, Sudha P (2017) Adsorption and kinetic studies on the removal of chromium and copper onto Chitosan-g-maliec anhydride-g-ethylene dimethacrylate. Int J Biol Macromol 104:1578–1585. https://doi.org/10.1016/j.ijbiomac.2017.01.142

Grant SB, Saphores JD, Feldman DL, Hamilton AJ, Fletcher TD, Cook PLM, Stewardson M, Sanders BF, Levin LA, Ambrose RF, Deletic A, Brown R, Jiang SC, Rosso D, Cooper WJ, Marusic I (2012) Taking the "waste" out of "wastewater" for human water security and ecosystem sustainability. Science 337(6095):681–686. https://doi.org/10.1126/science.1216852

HPS AK, Saurabh CK, Adnan AS, Fazita MRN, Syakir M, Davoudpour Y, Rafatullah M, Abdullah CM, Haafiz M, Dungani R (2016) A review on chitosan-cellulose blends and nanocellulose reinforced chitosan biocomposites: properties and their applications. Carbohyd Polym 150:216–226. https://doi.org/10.1016/j.carbpol.2016.05.028

Hande PE, Kamble S, Samui AB, Kulkarni PS (2016) Chitosan-based lead ion-imprinted interpenetrating polymer network by simultaneous polymerization for selective extraction of Lead(II). Ind Eng Chem Res 55(12):3668–3678. https://doi.org/10.1021/acs.iecr.5b04889

Hargreaves AJ, Constantino C, Dotro G, Cartmell E, Campo P (2018) Fate and removal of metals in municipal wastewater treatment: a review. Environ Technol Rev 7(1):1–18. https://doi.org/10.1080/21622515.2017.1423398

Herath I, Vithanage M, Bundschuh J, Maity JP, Bhattacharya P (2016) Natural arsenic in global groundwaters: distribution and geochemical triggers for mobilization. Current Pollut Rep 2(1):68–89. https://doi.org/10.1007/s40726-016-0028-2

Hossain M, Ngo H, Guo W, Setiadi T (2012) Adsorption and desorption of copper(II) ions onto garden grass. Biores Technol 121:386–395. https://doi.org/10.1016/j.biortech.2012.06.119

Ibrahim AG, Saleh AS, Elsharma EM, Metwally E, Siyam T (2019) Chitosan-g-maleic acid for effective removal of copper and nickel ions from their solutions. Int J Biol Macromol 121:1287–1294. https://doi.org/10.1016/j.ijbiomac.2018.10.107

Imam EA, El-Tantawy El-Sayed I, Mahfouz MG, Tolba AA, Akashi T, Galhoum AA, Guibal E (2018) Synthesis of α-aminophosphonate functionalized chitosan sorbents: effect of methyl vs phenyl group on uranium sorption. Chem Eng J 352:1022–1034. https://doi.org/10.1016/j.cej.2018.06.003

Jung WJ, Song YS (2019) Expression pattern of β-glucanase and production of water-soluble polysaccharide in Pedobacter sp. PR-M6. J Chitin Chitosan 24(1):16–23. https://doi.org/10.17642/jcc.24.1.3

Sardar UR, Bhargavi E, Devi I, Bhunia B, Tiwari ON (2018) Advances in exopolysaccharides based bioremediation of heavy metals in soil and water: a critical review. Carbohyd Polym 199:353-364. https://doi.org/10.1016/j.carbpol.2018.07.037

Kahraman HT (2017) Development of an adsorbent via chitosan nano-organoclay assembly to remove hexavalent chromium from wastewater. Int J Biol Macromol 94:202–209. https://doi.org/10.1016/j.ijbiomac.2016.09.111

Kanmani P, Aravind J, Kamaraj M, Sureshbabu P, Karthikeyan S (2017) Environmental applications of chitosan and cellulosic biopolymers: a comprehensive outlook. Biores Technol 242:295–303. https://doi.org/10.1016/j.biortech.2017.03.119

Khoramzadeh E, Nasernejad B, Halladj R (2013) Mercury biosorption from aqueous solutions by Sugarcane Bagasse. J Taiwan Inst Chem Eng 44(2):266–269. https://doi.org/10.1016/j.jtice.2012.09.004

Krishnani K, Meng X, Christodoulatos C, Boddu V (2008) Biosorption mechanism of nine different heavy metals onto biomatrix from rice husk. J Hazard Mater 153(3):1222–1234. https://doi.org/10.1016/j.jhazmat.2007.09.113

Kumar PS, Ramalingam S, Kirupha SD, Murugesan A, Vidhyadevi T, Sivanesan S (2011) Adsorption behavior of nickel(II) onto cashew nut shell: Equilibrium, thermodynamics, kinetics, mechanism and process design. Chem Eng J 167(1):122–131. https://doi.org/10.1016/j.cej.2010.12.010

Lagoa R, Rodrigues J (2009) Kinetic analysis of metal uptake by dry and gel alginate particles. Biochem Eng J 46(3):320–326. https://doi.org/10.1016/j.bej.2009.06.007

Leyva-Ramos R, Landin-Rodriguez L, Leyva-Ramos S, Medellin-Castillo N (2012) Modification of corncob with citric acid to enhance its capacity for adsorbing cadmium(II) from water solution. Chem Eng J 180:113–120. https://doi.org/10.1016/j.cej.2011.11.021

Lim AP, Aris AZ (2014) A review on economically adsorbents on heavy metals removal in water and wastewater. Rev Environ Sci Bio/technol 13(2):163–181. https://doi.org/10.1007/s11157-013-9330-2

Liu P, Borrell PF, Božič M, Kokol V, Oksman K, Mathew AP (2015) Nanocelluloses and their phosphorylated derivatives for selective adsorption of Ag+, Cu2+ and Fe3+ from industrial effluents. J Hazard Mater 294:177–185. https://doi.org/10.1016/j.jhazmat.2015.04.001

Luo W, Bai Z, Zhu Y (2018) Fast removal of Co(ii) from aqueous solution using porous carboxymethyl chitosan beads and its adsorption mechanism. RSC Adv 8(24):13370–13387. https://doi.org/10.1039/c7ra13064c

Luo X, Yuan J, Liu Y, Liu C, Zhu X, Dai X, Ma Z, Wang F (2017) Improved solid-phase synthesis of phosphorylated cellulose microsphere adsorbents for highly effective Pb2+ removal from water: batch and fixed-bed column performance and adsorption mechanism. ACS Sustain Chem Eng 5(6):5108–5117. https://doi.org/10.1021/acssuschemeng.7b00472

Ma X, Liu X, Anderson DP, Chang PR (2015) Modification of porous starch for the adsorption of heavy metal ions from aqueous solution. Food Chem 181:133–139. https://doi.org/10.1016/j.foodchem.2015.02.089

Mogoşanu GD, Grumezescu AM (2014) Natural and synthetic polymers for wounds and burns dressing. Int J Pharm 463(2):127–136. https://doi.org/10.1016/j.ijpharm.2013.12.015

Mohammad N, Atassi Y, Tally M (2017) Synthesis and swelling behavior of metal-chelating superabsorbent hydrogels based on sodium alginate-g-poly(AMPS-co-AA-co-AM) obtained under microwave irradiation. Polym Bull 74(11):4453–4481. https://doi.org/10.1007/s00289-017-1967-5

Mohammadzadeh Pakdel P, Peighambardoust SJ (2018) Review on recent progress in chitosan-based hydrogels for wastewater treatment application. Carbohyd Polym 201:264–279. https://doi.org/10.1016/j.carbpol.2018.08.070

Mohammed N, Grishkewich N, Tam KC (2018) Cellulose nanomaterials: promising sustainable nanomaterials for application in water/wastewater treatment processes. Environ Sci Nano 5(3):623–658. https://doi.org/10.1039/c7en01029j

Monier M, Ayad D, Wei Y, Sarhan A (2010) Adsorption of Cu(II), Co(II), and Ni(II) ions by modified magnetic chitosan chelating resin. J Hazard Mater 177(1–3):962–970. https://doi.org/10.1016/j.jhazmat.2010.01.012

Muzzarelli RA (2011) Potential of chitin/chitosan-bearing materials for uranium recovery: an interdisciplinary review. Carbohyd Polym 84(1):54–63. https://doi.org/10.1016/j.carbpol.2010.12.025

Naushad M, Ahamad T, Sharma G, Al-Muhtaseb AH, Albadarin AB, Alam MM, ALOthman ZA, Alshehri SM, Ghfar AA (2016) Synthesis and characterization of a new starch/SnO2 nanocomposite for efficient adsorption of toxic Hg2+ metal ion. Chem Eng J 300:306-316. https://doi.org/10.1016/j.cej.2016.04.084

Negm NA, El Sheikh R, El-Farargy AF, Hefni HH, Bekhit M (2015) Treatment of industrial wastewater containing copper and cobalt ions using modified chitosan. J Ind Eng Chem 21:526–534. https://doi.org/10.1016/j.jiec.2014.03.015

Nussinovitch A, Dagan O (2015) Hydrocolloid liquid-core capsules for the removal of heavy-metal cations from water. J Hazard Mater 299:122–131. https://doi.org/10.1016/j.jhazmat.2015.06.013

Olivera S, Muralidhara HB, Venkatesh K, Guna VK, Gopalakrishna K, Kumar K (2016) Potential applications of cellulose and chitosan nanoparticles/composites in wastewater treatment: a review. Carbohyd Polym 153:600–618. https://doi.org/10.1016/j.carbpol.2016.08.017

Pehlivan E, Altun T, Parlayici (2012) Modified barley straw as a potential biosorbent for removal of copper ions from aqueous solution. Food Chem 135(4):2229–2234. https://doi.org/10.1016/j.foodchem.2012.07.017

Pereira FV, Gurgel LVA, Gil LF (2010) Removal of Zn2+ from aqueous single metal solutions and electroplating wastewater with wood sawdust and sugarcane bagasse modified with EDTA dianhydride (EDTAD). J Hazard Mater 176(1–3):856–863. https://doi.org/10.1016/j.jhazmat.2009.11.115

Rajamani M, Rajendrakumar K (2019) Chitosan-boehmite desiccant composite as a promising adsorbent towards heavy metal removal. J Environ Manage 244:257–264. https://doi.org/10.1016/j.jenvman.2019.05.056

Ren Y, Zhang M, Zhao D (2008) Synthesis and properties of magnetic Cu(II) ion imprinted composite adsorbent for selective removal of copper. Desalination 228(1–3):135–149. https://doi.org/10.1016/j.desal.2007.08.013

Riaz T, Ahmad A, Saleemi S, Adrees M, Jamshed F, Hai AM, Jamil T (2016) Synthesis and characterization of polyurethane-cellulose acetate blend membrane for chromium (VI) removal. Carbohyd Polym 153:582–591. https://doi.org/10.1016/j.carbpol.2016.08.011

Rocha CG, Zaia DAM, Alfaya RVDS, Alfaya AADS (2009) Use of rice straw as biosorbent for removal of Cu(II), Zn(II), Cd(II) and Hg(II) ions in industrial effluents. J Hazard Mater 166(1):383–388. https://doi.org/10.1016/j.jhazmat.2008.11.074

Rojas S, Horcajada P (2020) Metal-organic frameworks for the removal of emerging organic contaminants in water. Chem Rev 120(16):8378–8415. https://doi.org/10.1021/acs.chemrev.9b00797

Sargın D, Arslan G (2015) Chitosan/sporopollenin microcapsules: Preparation, characterisation and application in heavy metal removal. Int J Biol Macromol 75:230–238. https://doi.org/10.1016/j.ijbiomac.2015.01.039

Schleuter D, Günther A, Paasch S, Ehrlich H, Kljajić Z, Hanke T, Bernhard G, Brunner E (2013) Chitin-based renewable materials from marine sponges for uranium adsorption. Carbohyd Polym 92(1):712–718. https://doi.org/10.1016/j.carbpol.2012.08.090

Sha L, Xueyi G, Ningchuan F, Qinghua T (2009) Adsorption of Cu2+ and Cd2+ from aqueous solution by mercapto-acetic acid modified orange peel. Colloids Surf B 73(1):10–14. https://doi.org/10.1016/j.colsurfb.2009.04.021

Sharma PR, Chattopadhyay A, Zhan C, Sharma SK, Geng L, Hsiao BS (2018) Lead removal from water using carboxycellulose nanofibers prepared by nitro-oxidation method. Cellulose 25(3):1961–1973. https://doi.org/10.1007/s10570-018-1659-9

Sousa FW, Oliveira AG, Ribeiro JP, Rosa MF, Keukeleire D, Nascimento RF (2010) Green coconut shells applied as adsorbent for removal of toxic metal ions using fixed-bed column technology. J Environ Manage 91(8):1634–1640. https://doi.org/10.1016/j.jenvman.2010.02.011

Stela DE, Valentina DM, Shankar G (2015) Recent developments in composite biosorbents and their applications for wastewater treatment. J Environ Chem Eng 19:42–58

Gupta VK, Carrott PJ, Singh R, Chaudhary M, Kushwaha S (2016) Cellulose: a review as natural, modified and activated carbon adsorbent. Biores Technol 216:1066-1076. https://doi.org/10.1016/j.biortech.2016.05.106

Tabasum S, Noreen A, Maqsood MF, Umar H, Akram N, Nazli ZIH, Chatha SAS, Zia KM (2018) A review on versatile applications of blends and composites of pullulan with natural and synthetic polymers. Int J Biol Macromol 120:603–632. https://doi.org/10.1016/j.ijbiomac.2018.07.154

Tan G, Xiao D (2009) Adsorption of cadmium ion from aqueous solution by ground wheat stems. J Hazard Mater 164(2–3):1359–1363. https://doi.org/10.1016/j.jhazmat.2008.09.082

Tan G, Yuan H, Liu Y, Xiao D (2010) Removal of lead from aqueous solution with native and chemically modified corncobs. J Hazard Mater 174(1–3):740–745. https://doi.org/10.1016/j.jhazmat.2009.09.114

Torab-Mostaedi M, Asadollahzadeh M, Hemmati A, Khosravi A (2013) Equilibrium, kinetic, and thermodynamic studies for biosorption of cadmium and nickel on grapefruit peel. J Taiwan Inst Chem Eng 44(2):295–302. https://doi.org/10.1016/j.jtice.2012.11.001

Vakili M, Rafatullah M, Salamatinia B, Abdullah AZ, Ibrahim MH, Tan KB, Gholami Z, Amouzgar P (2014) Application of chitosan and its derivatives as adsorbents for dye removal from water and wastewater: A review. Carbohyd Polym 113:115–130. https://doi.org/10.1016/j.carbpol.2014.07.007

van Loosdrecht MCM, Brdjanovic D (2014) Anticipating the next century of wastewater treatment. Science 344(6191):1452–1453. https://doi.org/10.1126/science.1255183

Vandenbossche M, Jimenez M, Casetta M, Traisnel M (2014) Remediation of heavy metals by biomolecules: a review. Crit Rev Environ Sci Technol 45(15):1644–1704. https://doi.org/10.1080/10643389.2014.966425

Varma A, Deshpande S, Kennedy J (2004) Metal complexation by chitosan and its derivatives: a review. Carbohyd Polym 55(1):77–93. https://doi.org/10.1016/j.carbpol.2003.08.005

Velazquez-Jimenez LH, Pavlick A, Rangel-Mendez JR (2013) Chemical characterization of raw and treated agave bagasse and its potential as adsorbent of metal cations from water. Ind Crops Prod 43:200–206. https://doi.org/10.1016/j.indcrop.2012.06.049

Wang G, Xin Y, Uyama H (2015) Facile fabrication of mesoporous poly(ethylene- co -vinyl alcohol)/chitosan blend monoliths. Carbohyd Polym 132:345–350. https://doi.org/10.1016/j.carbpol.2015.06.040

Wang J, Liu M, Duan C, Sun J, Xu Y (2019) Preparation and characterization of cellulose-based adsorbent and its application in heavy metal ions removal. Carbohyd Polym 206:837–843. https://doi.org/10.1016/j.carbpol.2018.11.059

Wang J, Wei L, Ma Y, Li K, Li M, Yu Y, Wang L, Qiu H (2013) Collagen/cellulose hydrogel beads reconstituted from ionic liquid solution for Cu(II) adsorption. Carbohyd Polym 98(1):736–743. https://doi.org/10.1016/j.carbpol.2013.06.001

Werber JR, Osuji CO, Elimelech M (2016) Erratum: Materials for next-generation desalination and water purification membranes. Nat Rev Mater. https://doi.org/10.1038/natrevmats.2016.37

Witek-Krowiak A (2012) Analysis of temperature-dependent biosorption of Cu2+ ions on sunflower hulls: Kinetics, equilibrium and mechanism of the process. Chem Eng J 192:13–20. https://doi.org/10.1016/j.cej.2012.03.075

Wu MT, Tsai YL, Chiu CW, Cheng CC (2016) Synthesis, characterization, and highly acid-resistant properties of crosslinking β-chitosan with polyamines for heavy metal ion adsorption. RSC Adv 6(106):104754–104762. https://doi.org/10.1039/c6ra21993d

Xiang B, Fan W, Yi X, Wang Z, Gao F, Li Y, Gu H (2016) Dithiocarbamate-modified starch derivatives with high heavy metal adsorption performance. Carbohyd Polym 136:30–37. https://doi.org/10.1016/j.carbpol.2015.08.065

Xu SM, Feng S, Peng G, Wang JD, Yushan A (2005) Removal of Pb (II) by crosslinked amphoteric starch containing the carboxymethyl group. Carbohyd Polym 60(3):301–305. https://doi.org/10.1016/j.carbpol.2005.01.018

Yang D, Li L, Chen B, Shi S, Nie J, Ma G (2019) Functionalized chitosan electrospun nanofiber membranes for heavy-metal removal. Polymer 163:74–85. https://doi.org/10.1016/j.polymer.2018.12.046

Yang R, Su Y, Aubrecht KB, Wang X, Ma H, Grubbs RB, Hsiao BS, Chu B (2015) Thiol-functionalized chitin nanofibers for As (III) adsorption. Polymer 60:9–17. https://doi.org/10.1016/j.polymer.2015.01.025

Yule B, Roberts S, Marshall J (2000) The thermal evolution of sporopollenin. Org Geochem 31(9):859–870. https://doi.org/10.1016/s0146-6380(00)00058-9

Zaman CZ, Pal K, Yehye WA, Sagadevan S, Shah ST, Adebisi GA, Johan RB (2017) Pyrolysis: a sustainable way to generate energy from waste. INTECH. https://doi.org/10.1016/j.colsurfa.2011.12.014

Zhang X, Shen J, Pan S, Qian J, Pan B (2020) Metastable zirconium phosphate under nanoconfinement with superior adsorption capability for water treatment. Adv Func Mater 30(12):1909014. https://doi.org/10.1002/adfm.201909014

Zhao F, Repo E, Yin D, Sillanpää ME (2013) Adsorption of Cd(II) and Pb(II) by a novel EGTA-modified chitosan material: Kinetics and isotherms. J Colloid Interface Sci 409:174–182. https://doi.org/10.1016/j.jcis.2013.07.062

Zhao X, Yu X, Wang X, Lai S, Sun Y, Yang D (2021) Recent advances in metal-organic frameworks for the removal of heavy metal oxoanions from water. Chem Eng J 407:127221. https://doi.org/10.1016/j.cej.2020.127221

Chapter 10
Impact of Industrial Wastewater on Environment and Human Health

Shivani Garg, Zaira Zaman Chowdhury, Abu Nasser Mohammad Faisal, Nelson Pynadathu Rumjit, and Paul Thomas

Abstract The wastewater generated from different industries is discharged into adjoining environment and water body. Sometimes this wastewater discharged untreated or partially treated. Due to the industrialization in recent years, the environment deterioration is the major issues for consideration in different countries. Wastewater also contains several microorganisms such as virus, bacteria, protozoans, algae that have major public health concerns as these are cause of many waters borne diseases. The untreated wastewaters affect the quality of water in water bodies and human health to entering into trophic levels of food chain. The wastewater effluent may contain certain type of emerging contaminants that have the endocrine disruptive characteristics. Sometimes the discharged water contains certain contaminants which are not removed by the techniques used in that industry also cause harm to environment and public health. The chapter focuses on the impact of industrial wastewater on environment and human health with details.

Keywords Pollutants · Ground water quality · Eutrophication · Pathogens · Diseases by toxicants

Abbreviations

BOD	Biochemical Oxygen Demand
COD	Chemical Oxygen Demand
PCP	Personal Care Products
PhCs	Pharmaceuticals
PAH	Polycyclic Aromatic Hydrocarbon
SS	Suspended Solids

S. Garg (✉)
Institute of Environmental Studies, Kurukshetra University, Kurukshetra 136119, Haryana, India

Z. Z. Chowdhury · A. N. M. Faisal · N. P. Rumjit · P. Thomas
Nanotechnology & Catalysis Research Centre (NANOCAT), Institute for Advanced Studies (IAS), University of Malaya (UM), 50603 Kuala Lumpur, Malaysia

S. Roy et al. (eds.), *Advanced Industrial Wastewater Treatment and Reclamation of Water*, Environmental Science and Engineering,
https://doi.org/10.1007/978-3-030-83811-9_10

USEPA United State Environmental Protection Agency

10.1 Introduction

The water resources pollution become major problem due to human activities from industrialization and urbanization now a day (Molden 2013). The higher the population, the higher the demand of water for different activity that in turn gives the idea of reuse of wastewater after treatment (Scheierling et al. 2011). The industries use generally fresh water as a constituent in its operations. As fresh water resources are limited on Earth. Pollutants such as oils, heavy metals, radioactive substances, pesticides, hydrocarbons etc. discharge into environment directly from agriculture lands, domestic and industries (Fathi et al. 2008). Scientists and engineers are working on discovering new methods to make this industrial wastewater reusable. The wastewater from industries is the only source of irrigation in some developing countries for agriculture (Ensink et al. 2004). But this usage leads to environmental and human health issues due to lack of knowledge just for some financial benefits (Wichelns and Drechsel 2011). The discharge of contaminated wastewater without treatment in environment cause several diseases such as cancer, mutagenic changes, nervous responses delay etc. by entering the food chain (Balaji et al. 2005). Agricultural practices involve mineral fertilizers, pesticides and other chemical for growth and yield. Alone in India, maximum increasing rate of consumption of pesticides is 5% currently (Kanekar et al. 2004). Added to these chemicals, if industrial wastewater is used to watering the crops and soil of agricultural field can increase the contaminants level to high level. These chemicals cause soil acidification, precipitation of pollutants, deficiency of essential nutrients, loss of physiochemical characteristics and complexation. The migration of these chemical leads to phytoaccumulation in food crops, hence leading to exposure to diet (Prasad 2011). The chapter discussed the different sources of industrial wastewater discharge along with characteristics and the ill-effect of treated or untreated wastewater discharge and reuse by domestic life, agricultural crops and soil, water sources and human health along with aquatic flora and fauna.

10.2 Physico-chemical Characteristics of Wastewater from Different Industries

The physico-chemical characteristics constitute physical, chemical and biological components. The physical components are color, temperature, odor and total solids. The chemical components represent organic and inorganic compounds. These components after reaction in certain conditions produce harmful toxic and pollutants. The biological components constitute microorganisms such as bacteria, viruses,

Table 10.1 The wastewater pollutant factors by different industrial sectors

Industrial sectors	Pollutant factors
Textile and leather	Total solids, BOD, surfactants, dyes
Food industries	BOD, COD, suspended solids, micro and macro nutrients
Iron and steel	COD, BOD, metals, cyanides, acids and ores
Petrochemical industries	Oil, phenols, BOD, COD, chromium
Mining industries	Suspended solids, metals, acids and salts
Electronic industries	Organic chemicals, metals and metal ores, COD

pathogens. The wastewater generates from different industries have various types of physico-chemical compositions. Table 10.1 enlists the different industrial sectors along with the different physico-chemical and biological pollutant factors possess by wastewater.

There are four types of effluent such as specific effluent, general manufacturing effluent, general services effluent and intermittent effluent from industries which considered as harmful for human health and environment. The different type of wastewater on the basis of types and sources (industries) generates are explained as follow.

10.2.1 Agricultural and Food Industrial Wastewater

According to National Water Quality inventory conducted by EPA (2000), the agricultural industry wastewater accounts for steam, river water and lake water pollution by 41–48%. The most of agricultural industry wastewater generated animal waste, industrial livestock waste, fertilizers and pesticides industries along with other chemical used to preserve and process agricultural raw material wastewater etc. Pesticides and fertilizer wastewater have high concentration of pollutant and nutrients, COD, high toxicity (phenol, arsenic, mercury), foul odor and many biodegradable substances. The common source of water pollution is nitrogen and phosphorus concentration in water by agricultural runoff or from agriculture-based industries. Water pollution from these sources can cause extreme damage to environment and human health. Due to the difference in pH and BOD in wastewater from agriculture and food industries such as fruits, vegetable and meat products along with raw material packaging and preservation, it is very difficult to analyze the components of wastewater. The dairy industry wastewater contains conventional pollutants such as suspended solids, BOD (USEPA 2018). Meat industry generates organic waste (body fluid such as blood and pollutants produce BOD, organic nitrogen, oil and grease, ammonia,

SSetc. Processing and packaging the food produces coloring agents, salts, flavoring and preservative chemicals.

10.2.2 Personal Care Products and Pharmaceutical Industries Wastewater

Wastewater from these industries have recalcitrant substances such as emerging contaminants (painkillers, steroids, hormones, antibiotics) (Lateef 2004) and insufficient carbon and nutrient sources. Problems caused by these pollutants include inhibition of microbial and other useful organism activities and development, flotation and development of sludge and algae in water sources etc. There are many treatment methods such as advanced oxidation processes, coagulation—flocculation, Ozonation, Ultraviolet irradiation etc. are used to remove these contaminants still some amount of contaminants discharge into environment. The pretreatment of wastewater from pharmaceutical and PCPs effluent and bio-augmentation may enhance the biodegradation process.

10.2.3 Iron and Steel Industry Wastewater

Iron productions from its ores involve strong reduction reaction performed in the blast furnace. The water used to cool the furnace contaminated with the ores along with ammonia and cyanide. The wastewater steam contains gasification products such as anthracene, cyanide, benzene, phenols, ammonia, and cresols along with organic compounds commonly known as PAH. The iron used in the form of sheets, rods, wires and this conversion also used water at every stages. The wastewater also contains lubricants, coolants, hydraulic oil, particulate solids, ores particles, paint residue used for plating and galvanization. Hydrochloric and sulfuric acids are also used in the iron and steel industries. The hydraulic oil contaminated wastewater from steel industry is known as soluble oil.

10.2.4 Mining and Quarries Industry Wastewater

The wastewater from mines and quarries contains slurries of rocks. The wastewater generated from grading, washing of rocks and washing of tow roads and exposed surfaces. The wastewater from coal mines contains hematite, hydraulic oils and surfactants. The metal mines discharges wastewater with minerals. Crushing, drilling and extraction of materials may become contaminants in wastewater. The gold and silver extraction from high value metal constitutes fine particles of metals and ores.

10.2.5 Paper Mill Industry Wastewater

The wastewater from the paper industry constitutes sugar and lignocellulose which increase the development of bacteria, fungi and yeast. The other components in wastewater are chlorinated resin acids, phenols, hydrocarbons along with 500 different chlorinated organic compounds such as furans, dioxins, chloroforms, Syringols etc. The colored and adsorbable organic halogens compounds from pulp and paper industries cause numerous environmental problems. The bacterial growth (*Klebsiella sp.* and *Enterobacter*) in the wastewater disturbs the ecological equilibrium of the nearby water body by demanding excessive oxygen for their functioning. This will lead to loss of aquatic life and environmental issues such as air, soil and water pollution (Calamari 1985).

10.2.6 Textile Industry Wastewater

The textile industry wastewater constitutes heavy metals, high pH, BOD, COD, alkalinity, suspended solids, dyes, colors, oils, odor. The textile industry used raw material from direct source and synthetic components for manufacturing. The water used for grading and washing purposed during production have natural contaminants, fiber lint, animal oil and chemicals used to improve the structure of the raw material.

10.2.7 Complex Organic Chemical Industries Wastewater

Some industries such as dyes, paints, petrochemical, detergents, plastics etc. manufacture and utilize complex organic chemical during its operations. Wastewater from these industries contaminated by soluble particulate and solids matter, by-products, feed-stock materials, solvents, washing and cleaning agents.

10.2.8 Soap and Detergent Industries Wastewater

The wastewater from these industries contains toxic compounds, high COD, BOD (El-Gohary et al. 1987). These toxics are capable of damaging the microbial habitat. The increase in production of surfactants and consumption of detergent is the main cause of foam formation in river, lake and treatment plants. The other sources of surfactants in water bodies from the use of surfactants in cosmetics, leather tanning, textiles, dyes and paints, laundries and dry-cleaning industries. The presence of surfactants in water bodies inhibit the self- purification processes and enhance the anaerobic conditions.

10.2.9 Battery Manufacturing Industry Wastewater

Fabricating devices for portable equipment such as power tools and electronic or highly powered larger units for automobiles and motorized vehicles are the battery manufacturing industries. The pollutants produce from this industry are lead, nickel, chromium, copper, cadmium, iron, cyanide, manganese, silver, zinc and oil and grease (U.S. Environmental Protection Agency 2017).

10.3 Impact of Industrial Wastewater

The severity of effects of industrial effluent on human health and environment depend upon the physico-chemical characteristics of wastewater generated from particular industries and the size and type of effluent treatment plant in industries. The different world organizations give standard and guidelines the for discharge, recycling and reuse of treated or untreated wastewater of domestic and industrial sources along with sludge production and usage and air pollutant emission. The majority of impact on environment and human health are discussed as follows.

10.3.1 Environmental Hazards of Industrial Waste Effluents

The consequences of waste effluent generated from industry can reduce the concentration of dissolved oxygen; modify the physical properties of surface, ground and marine water. It can further release hazardous chemicals which can be bioaccumulated in aquatic species and can lead to enhanced nutrient loads of aqueous stream (Environmental Canada 1997). Wastewater can be utilized as a dynamic resource having number of benefits for enhanced crop growth. Wastewater and its nutritional content can be used to cultivate crops. However, application of wastewater or its improper discharge can have a detrimental impact on populations and overall ecosystem as a whole.

The extensive wastewater usage releasing toxic contaminants and an absence of funds to ensure proper disposal will likely to result in an increase in the prevalence of wastewater-borne pathogens and a further fast deterioration of the environment. Although the negative impacts of using polluted wastewater can be deferred for many years by intense irrigation, it has a detrimental impact on groundwater quality as nutrients usually will percolate down the soil (Mahmood and Maqbool 2006). Eutrophication, takes place when nutrients are added in excess quantity. This will result in the loss of dissolved oxygen. Solids (organic) in wastewater can be broken down by bacterial colony. Consequently, chemical oxidation will take place consuming a significant amount of dissolved oxygen in aqueous stream (Borchardt and Statzner 1990). These impacts can be instantaneous and temporary, or they may last over months or even years due to the deposition or sedimentation of oxygen-consuming content in the lowest part of the soils (Environmental Canada 1999). Low DO (dissolved oxygen) levels have a detrimental impact on ability of fish survival by decreasing their immunity to disease, inhibiting their growth, impeding swimming capacity, changing their feed intake and migration aptitude and in life-threatening circumstances, resulting in rapid death. Reduced dissolved oxygen concentrations over time can alter the composition of organisms (Environmental Canada 1997; Chambers and Mills 1996). Additionally, improperly handled effluent may change physical properties of water. Each type of aquatic life prefers specific temperature or their survival. Furthermore, their tolerance limit is also different. Any increase in the water surface temperature can have significant ecological consequences. Additionally, the discharge of suspended solids into aquatic ecosystems has several active and passive environmental consequences, including decreased penetration of sun-light (limited photosynthesis rate), physical harm to fish, and adverse impacts of toxins (Horner et al. 1994).

Another potential health hazard associated with untreated wastewater is biomagnification of pollutants. Certain trace compounds are hardly detectable in water can indeed be identified in massive quantities in plant and animal tissues because of bioaccumulation mechanism. These compounds are generally non-degradable, have a long chemical life, and are resistant to degradation (Environmental Canada 1997, 1999). In certain circumstances, the toxins can transfer throughout the food chain from lowest member to the highest one (Chambers and Mills 1996). Examples of such contaminants are heavy metals and organo-chlorine pesticides. Toxic compounds released from wastewater have a direct harmful effect on aquatic flora and fauna. Toxic effects may be immediate or chronic and persistent. Critical effects from effluents are typically caused by high concentration of chlorine and ammonia, excessive quantities of oxygen-demanding chemicals, or high concentrations of organic pollutants and heavy metals. Cumulative effects occur as a result of toxins gradually accumulating in consuming water and being visible only after a certain level is exceeded (Chambers et al. 1997; Welch 1992). Additionally, eutrophication may have an enrichment effect on nutrients. The following negative effects result from nutrient-induced eutrophication process:

- Accumulation of algae, unpleasant odors and water discoloration, can affect the aesthetic beauty and safe consumption of water.
- A large growth of rooted aquatic life disrupts shipping, oxygenation and channel capability
- Dying macrophytes and phytoplankton settle down at the ground of the aquatic body. This will trigger microbial activities which require more oxygen leading to depletion of oxygen. Severe loss of oxygen can destroy the optimal aquatic life.
- Filamentous algae and certain types of diatoms may block filters of wastewater treatment plants and result in reduced backwashing.
- Excessive algal growth can prevent sunlight to penetrate inside reducing the rate of photosynthesis required for submerged or merged aquatic plants. This will reduce the productivity of aquatic plants.

- Where major algal blooms occur in aquatic streams, the overwhelming numbers of phytoplankton death takes place. The degradation of these materials will cause reduction of dissolved oxygen, endangering the survival of numerous fish and other life forms of water (Environmental Canada 2003). The net impact of eutrophication is typically an increment in the amount and diversity of a certain plant species and a decrease in the type and abundance of many other species of plants and animals (Environmental Canada 1999, 2003). Total ammonia concentrations greater than 2 mg/L are harmful to marine organisms, although this differs by specific habitat and their life cycle. The toxicity effect of ammonia was carried out for freshwater plants. The concentration of ammonia greater than 2.4 mg/L can disrupt photosynthesis process (Chambers et al. 1997). Inappropriate amount of nitrate can decline the growth of amphibian species. The larva growth will be disrupted and the swimming ability will be reduced extensively for nitrate ions (Environmental Canada 1999).

10.3.2 Impact of Industrial Waste Water on Health

The major health risks connected with contaminated drinking water and other water sources are infections by viruses, bacteria and protozoa. The waste generated from animal and human are the major source of these microbes in wastewater (WHO 2006). This waste comprises a diverse array of bacteria, viruses and protozoa that can contaminate drinking water sources and receiving water bodies (Okereke et al. 2016). Microbial infections are known to be important contributors to various outbreaks of water-borne disease. Numerous microbial pathogens present in wastewater have the potential to cause chronic diseases with substantial lengthy consequences like degenerative cardiovascular disease and stomach ulcers. The diversity and concentration of these toxins varies with the severity and incidence of infection. Detecting, isolating, and identifying the various types of microbial toxins present in wastewater is often complex, costly, and consume so much time. To mitigate this, indicator species are often used to evaluate the potential hazard of a single pathogen being present in industrial wastewater (Paillard et al. 2005).

Viruses are one of the most prevalent and highly dangerous contaminants in waste effluents. They are often more resistant to medication, more contagious, harder to diagnose, and even they require a very small dose to cause an infection. Owing to the difficulties of identifying viruses due to their small population, bacteriophages (bacterial viruses) have been investigated for their potential role in faecal contamination (Okoh et al. 2007). Bacteria are the most prevalent microorganism in wastewater. They are responsible for a wide variety of diseases, including dysentery, diarrhea, skin and tissue infections, and others. Bacteria that cause infections can be found in water in a variety of forms, including *Leptosporosis, E. coli* O157:H7, *Vibrio, Listeria, Campylobacter, and Salmonella* (Absar 2005). Effluent contains a large number of bacteria, the majority of which are dangerous to people that can cause diseases such as dysentery, typhoid, and other intestinal disorders. The complete coliform and faecal coliform non-pathogenic bacteria tests are used to determine the existence of dangerous microorganisms (APHA 2001).

Due to the ease at which coliforms can be detected, faecal coliform screening has been recognized as the most accurate measure of faecal infections. In raw domestic wastewater, faecal coliform amounts up to 100 million per 100 ml. At levels of 2300–2400 total coliforms per 100 ml in water bodies, measurable health effects have been reported. To eliminate the harmful bacteria disinfection process is used and most common disinfection process used is chlorination (Absar 2005).

With bacteria as the susceptible agent, water-borne gastroenteritis of unknown origin is commonly identified. Certain strains of *Pseudomonas* and *E. coli* are two potential causes of this disorder, as they can affect newborn and have been linked to digestive disease epidemics (Metcalf and Eddy 2003). Additionally, extremely adaptable protozoa are abundant in natural waters, while only a small number of marine protozoa are pathogenic. Protozoal infections are often associated with relatively mild gastrointestinal symptoms than bacterial contagions. Among the pathogens, the protozoans *Giardia lamblia, Cryptosporidium parvum*, and *Cyclospora* are of particular interest due to their severe effects on persons with weakened immune systems, such as infants and the elderly people. Numerous *Giardiaoocysts* and *Cryptosporidium* are found in fresh water, but not each protozoan is pathogenic (Ingraham and Ingraham 1995; Metcalf and Eddy 2003). In drainage, the most resistant oocysts are those of *Giardia lamblia* and *Cryptosporidium parvum.* They are of significant relevance since they are typically present in all wastewaters. Due to traditional chlorine disinfection methods, these things can't be successfully removed by inactivating or destroying them under UV light or simple chlorination process. However, in subsequent years it has been demonstrated that UV disinfection is successful at inactivating *C. parvum*, Cysts of *Giardia lamblia* (Absar 2005; Metcalf and Eddy 2003).

The majority of Helminthes are classified into Annelida (roundworms) (segmented worms), Nematoda (roundworms) and Platyhelminthes (flatworms) three phyla. Helminth's life cycles often include two or more animal hosts, one of which may be helminth-infested animal or human waste. Additionally, pollution may occur through aquatic life of other species, along with insects or snails. While aquatic systems can act as a vector for the transmission of helminthic pathogens, new water disinfection and purification methods such as chlorination, sand filtration sedimentation, and chemical precipitation are extremely useful at eliminating helminthic pathogens (Absar 2005). Humans' excreta over hundreds distinct enteric viruses that are capable of causing infection or disease. These enteric viruses replicate in the digestive system and are excreted in the excreta of patients. The entero-viruses, adenoviruses, rotaviruses, reoviruses, Norwalk viruses, hepatitis A and calici-viruses virus seem to be the most significant human enteric pathogens from the health perspective (Absar 2005).

When used effectively filtration, sedimentation and disinfection generally have an appropriate level of virus elimination (EPA 1996). As mentioned previously, nutrients, particularly nitrogenous compounds with phosphorous, support the development of toxic phytoplankton species in both surface and sea waters. Consumption of radioactive algae or the creatures that consume them may be deadly for mankind and other terrestrial species. Toxins produced as a result of this process can cause liver damage, skin irritation, gastroenteritis, and neurological disability. Cyanotoxins have been linked to health conditions in a range of international, including the United Kingdom, Brazil, Canada, the United States of America, Australia, Zimbabwe and China (Department of Natural Science 2006). In certain cases, liver cancer in humans is believed to be caused by susceptibility to cyanobacterial toxins via the potable water line, with ingestion typically occurring by polluted drinking water or recreational water interaction (Runion 2008).

Toxins formed by microscopic algae can accumulate to dangerous levels during eutrophication. When shellfish and other sea species eat the algae, the toxins are concentrated higher up the food chain. Infections caused by toxic algae include paralytic shellfish toxicity, diarrheal diseases shellfish poisoning, and amnesic shellfish infecting (WHO 2006). Along with the health threats involved with contaminated effluent, populations and people can face odor and taste issues caused by significant algal growths. Although extra purification can be beneficial, it comes with a price (WHO 2006).

Regardless of the fact that nitrate is not toxic in and of itself; about a portion of consumed nitrate is transformed to nitrite by salivary microbes. Nitrites, once ingested, hinder the blood's capacity to transport oxygen by transforming hemoglobin to methemoglobin. Consumption of excessive nitrate or nitrite by infants and sensitive individuals can lead to methemoglobinemia (Wigle et al. 2008). Nitrites react with amino acids to produce nitrosamines in the stomach, which have been potent carcinogens for living creatures (Runion 2008). Another major health hazard caused by wastewater effluents stems from usage of chlorine as a disinfectant during the treatment processes. Although chlorination is efficient at eradicating cholera, typhoid fever, and other water-borne diseases, chlorine's high oxidizing ability can combine with naturally occurring organic matter in raw wastewater effluent, forming thousands of toxic chemicals such as bromodichloromethane, trihalomethanes and chloroform (Wigle et al. 2008).

Effluents contain a number of humans causedemerging contaminants, the majority of them cause endocrine disruption. Susceptibility to waste disposal effluents containing estrogenic compounds which has been found to impair marine life's endocrine activity, resulting in irreversible changes to the hormonal system's structure and function (Liney et al. 2006). Numerous laboratory experiments have established that many environmental chemicals have the ability to damage the endocrine system at environmentally relevant exposure levels. Such effects have apparently been found in marine environments in mollusks, fish, reptiles, birds, and rodents from North America, Europe and other regions. Heavy metal concentrations in soft tissues of Periwinkles from various rivers in Nigeria's Niger Delta region have been reported (Obasi et al. 2015). Although numerous laboratory experiments have shown the influence of specific endocrine disruptors at increased levels, no research has aimed to describe the impacts of external exposure to blends of endocrine disruptors (Sowers 2009).

Chemicals in wastewater can endanger people in a variety of ways. They can consume trace amounts of contaminants by absorbing through their skin by bathing/swimming/drinking with air during inhalation of airborne condensations and also by consumption of food that has been tainted by waterborne toxins, such as seafood (Vos et al. 2000). While ammonia is not a threat to human health at levels found in the atmosphere, but if expose to it especially in aquatic environment, may have a number of negative effects on human health. The utmost harmful effect of ammonia toxicity is pulmonary edema, which is accompanied by extreme inflammation of wet skin (WHO 2006).

10.4 Conclusion and Future Prospects

The wastewater discharged from different industries is major threat to environment and human health due to the presence of various kind of inorganic and organic toxic material. The effluent discharge into the water bodies or land causes extensive environment pollution and hazardous to human, animals, aquatic plants and animals as well as microbes. The main objective of industrial wastewater treatment is to protect health and environment. The wastewater characterization and treatment are also a big challenge to ensure the contaminants removal. The treated wastewater used in agricultural or other domestic uses cause salinity, metal-toxicity and nutrient imbalance in soil. Salinity and imbalance in nutrient in soil can alter soil infiltration, structure and permeability rate. The toxicity in surface water can also pollute the groundwater bodies through leachate formation. Drinking water pollution can cause neurological disorder, physiological and physical inability, retarded growth and development of human etc. No proper regulations are there for water reclamation. The main objective of water reclamation was to reuse and conserve the water. But, a very little consideration is given to effects of components of wastewater on plants, animals, environment and human health. There is the requirement of proper standards and guidelines to limit the reuse of specific types industry treated water. The toxicity and hazardous level should be assessed and analyzed before use in irrigation as this can bio accumulated and affect human internal organ directly. More researches are required to environmentally friendly alternatives to pesticides such as bio pesticides with improved genetic coding and their impact on other species in soil and environment.

References

Absar AK (2005) Water and wastewater properties and characteristics. In: Lehr JH, Keeley J (eds) Water encyclopedia: domestic, municipal and industrial water supply and waste disposal. Wiley, Inc., New Jersey, pp 903–905

APHA (2001) Standard methods for the examination of water and wastewater, 20th edn. APHA, Washington, D.C.

Balaji V, Datta S, Bhattacharjee C (2005) Evaluation on biological treatment for industrial wastewater. Biomed Environ Sci 85(18):320–326

Borchardt D, Statzner B (1990) Ecological impact of urban stormwater runoff studied in experimental flumes: population loss by drift and availability of refugial space. Aquat Sci 52(4):299–314

Calamari D (1985) Review of the state of aquatic pollution of West and Central African inland waters. CIFA Occasional Paper (FAO)

Chambers PA, Mills T (1996) Dissolved oxygen conditions and fish requirements in the Athabasca, Peace and Slave rivers: assessment of present conditions and future trends. Northern River Basins Study Synthesis Report No. 5. Environment Canada/Alberta Environmental Protection, Edmonton, AB

Chambers PA, Allard M, Walker SL, Marsalek J, Lawrence J, Servos M, Busnarda J, Munger KS, Adare K, Jefferson C, Kent RA, Wong MP (1997) Impacts of municipal effluents on Canadian waters: a review. Water Qual Res J Can 32(4):659–713

Department of Natural Science (2006) Wastewater characterization for evaluation of biological phosphorus removal

El-Gohary FA, Abo-Elela SI, Ali HI (1987) Management of wastewater from soap and food industries: a case study. Sci Total Environ 66:203–212

Ensink JH, Mahmood T, Van der Hoek W, Raschid-Sally L, Amerasinghe FP (2004) A nationwide assessment of wastewater use in Pakistan: an obscure activity or a vitally important one? Water Policy 6(3):197–206

Environmental Canada (1997) Review of the impacts of municipal wastewater effluents on Canadian waters and human health. Ecosystem Science Directorate, Environmental Conservation Service, Environmental Canada, p 25

Environmental Canada (1999) State of the Great Lakes. Environmental Canada and the US Environmental Protection Agency, p 8

Environmental Canada (2003) Environmental indicator series. http://www.ec.gc.ca/soer.ree. Accessed 12 Dec 2015

EPA (1996) U.S. Environmental Protection Agency, American Society of Civil Engineers, and American Water Works Association. Technology transfer handbook: management of water treatment plan residuals. EPA/625/R-95/008. Washington, D.C.

EPA (2000) Nutrient criteria technical guidance manual-rivers and streams. EPA-822-B-00-002. Washington, D.C.

Fathi AA, El-Shahed AM, Shoulkamy MA, Ibraheim HA, Rahman OA (2008) Response of Nile water phytoplankton to the toxicity Cobalt, Copper and Zinc. Res J Environ Toxicol 2:67–76

Horner RR, Skupien JJ, Livingstone EH, Shaver HE (1994) Fundamentals of urban runoff management. Technical and institutional issues, Terrene Institute, Washington, D.C.

Ingraham JL, Ingraham CA (1995) Introduction to microbiology. Wadworth Publishing Company, Belmont, CA

Kanekar PP, Bhadbhade BJ, Deshpande NM, Sarnaik SS (2004) Biodegradation of organophosphorus pesticides. Proc Indian Natl Sci Acad Part B 70(1):57–70

Lateef A (2004) The microbiology of a pharmaceutical effluent and its public health implications. World J Microbiol Biotechnol 20(2):167–171

Liney KE, Hagger JA, Tyler CR, Depledge MH, Galloway TS, Jobling S (2006) Health effects in fish of long-term exposure to effluents from wastewater treatment works. Environ Health Perspect 114(Suppl 1):81–89

Mahmood S, Maqbool A (2006) Impacts of wastewater irrigation on water quality and on the health of local community in Faisalabad [Pakistan]. Pak J Water Resour (Pakistan)

Metcalf X, Eddy X (2003) Wastewater engineering: treatment and reuse. In: Techobanoglous G, Burton FL, Stensel HD (eds) Wastewater engineering, treatment, disposal and reuse, 4th edn. Tata McGraw-Hill Publishing Company Limited, New Delhi, India

Molden D (ed) (2013) Water for food water for life: a comprehensive assessment of water management in agriculture. Routledge

Obasi KO, Chinedu K, Udebuani AC, Okereke JN, Ezeji EU, Anyadoh SO (2015) Study on concentrations of selected heavy metals: Cadmium, Lead, Arsenic and Mercury in the soft tissue of

Periwinkle (Tympanotonos fuscatus Var Radula) from Brass Island River, Bayelsa State, Nigeria. Int J Sci Technol 4(10):1–7

Okereke JN, Ogidi OI, Obasi KO (2016) Environmental and health impact of industrial wastewater effluents in Nigeria—a review. Int J Adv Res Biol Sci 3(6):55–67

Okoh AT, Odjadjare EE, Igbinosa EO, Osode AN (2007) Wastewater treatment plants as a source of microbial pathogens in receiving water sheds. Afr J Biotechnol 6(25):2932–2944

Paillard D, Dubois V, Thiebaut R, Nathier F, Hoogland E, Caumette P, Quentin C (2005) Occurrence of Listeria spp. in effluents of French urban wastewater treatment plants. Appl Environ Microbiol 71(11):7562–7566

Prasad MNP (2011) State-of-the-art report on bioremediation, its applications to contaminated sites in India. Ministry of Environments and Forests, Government of India

Runion R (2008) Factors to consider in wastewater-2. Ezine Articles, 14 Apr 2008

Scheierling SM, Bartone CR, Mara DD, Drechsel P (2011) Towards an agenda for improving wastewater use in agriculture. Water Int 36(4):420–440

Sowers A (2009) The effects of wastewater effluent on a fish and amphibian species

U.S. Environmental Protection Agency (2017) Battery manufacturing effluent guidelines. The U.S. EPA, Washington, D.C., 12 June 2017

U.S. Environmental Protection Agency (2018) Dairy products processing effluent guidelines. EPA, 30 Nov 2018

Vos JG, Dybing E, Greim HA, Ladefoged O, Lambre C, Tarazona JV, Brandt I, Vethaak AD (2000) Health effects of endocrine-disrupting chemicals on wildlife, with special reference to the European situation. Crit Rev Toxicol 30(1):71–133

Welch EB (1992). Ecological effects of wastewater. Applied limnology and pollutant effects, 2nd edn. Chapman and Hall, New York, p 45

Wichelns D, Drechsel P (2011) Meeting the challenge of wastewater irrigation: economics, finance, business opportunities and methodological constraints. Water Int 36(4):415–419

Wigle DT, Arbuckle TE, Turner MC, Bérubé A, Yang Q, Liu S, Krewski D (2008) Epidemiologic evidence of relationships between reproductive and child health outcomes and environmental chemical contaminants. J Toxicol Environ Health B 11(5–6):373–517

World Health Organization (2006) WHO guidelines for the safe use of wastewater excreta and greywater, vol 1. World Health Organization

Chapter 11
Fundamentals of Adsorption Process onto Carbon, Integration with Biological Process for Treating Industrial Waste Water: Future Perspectives and Challenges

Arnab Barua, Zaira Zaman Chowdhury, Ahmed Elsayid Ali, Rabia Ikram, Abu Nasser Mohammad Faisal, Shah Jalal Mohammad Shibly, Rahman F. Rafique, Rafie Bin Johan, and Suresh K. Bhargava

Abstract Presence of various types of micro-pollutants in domestic, industrial and municipal wastewater and landfill leachates are hazardous for living organisms. Thus, industries need to follow stringent requirements for discharging the chemicals in aquatic system. Organic pollutants, including dyes, hormones, phenols, pharmaceuticals, aromatic compounds, dyes, herbicides, pesticides, bromine containing retardants, industrial by-products, domestic products, sterilisers, and antiseptics, are referred as xenobiotic organic compounds (XOCs). Traditional treatment techniques used for organic pollutant removal is not sufficient. The porous carbonaceous structure of granular (GAC) and powdered (PAC) activate carbon can adsorb targeted pollutants from water. The surface area of activated carbon can be varied from (500–1500 m^2g^{-1}). The first part of the chapter illustrated the basic mechanism for adsorptive removal of contaminants over powdered and granular activated carbon (PAC or GAC). Now a days, adsorption onto activated carbon is integrated with biological process for treating industrial effluents. Adsorption integrated with biological treatment of waste water can be carried out using powdered activated carbon membrane bioreactor (PAC-MBR) and biological activated carbon reactor

A. Barua · Z. Z. Chowdhury (✉) · A. E. Ali · A. N. M. Faisal · S. J. M. Shibly · R. B. Johan
Nanotechnology and Catalysis Research Centre (NANOCAT), University of Malaya, 50603 Kuala Lumpur, Malaysia
e-mail: dr.zaira.chowdhury@um.edu.my

R. Ikram
Department of Chemical Engineering, University of Malaya, Kuala Lumpur, Malaysia

R. F. Rafique
John A. Reif, Jr. Department of Civil & Environmental Engineering, New Jersey Institute of Technology (NJIT), 323 MLK Blvd., New York, NJ 07102, USA

S. K. Bhargava
Centre for Advanced Materials and Industrial Chemistry, RMIT University, Melbourne, Australia

S. Roy et al. (eds.), *Advanced Industrial Wastewater Treatment and Reclamation of Water*, Environmental Science and Engineering,
https://doi.org/10.1007/978-3-030-83811-9_11

(BAC). The fundamentals of using activated carbon adsorption process integrated with biological treatment is discussed in the subsequent section of the chapter.

Keywords Adsorption · Powdered activated carbon · Granular activated carbon · Membrane bioreactor · Biological treatment

11.1 Introduction

Population of the world is growing day by day resulting 1% increment in consumption of water every year (WWAP/UN-Water 2018). Urbanization as well as growth of industrial sector is releasing lot of harmful chemicals, agricultural and industrial pollutants, into the aqueous stream (Razi et al. 2018; Oladipo et al. 2017). A recent statistics showed that, around 22 lacks population was suffering from diarrhoea and passed away due to consumption of polluted water. Furthermore, until now; around six hundred sixty three million residents don't have access to pure water for drinking (Huang et al. 2018; Zbair et al. 2017; Akinpelu et al. 2021). Consumption of polluted water supply has severe toxicological impact on living organisms (Chowdhury et al. 2018; Adebesi et al. 2017a, b; Ahmed et al. 2018). Thus, it is important to find out, eco-friendly, sustainable and cost effective protocols to treat industrial waste effluents containing multiple types of complex pollutants. Numerous technological approaches have been devised and implemented to detoxify impure, hazardous water. The list of most commonly utilised methods are Illustrated by Fig. 11.1.

Among the aforementioned strategies adsorption or sorption which is a combination of adsorption and absorption is the most prevalently used method for contaminant extraction from aqueous stream due to its reduced cost, easy operational mode, good biocompatibility and almost close to zero health hazards (Balasundram et al. 2017; Oladipo and Gazi 2015). The incorporation of activated carbon (AC) for the sorption

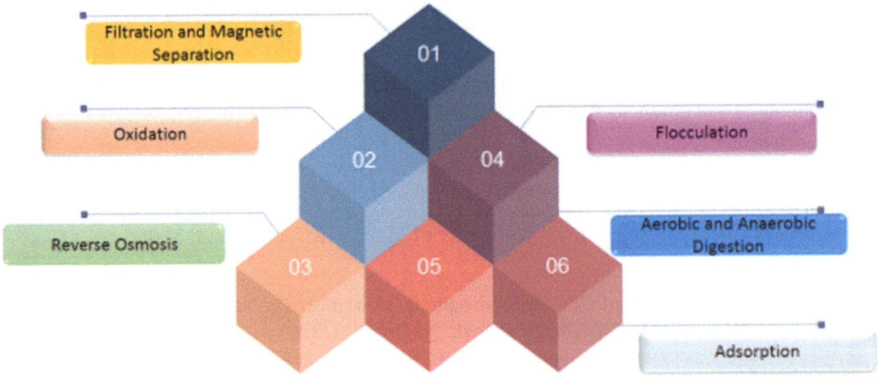

Fig. 11.1 Strategies used for industrial waste water treatment

process facilitates the eradication of a wide variety of impurities and carcinogenic chemicals from aqueous solutions-including medicines, hormones, heavy metallic cations, inorganic negative radicals, colour, odour or even taste to some extent (Din et al. 2017; Chowdhury et al. 2017).

In contrast to several other adsorbent materials such as clay, betonite, zeolite, resins, and selected polymers, AC has superior sorption capability endurance and stability (Regti et al. 2017). Consequently, it has been recognised that pollutant gas adsorption onto AC is also a promising technique for the coupling interaction between the contaminant and the functional active surface active groups (Le-Minh et al. 2018; Akinpelu et al. 2019, 2020). The generation of AC from organic matter can be viewed as beneficial in 2 ways: initially, it can help to inhibit the generation of CO_2 by fixing the carbon; and subsequently, it can also eventually find its route into the ground naturally. Advantage and disadvantages of adsorption process is illustrated by Fig. 11.2.

Activated carbon (AC) is most extensively utilized versatile amorphous phase adsorbent which is porous in texture containing lot of surface active sites-often referred as surface functional/chemical groups. Adsorption onto activated carbon can be used to remove a numerous variety of organic and inorganic solutes from wastewater. Commercially available precursors-which are used to manufacture AC with their chemical constituents are listed in Table 11.1. The highly permeable surfaces of activated carbon binds and holds the pollutant solutes and gases, and the amount of pollutant adsorb is significant due to its extensive internal surface (Din et al. 2017;

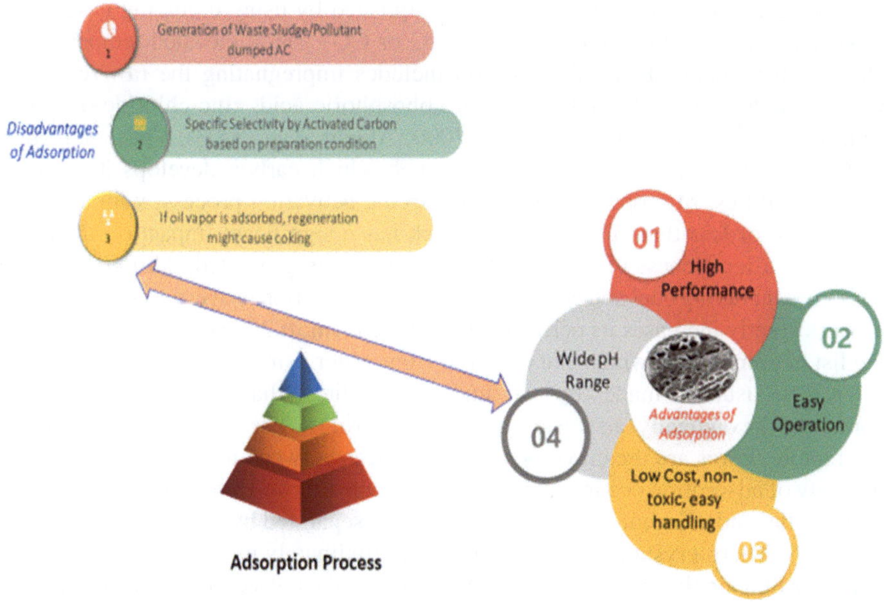

Fig. 11.2 Advantages and disadvantages of adsorption process

Table 11.1 Chemical constituent used generally for production of AC (Inglezakis and Poulopoulos 2006; Hung et al. 2005)

Starting precursors	Percentage carbon (wt%)	Percentage ash (wt%)	Percentage volatile constituents (wt%)
Coke (petroleum)	55–70	0.5–0.7	15–20
Coal	65–90	2–15	5–30
Lignite	70–85	5–6	25–40
Wood	40–45	–	55–60
Shell of nuts	40–45	0.3–1.1	55–60

Danish et al. 2011, 2013). Ac has a substantial adsorptive surface area of 500–1500 m^2/g and its' pore volume can ranges from 0.7–1.8 cm^3/g. It is utilised mostly as a powdered (PAC) or granular (GAC) activated carbon (Adebisi et al. 2016; Chowdhury et al. 2015, 2016).

Commercially distributed activated carbons are made from high-carbonaceous sources such as wood, coal, coke, lignite, nuts, peat, coconut fibre, shell, husk, lignin, and synthetic polymeric materials. Production of activated carbon mainly involves two stages: carbonization and activation. Carbonization is a drying and heating procedure that removes unwanted residues; as for example, volatile hydrocarbons or tar. The carbonized materials are subsequently pyrolyzed and activated in an oxygen free environment at temperatures ranging from 400 to 1000 °C. This eliminates the volatile less-molecular-weight component and activates the substance. Thermal activation can be accomplished by using oxidizing gases such as steam at temperatures more than 800 °C or CO_2 gas at elevated temperatures. On the other hand, chemical activation includes impregnating the raw resources with reagents like sulphuric/hydrochloric/phosphoric acid, zinc chloride or potassium/sodium hydroxide (Khalili et al. 2000; Ahnert et al. 2003). Thus, activation is a phrase that refers to the process through which carbon develops its adsorption capabilities. Micropores develop due to the activation process, with a typical throughput of less than 50%. The feedstock has a significant impact on the properties and effectiveness of the activated carbon that is generated. Although natural resources including coal and charcoal have these adsorption properties, the activation process greatly increases its performance (Cheremisinoff 2002). Table 11.2 outlines the list of compounds which can removed from water using activated carbon (AC).

AC is usually made up of micro-crystallites that contains fused and connected hexagonally arranged carbon atoms rings. The structure is analogous to graphite. Pores are the crevices and void spaces between micro-crystallites. The majority of adsorption occurs in micropores, which are two-dimensional gaps within two parallel and adjacent crystallographic planes separated by an inter-layer spacing of 3–35A. Micro-crystallites have diameter approximately nine times more than the diameter of a single carbon hexagon. Adsorption takes place on the flat surfaces of the micro-crystallites and over the functional active groups attached at the boundaries (Yonge et al. 1985; Yonge and Keinath 1986). Adsorption on the planes of

Table 11.2 Major categories of organic compounds which can be removed using AC (Cheremisinoff 2002)

Categories of organic chemicals	Example
Benzene and its derivatives	Benzene, xylene toluene, ethyl/methyl benzene
Polynucleated Benzene/aromatic compounds	Poly-aromatic hydrocarbon, Anthracenes, naphthalene, pyrene, biphenyls
Chlorine substituted aromatic compound	Chloro-benzene
Phenols	Phenol, nitrophenols, cresol, chloro-phenols
Hydrocarbon with high molecular weight	Kerosene, gasoline
Chlorine substituted aliphatic compounds	Carbon tetrachloride, trichloroethylene,
Aromatic and aliphatic acids	Benzoic and tar acids
Ketones, esters, ethers, and alcohols	Polyethylene glycol, Hydroquinone,
Surfactants	Alkyl benzene sulfonates
Dye, pharmaceutical products, pesticides, herbicides, hormones, caffeine, pharmaceutical products, inorganic chemicals etc.	Acidic dyes, basic dyes, bentazon, paracetamol, methylene blue, indigo carmine, heavy metals, inorganic ions

micro-crystallites happens due to the van der Waals forces. Adsorption somewhere along the micro-crystallite's edges, proceeds due to chemical bonding.

Generally, AC surfaces involve a lot of oxygen complexes. These complexes are formed throughout the activation phase due to chemical adsorption of oxygen. On AC surface, oxygen complexes primarily form four distinct acidic oxide groups, including weak –COOH (carboxylic), carbonyl and –C–OH (phenolic) groups. Additionally, there are cyclic ethers which are basic in nature and usually forms when activation takes place at elevated temperatures. The existence of oxide layer endows AC with polar character. Carbons can be thermally treated in an oxygen deficient, inert environment or under vacuum to eliminate these oxide surface active groups (Yonge and Keinath 1986). Surface functional groups are crucial for the adsorption of many inorganic and organic pollutants from water. Aromatic molecules, can be adsorbed due to presence of carbonyl oxygen groups over the surface of carbon via a donor acceptor donor complex formation mechanism. The carbonyl oxygen donates electrons, whereas the aromatic/benzene ring of the pollutants accepts electrons. Adsorption is also caused by phenolic protons which can form hydrogen bond with surface functional groups (Yonge and Keinath 1986). Different types of minerals, primarily sulphates, calcium, and phosphate ions, can be found in high concentrations inside the AC. Presence of minerals and the acidic or basic surface functional groups, have an impact on the surface characteristics of AC (Julien et al. 1998). AC also comprises of some fraction of ash produced from the starting material, with the amount varying from 1 to 12%. Iron oxides, silica, alumina, and alkali and group two-alkaline earth metals are the most common constituents of ash. The presence of ash in AC increases its hydrophilic nature. When powdered activated carbon (PAC) is used for water purification, this is helpful since PAC does not adhere to the walls of the reactor due to high content of ash.

This chapter illustrates the basic mechanism of adsorption used in industrial waste water treatment process. The mechanism of pollutant transportation during the sorption process is described in subsequent section of the chapter. The chapter provides a brief overview on using granular activated carbon (GAC) for fixed bed, column sorption process in terms of breakthrough curve analysis. The importance of integration of biological system with carbon adsorption process is described. The concluding section of the chapter describes about the safe handling protocols and regeneration process of the exhausted carbon used for waste effluents.

11.2 Classification of Activated Carbon Used as Adsorbent for Waste Water Treatment

For industrial waste water treatment; following two types of AC is extensively used:

11.2.1 Powdered Activated Carbon (PAC)

Based on the formulation proposed by the Water Works Association, America, PAC particles should be able to pass through a filter having diameter 0.297 mm. However, based on ASTM D5158 classification, 95% PAC should pass through or 0.177 mm mesh (Cheremisinoff 2002). PAC is typically manufactured from sawdust derived from wood. The mean particulate sizes of wood saw dust can vary from 15 to 25 mm. PAC is widely used for purification of both drinking water and industrial waste effluents (Chowdhury et al. 2012). It is injected to activated sludge or kept in a separate, special unit during the filtration process. PAC may also be used in the aqueous phase to coagulate the colloidal particles. Regeneration of powdered activated carbon (PAC) can be challenging due to the requirement to isolate the colloidal contaminants from water prior to the regeneration process (Cheremisinoff 2002).

11.2.2 Granular Activated Carbon (GAC)

Granular (GAC) activated carbon is often produced from the crushed coal or shell granules having more bulk density that PAC. Additionally, GAC can be made by granulating the crushed powders with binders like coal tar pitch (Oladipo et al. 2017). Like PAC adsorption, intraparticle diffusion process is frequently regarded as the rate-limiting step in casc of GAC sorption from the liquid state. The size of GAC particles ranges from 0.2 to 5 mm. GAC is classified according to its mesh size, with 8/20, 20/40, or 8/30 being used in aqueous phase usage and 4/6, 4/8, or 4/10 being used in gas phase application (Cheremisinoff 2002). GAC particle sizes between 12/42 mesh are favourable for aqueous stage application (Cheremisinoff 2002).

GAC column filters are frequently utilised as an advanced processing step for purification of drinking, ground, and wastewater, notably for the disposal of hazardous organic pollutants from water. Specific GAC usage in drinking water purification can result in the formation of a microbial layer over the particles. Thus, biological pollutant removal process is integrated with GAC sorption. The resulting substance is often termed as biological activated carbon (BAC).

11.3 Fundamental of Adsorption Process for Removal of Aqueous Pollutant from Industrial Effluents

Adsorption is an important aspect in the majority of spontaneous physical, ecological, and chemical procedures and AC is the most often employed as adsorbent in wastewater purification process. Adsorption is the process through which pollutants accumulate or concentrate onto the surface of adsorbent or interface between the liquid of solid phase. The adsorptive phase is referred as the adsorbent, whereas the component that is adsorbed is referred as the adsorbate (Karimi et al. 2019; Veil 2011). Adsorption can occur at the interfaces of two phases, such as liquid–liquid, gas–liquid, gas–solid, or liquid–solid (Hansen and Davies 1994; Yousef et al. 2020). AC is employed as the adsorptive solid phase. For AC based adsorption process, between the liquid phases- a hypothetical boundary layer is present.

In adsorption process, surface reactions are induced by strong forces within the layer/phases or boundaries at the surface. These forces provide distinctive energies at boundaries. The surface tension created at the liquid phase's surface is a result of the attraction forces between the liquid's molecules. The solvent molecules are more attracted to the solute particles than they are to each other. As a result, a solute with a low surface tension is accumulated at the surface. Adsorption is a term that refers to the phenomenon of an increment of a dissolved components at the boundaries or interfaces (Yousef et al. 2020). The term 'adsorption' relates to the procedure by which particles aggregate at the interfacial region, while the term 'desorption' refers to the mechanism by which the particles desorb (Thomas and Crittenden 1998).

The lyophobic (rejection towards solvents) force of the solute-pollutant is the primary driving force which enables adsorption to take place. The primary determinant of the adsorption intensity is the dissolution of the dissolved pollutant particles present in water. A hydrophilic molecule has a tendency to remain in the aqueous system. As a result, it has a lower adsorption capacity onto the solid phase.

However, a hydrophobic pollutant is more likely to be adsorbed over the solid adsorbent than to remain in water. Humic acids are regarded as complex organic pollutant which contains both hydrophilic and hydrophobic domains. During adsorption, hydrophobic portion of the molecule tends to be adsorbed over the solid surface of adsorbent while the hydrophilic region of the molecule seemed to remain inside the liquid.

The second possible energetic factor underpinning the sorption process is the pollutant-solute's attraction towards the solid adsorbent. The affinity between solute and solid surface takes place as a result of electrostatic attractions. Adsorption of this sort may happen due to consequence of Van der Waals forces or chemical interactions with the sorbent. The forces are generated as a result of the solute's strong affinity for the solid adsorbent under consideration (Worch 2009). Adsorption generated by van der Waals forces is commonly mentioned to as physical adsorption or physisorption, because it occurs as a result of intermolecular attraction between favourable energy sites. The adsorbate initiates chemical interactions with the solid adsorbent during chemical adsorption, or chemisorption. Chemisorption tends to occur when electrons are exchanged between certain surface active sites and pollutant-solute chemicals present in water, leading to the formation of a chemical bond. Adsorbates that have been chemically sorbed are not able to transfer across the surface or interfaces between the surfaces (Worch 2008).

11.4 Mechanism for Transportation of Pollutant-Solute (Adsorbate) Particles Inside the Carbon Matrix (Adsorbent) During Wastewater Treatment

11.4.1 Bulk Transportation of Solution

It is referred as Advection where the pollutant-adsorbate molecules essentially needs to transported and leave the bulk solution to proceed towards the liquid film of water or boundary layer (Fig. 11.3a, b). This transportation of pollutant-solute may occur by diffusion in stagnant water with dispersed AC placed inside it. Additionally, turbulent blending occurs with flow across a densely packed fixed bed of granular (GAC), or powdered (PAC) mixed inside the water purifier division.

11.4.2 External/Peripheral Diffusion

There might be a possible supplemental resistance to mass transportation of solute towards the surface of the carbon during adsorption.

This is linked with the mobility of sorbate from the solution bulk throughout the static layer of aqueous effluents-which is regarded as the hydrodynamic boundary

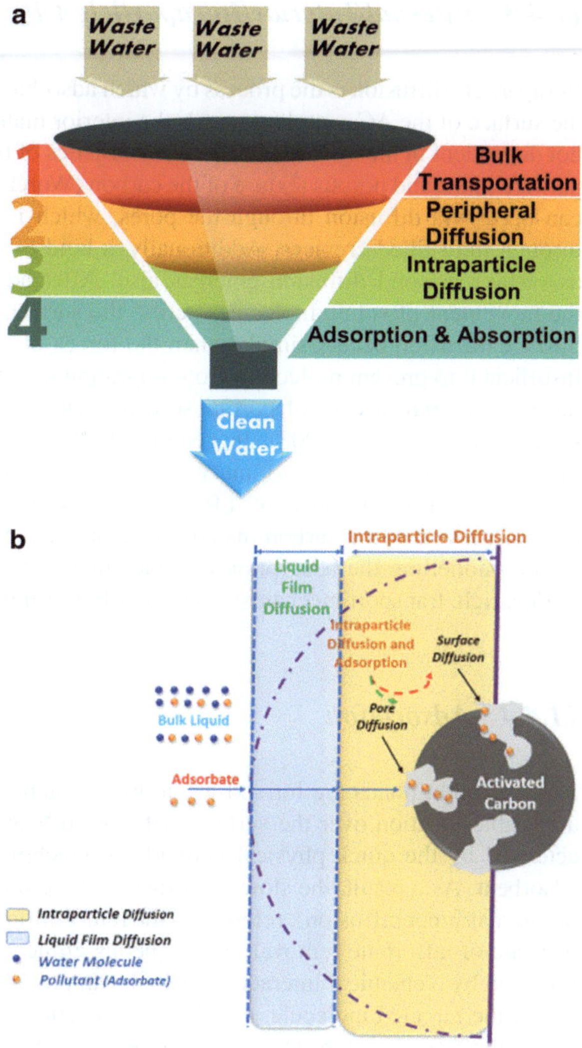

Fig. 11.3 **a** Stages and **b** mechanism for transportation of pollutant-solute (adsorbate) particles inside the carbon matrix (adsorbent) during wastewater treatment

film or layer. It is also known as liquid film, or exterior film, which mostly encircles the adsorbent carbon particles, as illustrated in Fig. 11.3. The concentration gradient provides the energy for mass transportation of pollutants through this layer. The hydrodynamic characteristics of the sorption system determine the rate at which the diffusion takes place (Worch 2008).

11.4.3 Internal/Interior/Intraparticle Diffusion

Intraparticle diffusion is the process by which adsorbate/pollutant is transferred from the surface of the AC sample, towards the interior matrix of the carbon particle. It is not dependent in the hydrodynamic characteristics of the system but is reliant on the particle's size and porous texture of the carbon (Worch 2008). Intraparticle diffusion can occur via diffusion through the pores, which is the movement of molecules through fluid-filled crevices. Additionally, a kind of intraparticle diffusion known as surface or solid diffusion emerges also. After adsorption, surface diffusion is the movement of solute molecules across the surface of the adsorbent (Fig. 11.3). Surface diffusion occurs simply when the forces of attraction over the surface are insufficient to prevent molecules from migrating over the surface. This effect is still most substantial in case of porous sorbents with a large active surface area with smaller pores (Worch 2008). In general, the terms 'pore diffusion' and 'surface diffusion' are used to refer to 'macropore' and 'micropore' diffusion, respectively.

Surface diffusion and Pore diffusion frequently occur concurrently within the interior surface of the carbon matrix (Snoeyink and Summers 1999). As discussed earlier, modelling the adsorption of contaminants from waste effluents, the major intraparticle transport mechanism is generally presumed to be surface diffusion.

11.4.4 Adsorption

An adsorption bonds are formed when the pollutant-adsorbate is transported to an accessible location over the surface of the adsorbent. Physical adsorption is characterised by the quick physical network of attachment by the adsorbate over the adsorbent. As a result, the slowest diffusion phase (peripheral diffusion, advection, or internal/inner diffusion), referred as the rate-limiting phase, will restrict the effective rate of adsorbate removal from aqueous effluents. Nevertheless, if sorption is followed by a chemical interactions which can modify the physio-chemical properties of the targeted molecule, the chemical reaction kinetics could be delayed than that of the diffusion step. Thus overall removal rate might be controlled by diffusion process and at the last stage-when the pollutant enters inside the carbon matrix-it is known as absorption (Sun et al. 2007; Tan and Hameed 2017). The concentration profile of a pollutant in liquid, interface and solid phases are illustrated by Fig. 11.4 (Hendricks 2006).

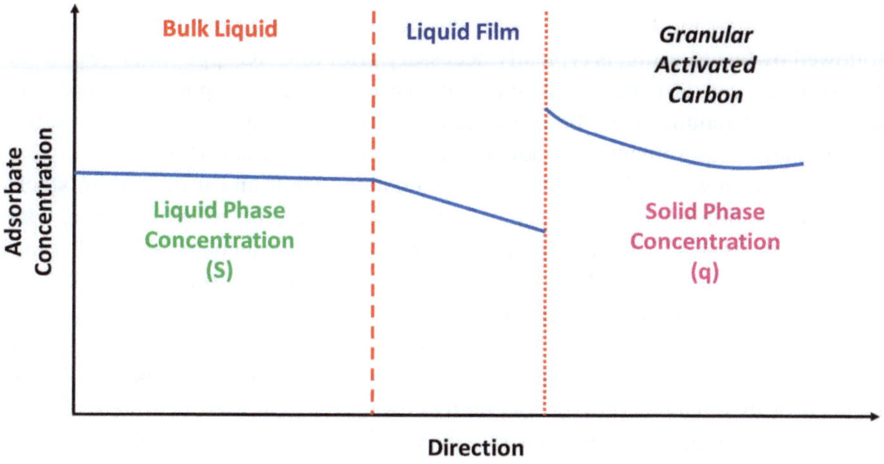

Fig. 11.4 Concentration profile of pollutant travelling during adsorption process

11.5 Types of Carbon Reactors Used for Industrial Waste Water Treatment

11.5.1 Reactors Using PAC

For drinking water supplies, PAC is utilised in one of three process variations multiple stage simultaneous flow, single stage continuous flow, or multi-stage concurrent flow (Danish et al. 2013). PAC is applied by contacting the sorbent with the liquid for a specified period of time. Then, the wastewater must be separated properly to produce a purified effluent. PAC could also be utilized in wastewater treatment, mostly for the removal of non-biodegradable contaminants subsequent after the biological pre-treatment. However, in comparison to GAC based reactors, the usage of PAC contactors is largely unfeasible for wastewater treatment.

11.5.2 Reactors Using GAC

11.5.2.1 Reason for Using GAC in Industrial Absorbers/Reactors for Wastewater Treatment

GAC adsorption is used as a third treatment step for industrial and municipal wastewaters (physico-chemical treatment following second phase of treatment) or as a phase in the physico-chemical pre-treatment step (flocculation, coagulation, settling, filtration, adsorption over GAC) to replace biological treatment. When utilised in tertiary phase of treatment, GAC is primarily utilised to adsorb organic compounds that remain after biological treatment. Pre-treatment, using precipitation with lime followed by fast filtering, is typically necessary prior to GAC application (Suthersan 1999). GAC can also be employed in industrial wastewater purification to meet pre-treatment requirements before release to urban sewerage system or to satisfy the discharge requirements into aquatic ecosystems (Suthersan 1999).

Continual flow GAC system/adsorbers contains fresh and used carbon stocks, carbon transportation facilities, and carbon regeneration units to initiate waste water treatment process. Carbon adsorbers/reactors are made of a coated steel column. It can be rectangular tank made of steel or concrete into which certain volume of GAC is deposited and known as 'filter' bed.

A down flow from a fixed-bed GAC is sometimes used in conjunction with a GAC adsorber to adsorb pollutant from water or wastewater. Carbon adsorbers with fixed beds can be driven under pressure or gravity flow in case of down flow system of water. Water is introduced at the highest part of the GAC column, whereby water flows down throughout the filter bed. At the end of the process, it is withdrawn from the bottom. However, in most of the literature, upward flow of waste water is encouraged to initiate sufficient contact time of waste water with the carbon (Chowdhury et al. 2013; Al-Hameed 2009; Pam et al. 2018). Contaminants are adsorbed as the wastewater runs through the column. A drainage channel at the base of the contactor keeps the carbon in place. Backwash and surface washing of the filter bed is intended to stop the aggregation of significant head loss induced by particles deposition and to preserve the surface of the bed from blockage.

For the adsorption, using GAC, there are three different reactors configurations. These are known as the fixed bed, extended/expanded bed, and fluidized bed reactors. A down flow adsorber with slow fluid velocity functions similarly to a fixed-bed reactor. Column packed with fixed-bed of carbon offers both filtration and adsorption. The fixed-bed arrangement is appropriate for wastewaters with a low concentration of suspended particles or no suspended particles (Fig. 11.5).

Fixed-bed adsorption is frequently used in the purification of water used for drinking purposes since it encompasses relatively few contaminants. In these instances, fixed-bed GAC reactors are mostly operated in down flow operation mode and together with gravitational force usually (Fig. 11.5).

On the contrary, if fixed beds are used for treating industrial and municipal waste water, containing relatively large number of suspended pollutants; the possibility of biological growth, might create complications. Raising wastewater from the lower part and permitting it to ascend into the treatment plant is a technique used to reduce fouling, scaling, blockage, and pressure drop during treating industrial waste water. As a result, methods that permit the bed to expand, move, or become fluidized may be used. For upward transportation of water, the bed expands as this fluid flows upward. Carbon that has been depleted throughout the process of moving towards the subsequent bed is replaced on a regular basis to avoid the accumulation of head loss. A third option for waste effluent treatment is to use fluidized-bed reactors (FBRs), wherein the up flowing solution flows through the FBR, suspending the tiny solute elements that persist in the reactor. Thus, fluidized beds aid in the prevention of blockage and accidental filtering (Karri et al. 2017; Chang et al. 2006; Jones et al.

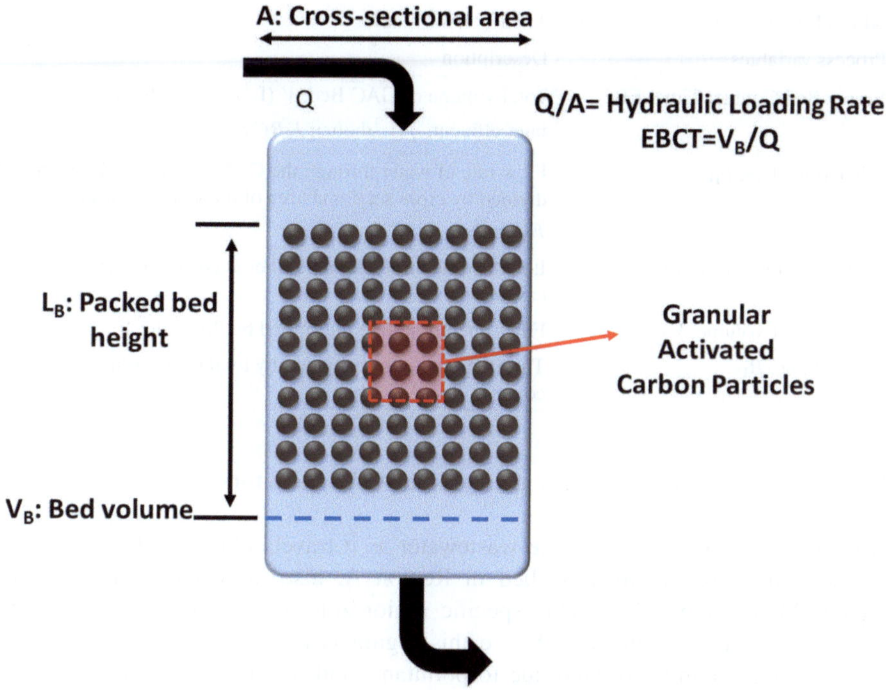

Fig. 11.5 Fixed bed GAC adsorber/reactor with design/operational variables

2010). To remove carbon from the reactor, individuals can either add fresh carbon at a constant rate or utilise an automated mechanism that continuously adds fresh carbon to the reactor's top. In contrast to fixed-bed operations, which demands reactor shutdown upon exhaustion, this approach will not entail the reactor shutdown after exhaustion.

11.5.2.2 Process Variables in GAC Reactors and Its Impact on Water Treatment

The following criteria are primarily considered while designing GAC columns: hydraulic loading rate, empty-bed contact time (EBCT), depthless of carbon bed, and the number of columns (Jones et al. 2010). GAC column filters can be configured as up- or down-flow arrangements, with more than one unit connected in parallel or series layout, depending on the kind and amount of pollutant in water to be purified (Jones et al. 2010). The following Table 11.3 listed the primary parameters that specify the GAC operation.

Table 11.3 Process variables for fixed bed-column studies

Process variables	Description
Empty Bed Contact Time, EBCT	Total volume of GAC Bed, V (L^3) divided by flow rate of inlet effluent, q (L^3/hour); $EBCT = \frac{V}{q}$
Filter velocities, F_B	Flow rate of water through the GAC column, q (L^3/hour) divided by cross sectional area of the filter bed used, A; $F_B = \frac{q}{A}$
Operating time of filter t_F	It is the time until when the filter needs no replacement of carbon
Through-put volume, V_L	Filter operation time multiplied by flow rate; $V_L = q \times t_F$
Bed volume, B_V	Through-put volume divided by total GAC volume in column; $B_V = V_L/V_B$

11.5.2.3 Breakthrough Curve Analysis for GAC Reactors

Pollutants are separated from the wastewater as it travels along the GAC reactor. Contaminants are mainly adsorbed in Region A, also known as mass transfer region (MTZ) (Fig. 11.6). This specific region/zone travels down as the GAC reactor continues to function. Above this region is known as "Zone-Saturated", which is packed and saturated due to pollutants and the inlet concentration S_0 is in equilibrium state with the saturated zone.

Continuous flow adsorption through fixed bed or column dynamics, in comparison to sorption kinetics, is a comprehensive phrase that relates to the dynamic responses of sorption process in the GAC adsorbers. Contaminant concentrations within a GAC

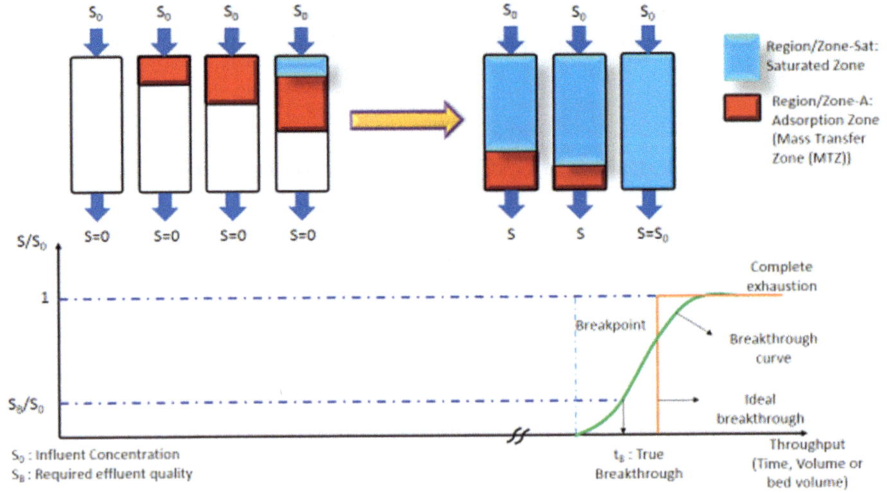

Fig. 11.6 Breakthrough curve analysis of a fixed bed GAC adsorber/reactor

adsorber vary with time and reactor length. The wave front of the GAC adsorber is the sector of substantial, quick changes in concentration along the length z. However, it is more important to monitor the concentration at the adsorber's exit or the concentration of the effluent, S than it is to monitor the concentrations within the bed. The 'breakthrough curve' is generated by plotting outlet/effluent concentration divided by inlet concentration (S/S_0) with respect to time interval, handled volume of water or the bed volume. The 'breakthrough point' is specified as the point at which a particular concentrations (SB) of the pollutants emerges. This point is somewhat arbitrarily chosen in accordance with the purification criteria (Jones et al. 2010).

Essentially, the breakthrough patterns of a GAC adsorber purifying the wastewater are determined by the equilibrium adsorption requirements and limitations in the mass transfer (Jones et al. 2010). In an idealistic situation, where mass transportation proceeds at an infinite speed, as can be seen in Fig. 11.1, the elevation of such mass transfer region approaches to zero. Under these circumstances, the appropriate breakthrough curve might resemble with Fig. 11.4 and can be simply predicted through using data from equilibrium state. In this situation, the breakthrough time will be equivalent to the GAC adsorber's saturation time.

However, as illustrated by Fig. 11.6, breakthrough curves appeared to be dispersed which vary from the standard one. This types of breakthrough curves appeared to be dispersed which vary from the standard one. This is because adsorption kinetics determines the altitude of the adsorption region (Zone-A) and the slope of the breakthrough curve. As discussed earlier, adsorption kinetics is dependent on both external and intraparticle diffusion rates (pore or surface diffusion). When mass transfer towards the internal layer of the adsorbent particles of the GAC reactor is slower, a flattened breakthrough curve which deviates considerably from the optimum breakthrough is evidenced. In this situation, the quantity of mass adsorbed till breakthrough is greater than the adsorber's saturation capacities, and a greater part of the adsorber's GAC continues to remain unusable. The gradient of concentration inside the reactor and the form of the breakthrough profiles are affected by the independent factors indicated in Fig. 11.7. For instance, the length of the adsorption Zone-A, or mass transfer region, is influenced by the chemical characteristics of the substance present in MTZ. The values for MTZ region for easily adsorbed chemicals are lower than those for complex and challenging substances. Additionally, a compound's concentration is critical since superior concentration lead to greater diffusion gradients which will improve the mass transfer. Furthermore, the form of an isotherm profile has a significant effect on the pattern of the breakthrough curve (Chowdhury et al. 2013; Çeçen and Aktaş 2011).

If single solute system is considered, it is quite simple to estimate breakthrough patterns using a GAC reactor. However, single solute adsorption occurs rarely, if ever. In comparison, when numerous chemicals are removed from the wastewater via adsorption, competitions for adsorption sites onto the activated carbon prevails. For the bisolute system, the inclusion of these kind of chemicals with varying absorbability has a significant effect on the geometry of the breakthrough curves in a GAC reactor. This results in the partitioning of less likely and highly adsorbing constituents into their corresponding mass transfer zones in a fixed-bed process. In comparison to

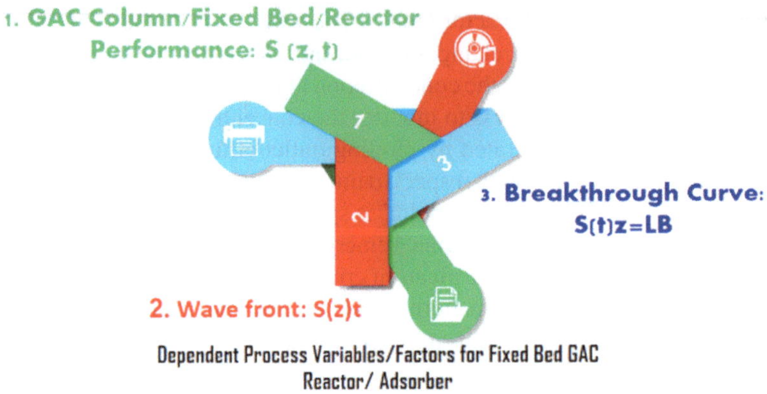

Fig. 11.7 Independent and dependent variables for GAC based column/fixed bed sorption

powerfully adsorbing chemicals, inadequately adsorbing compounds are retained in the column's deeper regions. Non-adsorbable chemicals which cannot be retained on activated carbon move fast across the filter carbon bed. Subsequently, it released into the outlet watercourse. As a result, breakthrough patterns might significantly differ from the contour depicted by Fig. 11.6.

11.6 Application of Activated Carbon (AC) in Secondary and Tertiary Phase Treatment for Purification of Industrial Waste Water

11.6.1 Application of PAC in Secondary and Tertiary Phase Treatment

Wastewaters having high concentrations of bio-degradable organic chemicals exhibit tendency for microbial growth. In that case, application of PAC suspended inside the aqueous stream is often more beneficial than fixed, immobile AC bed. In this context, PACT is directly suspended inside the biological units for concurrent adsorption and biological materials expulsion. PACT is frequently used for the secondary treatment method for industrial waste effluents and landfill residues. There are three types of PACT system: aerobic, anaerobic, and successive aerobic-anaerobic PACT reactor as illustrated by Fig. 11.8. PACT working under aerobic condition is mostly used to cleanse comparatively dilute waste stream, while anaerobic type PACT is used to address heavily contaminated wastewaters. The PACT using anaerobic process may occasionally pave the way for aerobic system of PACT, focusing on the quality of the waste effluents.

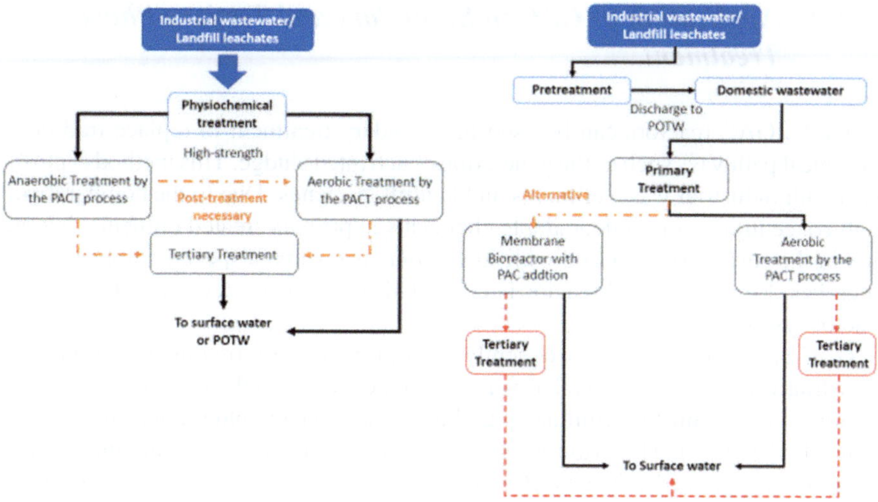

Fig. 11.8 Application of PACT in secondary and tertiary treatment of waste effluents

Additionally, PAC sorption can be used as a tertiary phase of pre-treatment to eradicate organics that remain in urban or industrial waste effluents following secondary filtration process. Separating PAC from treated water, on the other hand, is regarded as a difficult task. Separation can be accomplished through settling accompanied by sand filtering or through membrane processes, which takes extra energy (Çeçen and Aktaş 2011). Additionally, organics are eliminated more slowly through PAC adsorption than through other tertiary phase techniques, like Advanced Oxidation Techniques (AOPs) and ozonation. Thus, the unit containing PAC with a hydraulic retention time (HRT) in minutes or less, equilibrium between the surface of PAC and contaminants is generally not achieved, resulting in inefficient utilisation of PAC capacities.

To avoid these shortcomings, a better option is to incorporate PAC into secondary treatment concurrently. Powdered activated carbon (PAC) is kept inside the reactor for a certain period of contact time equivalent to sludge age also referred as sludge retention time (SRT). The capacity of PAC will be completely utilized due to the high rate of retention within the reactor. Additionally, the existence of microbes with PAC inside the system simultaneously results insynergistic outcomes. Furthermore, in a PACT technique, PAC exits the reactor with the discarded sludge, and additional separation step is not required.

11.6.2 Application of GAC in Secondary and Tertiary Phase Treatment

Similarly, GAC reactors can be used in secondary treatment to replace traditional biological pathways such as the generation of activated sludge. This method is mostly used with industrial waste effluents and landfill leachates. Due to the comparatively high concentration of biodegradable chemicals in primary treated effluents, interaction of the surface of the GAC with such effluents results in speedy colonisation by microbes. Finally, a GAC reactor is converted into a BAC reactor, that is, biofilm reactor system.

GAC based adsorption is frequently used in the tertiary treatment of urban and industrial waste stream, as well as waste leachates, where only a minimal amount of organic material must be eliminated by biodegradation and plugging difficulties are avoided (Fig. 11.9). The primary objective is to remove trace organic matter from the secondary treated effluents. If a GAC reactor receives secondary effluents that contain bacteria, conversion to a BAC reactor is quite feasible.

Wastewater passes over fixed, extended, or fluidized beds into the tertiary GAC sorption systems. AC segregation issues associated with PAC sorption are solved by employing these GAC beds. Simultaneously, the transit of wastewater via GAC beds enables maximum utilisation of the capacity of the carbon. After secondary phase of treatment of industrial or urban waste stream, adsorption as well as ozonation of contaminants are often used methods for removing remaining contaminants. These oxidative procedures, most notably ozonation, are routinely used prior to GAC adsorption. Generally, these procedures result in an increase in the biodegradable component in waste stream. As a result, they promote the transformation of GAC towards BAC.

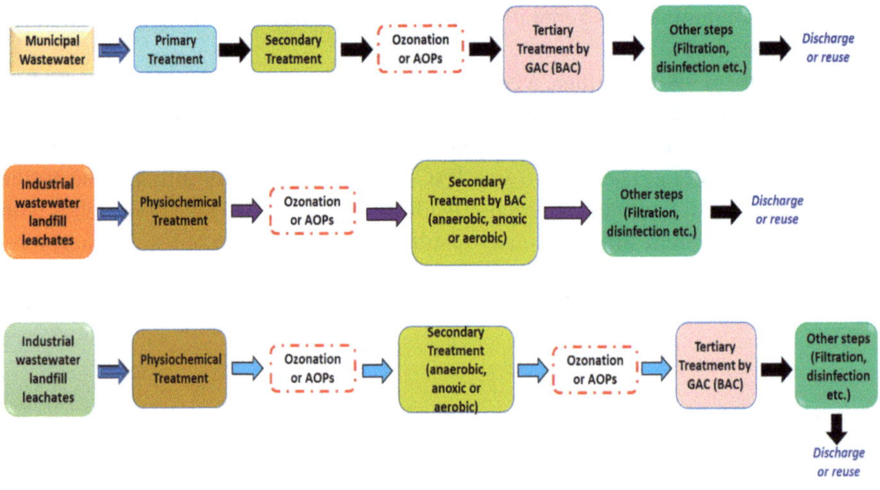

Fig. 11.9 Application of GAC in secondary and tertiary treatment of waste water

11.7 Integration of Biological System with Activated Carbon: Mechanism

11.7.1 Retention Properties of Non-biodegradable and Slow Biodegradable Organic Molecules Over the Surface of AC

In a biological system, the existence of AC offers porous texture with enhanced surface area for the sorption of organic chemicals. Hydrophobic chemicals having a log Kow <2 value are preferred for adsorption onto the surface of carbon. Surface phenomenon of adsorption is particularly advantageous for the biodegradation of organic materials that are gradually biodegradable or appear to be non-biodegradable. Numerous substances present in urban, industrial, and residential wastewaters, as well as landfill leachates, belong into such groups. Adsorption of these chemicals onto carbon would likely reduce their inhibitory properties in the feed liquid. Apart from the original organic products, AC is also capable of adsorbing the intermediates and/or the final yields/chemicals of biodegradation process (Yousef et al. 2020).

However, addition of too much organic material in waste effluent reduces activated carbon's adsorption capability for a particular contaminant. Additionally, it should be recognised that no pollutant can subsists in isolation in waste effluents. As a result, competition for adsorbent surface onto the AC may exist between different types of particular contaminants. Adsorption onto AC significantly increases the retention period of an organic molecule in a biological process. For instance, in the PACT process, dissolved organic substances adsorbed on AC surface will be held in the systems for around 10 days to 50 days, but in traditional biological systems, these can be retained approximately for 6–36 h (Lankford and Eckenfelder 1990).

11.7.2 Retaining Properties of Toxic Substances Over the Surface of AC

Certain organic and inorganic chemicals can be harmful or hinder biological activities. These negative impacts are frequently observed during treatment of wastewater, leachate, or even in case of POTWs that receive a considerable amount of such wastes. In general, a pollutant can hinder its very own elimination by substrate resistance or by competing and non-competitive inhibition effect on some other chemicals.

The inhibitory properties of a contaminant in a biological process cannot be predicted by a single variable alone; rather, studies must be undertaken to anticipate the nature of inhibition and the restrictive ranges. Additionally, the impact of a contaminant varies by system. For example, it is well established that the majority of organic contaminants have a greater detrimental effect on nitrification compared to organic carbon elimination. However, the majority of organic contaminants that

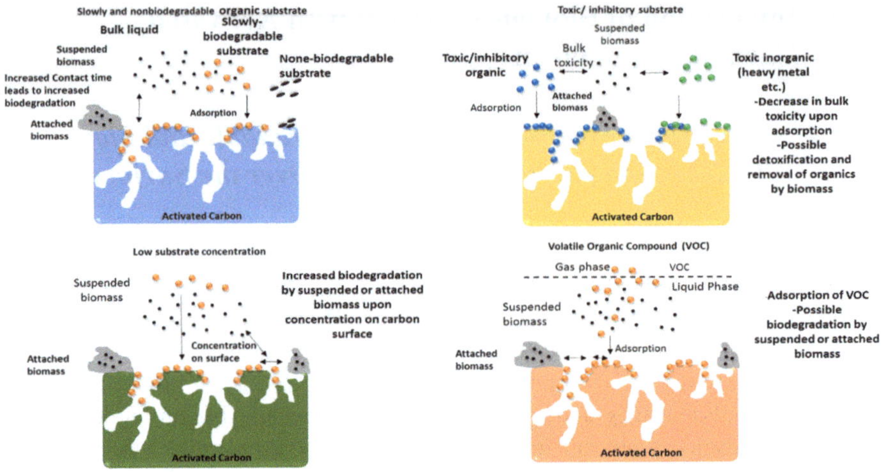

Fig. 11.10 Mechanistic evaluation of integrated adsorption and biological systems for **a** slowly and non-biodegradable substrate, **b** toxic/inhibitory substrate, **c** low substrate concentration, and **d** volatile organic compound (VOC)

are inhibiting or hazardous are susceptible to sorption onto AC (Fig. 11.10). Their sorption from bulk water over the surface of AC protects the biomass that remains suspended indefinitely or adhered inside the biological reactor. Additionally, microbiological processes can takes place onto the surface of AC, converting a particular hazardous organic contaminant to a harmless one. This positive impact of AC is amplified in BAC systems containing bacteria adhering to the surface of GAC.

Numerous approaches may be employed to determine a particular chemicals inhibitory function in the presence and absence of AC. Monitoring the Precise Uptake rate of Oxygen (*sour*) is considered as a standard approach for determining the inhibitory effect of inorganic and/or organic compounds in wastewater (Sher et al. 2000). Phenolic compounds are recognised for its inhibitory influence on biochemical procedures and has thus frequently being used as a typical molecule in studies of AC's influence on biochemical processes. Thus, for instance, adding activated sludge with PAC permitted uninterrupted supply at inlet having phenol concentrations around 1000 ppm (Meidl 1997). Metallic cations-mainly containing heavy metals and inorganic negative ions are the most abundant class of pollutants which have a detrimental impact on the biochemical activities. Calcium, lead, copper, cadmium, zinc, nickel, and chromium, for example, can impede or entirely block microbial metabolism and growth when they interact with enzymes (Aktaş and Çeçen 2007). Heavy metals, can also be trapped by AC surface, as illustrated in Fig. 11.10. Due to activated carbon's susceptibility to adsorb hazardous heavy metals, precipitation of metals before the biological treatment may not be necessary (Aktaş and Çeçen 2007).

11.7.3 Substrate Concentration Over the Surface of AC

The majority of organic contaminants detected in waste effluents are now present at extremely low concentrations. Most substrates can only be utilised as secondary feeds under such circumstances because microbes are incapable to exploit them as primary carbon and energy resources. In that perspective, the inclusion of AC inside the biological setup will be beneficial as it will provide substantial porous texture with enlarged surface area onto that the substances can be concentrated during the adsorption process (Fig. 11.10). As a result, AC is especially beneficial for diluted waste streams (Aktaş and Çeçen 2007). Moreover, based on the category of AC and the surroundings, AC can concentrate other components such as nutrients and oxygen over its surface. This abundance of diverse substrates facilitates biodegradation even more (Aktaş and Çeçen 2007).

11.7.4 Retaining Properties of Volatile (VOCs) Chemicals Over AC

Regardless the level of adsorption, VOC's retention onto the carbon surface prohibits it from being emitted to the air. Furthermore, if a VOC is biodegradable, its persistence on the surface of the carbon enhances the likelihood that it will be degraded by dispersed or associated biomass (Aktaş and Çeçen 2007).

11.7.5 Growth of Microbes Over the Surface of AC

One of the greatest significant properties of AC is that; it also offers an enlarged surface area for microbes to adhere with it. Surface of AC have reactive groups that promote microbe adhesion. Additionally, it protects microbes from hydrodynamic friction forces (Aktaş and Çeçen 2007). Stimulation of feedstock, food, and oxygen, for example, promotes microbial populations.

If adequate time is allocated for bacteria to become immobilised, biofilm formation takes place over the surface of AC, as illustrated by Fig. 11.10. In BAC reactors, biofilm development is very pronounced. Microbes can also populate over the PAC surface to a limit in suspended-growth setups as well.

11.7.6 Bio-regeneration of AC Surface

Bioregeneration is the process by which carbon's adsorptive potential is renewed by the activity of microbes to allow additional adsorption onto AC surface. Bioregeneration can be accomplished either through offsite blending or instantaneous mixing of bacteria and the exhausted/saturated AC (Silva 2004; Vinitnantharat et al. 2001). Initially, it was proposed that the elimination of inhibitory chemicals and the bioregeneration of AC were the two primary processes behind the synergistic effects detected in combined setups (Joss et al. 2006; Emanuelsson et al. 2006; Morgenroth 2008).

As illustrated in Fig. 11.10, AC can act as a storage facility for contaminant particles. Nevertheless, it is possible that bound molecules can desorb from carbon surface. This phenomenon takes place when solute concentration over the surface of the carbon and the aqueous media is reversed. Following desorption, the compounds may be eliminated via a biological procedure known as bioregeneration. Additionally, microbial enzymes released by microbes into the pores of carbon can promote external bioremediation of immobilized organic chemicals and initiate regeneration of AC. Thus, activated carbon bioregeneration is founded on three key mechanisms: adsorption, desorption, as well as biological elimination (Morgenroth 2008).

11.8 Conclusion

For environmental conservation and sustainable resource management, AC recycling or regeneration is essential (Chowdhury et al. 2011a, b; Bian et al. 2018; Bernhardt and Gyasi 2016). Once the contaminants have been extracted from the spent AC, it can be reused and regenerated approximately 80% with excellent efficiencies. Soft, powdered ACs have fewer particles after recycling as it becomes fragmented, whereas harder ACs could be reused several times because the rigid component is well preserved. Granular, powdered and activated carbon fibers are also used for water treatment (Rafique et al. 2015, 2018; Rafique and Lee 2014). Compared to granular activated carbon, activated carbon fiber (ACF) and powdered activated carbon is rather soft and difficult to regenerate. Thermal, chemical and steam treatments are used to reclaim and regenerate the exhausted AC (Shi et al. 2010; Mu'azu et al. 2017; El Gamal et al 2018; Li et al. 2015). Specifications about the procedure are as follows:

a. **Recycling using Thermal Treatment**:

The discharged AC is first rinsed and dried in an oven under vacuum. The contaminants are carbonised by heating the dehydrated AC at elevated temperature (Reza et al. 2020). The AC is subsequently renewed at temperatures ranging from 700 to 1000 °C to keep the pores intact and hollow (Radic et al. 2017; Ahmed 2017). Application of too high temperature can introduce additional proportion of ash residues in carbon-which is considered as impurities of carbon. However, modified ash is also used for purification of water containing heavy metals (Chowdhury et al. 2011a, b).

b. **Recycling using Steam Treatment**:

The AC filtration unit is constantly back washed with hot water. To eliminate the contaminants, the steam is then passed through the activated carbon at a temperature of 120–130 °C. Finally, the AC is re-flushed so that it can be utilized again.

c. **Recycling of AC using Chemical Treatment**:

To desorb or breakdown the sorbate from surface of AC, particular chemical solutions (acids and alkali) are employed. The efficiency is determined by the kinds of contaminants that is deposited on AC. Once the contaminants have been eliminated by solvents, the chemicals must be discarded in order to rejuvenate the AC.

The activation of carbon does not result in any substantial human health hazard when it is touched by the skin. Slight irritation is likely may cause due to interacting with AC. Since unused AC is benign, it can be discarded in the same manner as other harmless wastes. However, spent AC may exhibit the characteristics of a contaminant and consequently must be disposed in accordance with existing regulations. It is necessary to exercise caution when reusing AC. The application of reactivated ACs in water purification must adhere to particular rules and norms. Researchers who work with AC that generate dust may consider using protective eyewear and face masks with dust filtration to minimise dust inhalation. If the elevated levels of vapours are absorbed in the AC, the temperature may rise, necessitating either partial wetting with water or concentration reduction by passing more air. To ensure safety and to protect environment and resources; the processing, storage, dumping, reusing, and regeneration of AC should be complied with applicable standard regulations (Danish and Ahmad 2018; Azimi et al. 2017).

References

Adebisi GA, Chowdhury ZZ, Alaba PA (2017a) Equilibrium, kinetic, and thermodynamic studies of lead ion and zinc ion adsorption from aqueous solution onto activated carbon prepared from palm oil mill effluent. J Clean Prod 148:958–968. https://doi.org/10.1016/j.jclepro.2017.02.047

Adebisi GA, Chowdhury ZZ, Abd Hamid SB, Ali E (2017b) Equilibrium isotherm, kinetic, and thermodynamic studies of divalent cation adsorption onto Calamus gracilis sawdust-based activated carbon. BioResources 12(2). https://doi.org/10.15376/biores.12.2.2872-2898

Adebisi GA, Chowdhury ZZ, Abd Hamid SB, Ali E (2016) Hydrothermally treated banana empty fruit bunch fiber activated carbon for Pb(II) and Zn(II) removal. BioResources 11(4). https://doi.org/10.15376/biores.11.4.9686-9709

Ahnert F, Arafat HA, Pinto NG (2003) A study of the influence of hydrophobicity of activated carbon on the adsorption equilibrium of aromatics in non-aqueous media. Adsorption 9:311–319

Ahmed A, Abu Bakar MS, Azad AK, Sukri RS, Phusunti N (2018) Intermediate pyrolysis of Acacia cincinnata and Acacia holosericea species for bio-oil and biochar production. Energy Convers Manag 176:393–408. https://doi.org/10.1016/j.enconman.2018.09.041

Ahmed MJ (2017) Adsorption of non-steroidal anti-inflammatory drugs from aqueous solution using activated carbons: review. J Environ Manag 190:274–282. https://doi.org/10.1016/j.jenvman.2016.12.073

Akinpelu AA, Ali ME, Johan MR, Saidur R, Chowdhury ZZ, Shemsi AM, Saleh TA (2019) Effect of the oxidation process on the molecular interaction of polyaromatic hydrocarbons (PAH) with carbon nanotubes: adsorption kinetic and isotherm study. J Mol Liq 289:111107. https://doi.org/10.1016/j.molliq.2019.111107

Akinpelu AA, Chowdhury ZZ, Shibly SM, Faisal ANM, Badruddin IA, Rahman MM, Amin MA, Sagadevan S, Akbarzadeh O, Khan TMY, Kamangar S, Khalid K, Saidur R, Johan MR (2021) Adsorption studies of volatile organic compound (naphthalene) from aqueous effluents: chemical activation process using weak Lewis acid, equilibrium kinetics and isotherm modelling. Int J Mol Sci 22(4):2090. https://doi.org/10.3390/ijms22042090

Akinpelu AA, Ali ME, Owolabi TO, Johan MR, Saidur R, Olatunji SO, Chowdbury Z (2020) A support vector regression model for the prediction of total polyaromatic hydrocarbons in soil: an artificial intelligent system for mapping environmental pollution. Neural Comput Appl 32(18):14899–14908. https://doi.org/10.1007/s00521-020-04845-3

Aktaş Z, Çeçen F (2007) Bioregeneration of activated carbon: a review. Int Biodeterior Biodegrad 59(4):257–272. https://doi.org/10.1016/j.ibiod.2007.01.003

Alhamed YA (2009) Adsorption kinetics and performance of packed bed adsorber for phenol removal using activated carbon from dates' stones. J Hazard Mater 170(2–3):763–770. https://doi.org/10.1016/j.jhazmat.2009.05.002

Azimi A, Azari A, Rezakazemi M, Ansarpour M (2017) Removal of heavy metals from industrial wastewaters: a review. ChemBioEng Rev 4(1):37–59. https://doi.org/10.1002/cben.201600010

Balasundram V, Ibrahim N, Kasmani RM, Hamid MKA, Isha R, Hasbullah H, Ali RR (2017) Thermogravimetric catalytic pyrolysis and kinetic studies of coconut copra and rice husk for possible maximum production of pyrolysis oil. J Clean Prod 167:218–228. https://doi.org/10.1016/j.jclepro.2017.08.173

Bernhardt A, Gysi N (2016) World's worst pollution problems: the toxics beneath our feet. Fabrikstrasse 17, 8005, Zurich, Switzerland

Bian Y, Yuan Q, Zhu G, Ren B, Hursthouse A, Zhang P (2018) Recycling of waste sludge: preparation and application of sludge-based activated carbon. Int J Polym Sci 2018:1–17. https://doi.org/10.1155/2018/8320609

Chang H, Yuan XG, Tian H, Zeng AW (2006) Experiment and prediction of breakthrough curves for packed bed adsorption of water vapor on cornmeal. Chem Eng Process: Process Intensif 45(9):747–754. https://doi.org/10.1016/j.cep.2006.03.001

Çeçen F, Aktaş Ö (2011) Activated carbon for water and wastewater treatment. Wiley-VCH. ISBN: 978-3-527-32471-2

Cheremisinoff NP (2002) Handbook of water and wastewater treatment technologies. Butterworth-Heinemann, Woburn, MA, USA

Chowdhury Z, Krishnan B, Sagadevan S, Rafique R, Hamizi N, Abdul Wahab Y, Khan A, Johan R, Al-douri Y, Kazi S, Tawab Shah S (2018) Effect of temperature on the physical, electro-chemical and adsorption properties of carbon micro-spheres using hydrothermal carbonization process. Nanomaterials 8(8):597. https://doi.org/10.3390/nano8080597

Chowdhury ZZ, Pal K, Johan RB, Yehya Dabdawb WA, Ali ME, Faizur Rafique R (2017) Comparative evaluation of physiochemical properties of a solid fuel derived from Adansonia digitata trunk using torrefaction. BioResources 12(2). https://doi.org/10.15376/biores.12.2.3816-3833

Chowdhury ZZ, Yehye WA, Julkapli NM, Al Saadi MAH, Atieh MA (2016) Application of graphitic bio-carbon using two-level factorial design for microwave-assisted carbonization. BioResources 11(2). https://doi.org/10.15376/biores.11.2.3637-3659

Chowdhury ZZ, Hasan MR, Abd Hamid SB, MarlinaSamsudin E, Zain SM, Khalid K (2015) Catalytic pretreatment of biochar residues derived from lignocellulosic feedstock for equilibrium studies of manganese, Mn(II) cations from aqueous solution. RSC Adv 5(9):6345–6356. https://doi.org/10.1039/c4ra09709b

Chowdhury ZZ, Sharifuddin MZ, Rashid AK, Khalid K (2012) Application of Response Surface Methodology (RSM) for optimizing production condition for removal of Pb (II) and Cu (II) onto kenaf fiber based activated carbon. Res J Appl Sci Eng Technol (RJASET) 4(5):458–465

Chowdhury ZZ, Zain SM, Rashid AK, Rafique RF, Khalid K (2013) Breakthrough curve analysis for column dynamics sorption of Mn(II) ions from wastewater by using Mangostana garcinia peel-based granular-activated carbon. J Chem 1–8. https://doi.org/10.1155/2013/959761

Chowdhury ZZ, Zain SM, Rashid AK, Ahmed AA (2011a) Equilibrium kinetics and isotherm studies of Cu (II) adsorption from waste water onto alkali activated oil palm ash. Am J Appl Sci 8(3):230–237. https://doi.org/10.3844/ajassp.2011.230.237

Chowdhury ZZ, Zain SM, Rashid AK, Khalid K (2011b) Linear regression analysis for kinetics and isotherm studies of sorption of manganese (II) ions onto activated palm ash from waste water. Orient J Chem 27(2):405–415

Danish M, Hashim R, Ibrahim MNM, Rafatullah M, Ahmad T, Sulaiman O (2011) Characterization of Acacia mangium wood based activated carbons prepared in the presence of basic activating agents. BioResources 8(3):3019–3033

Danish M, Hashim R, Ibrahim MM, Sulaiman O (2013) Effect of acidic activating agents on surface area and surface functional groups of activated carbons produced from Acacia mangium wood. J Anal Appl Pyrolysis 104:418–425. https://doi.org/10.1016/j.jaap.2013.06.003

Danish M, Ahmad T (2018) A review on utilization of wood biomass as a sustainable precursor for activated carbon production and application. Renew Sustain Energy Rev 87:1–21. https://doi.org/10.1016/j.rser.2018.02.003

Din MI, Ashraf S, Intisar A (2017) Comparative study of different activation treatments for the preparation of activated carbon: a mini-review. Sci Prog 100(3):299–312. https://doi.org/10.3184/003685017X14967570531606

Emanuelsson MA, Henriques IS, Ferreira Jorge RM, Castro PM (2006) Biodegradation of 2-fluorobenzoate in upflow fixed bed bioreactors operated with different growth support materials. J Chem Technol Biotechnol 81(9):1577–1585. https://doi.org/10.1002/jctb.1565

El Gamal M, Mousa HA, El-Naas MH, Zacharia R, Judd S (2018) Bio-regeneration of activated carbon: a comprehensive review. Sep Purif Technol 197:345–359. https://doi.org/10.1016/j.seppur.2018.01.015

Hansen BR, Davies SR (1994) Review of potential technologies for the removal of dissolved components from produced water. Chem Eng Res Des 72:176–188

Hendricks D (2006) Water treatment unit processes: physical and chemical. CRC Press, USA

Huang T, Zhou R, Cui J, Zhang J, Tang X, Chen S, Feng J, Liu H (2018) Fast and cost-effective preparation of antimicrobial zinc oxide embedded in activated carbon composite for water purification applications. Mater Chem Phys 206:124–129. https://doi.org/10.1016/j.matchemphys.2017.11.044

Hung Y, Lo HH, Wang LK, Taricska JR, Li KH (2005) Granular activated carbon adsorption. In: Wang LK, Hung Y, Shammas NH (eds) Physicochemical treatment processes. Handbook of environmental engineering, vol 3. The Humana Press Inc., Totowa, New Jersey, USA

Inglezakis VJ, Poulopoulos SG (eds) (2006) Adsorption, ion exchange and catalysis: design of operations and environmental applications. Elsevier, Amsterdam The Netherlands

Jones RA, Thibault J, Tezel FH (2010) Simulation and validation of ethanol removal from water in an adsorption packed bed: isotherm and mass transfer parameter determination in batch studies. Can J Chem Eng. https://doi.org/10.1002/cjce.20337

Joss A, Zabczynski S, Göbel A, Hoffmann B, Löffler D, McArdell CS, Ternes TA, Thomsen A, Siegrist H (2006) Biological degradation of pharmaceuticals in municipal wastewater treatment: proposing a classification scheme. Water Res 40(8):1686–1696. https://doi.org/10.1016/j.watres.2006.02.014

Julien F, Baudu M, Mazet M (1998) Relationship between chemical and physical surface properties of activated carbon. Water Res 32(11):3414–3424. https://doi.org/10.1016/s0043-1354(98)00109-2

Karimi S, TavakkoliYaraki M, Karri RR (2019) A comprehensive review of the adsorption mechanisms and factors influencing the adsorption process from the perspective of bioethanol dehydration. Renew Sustain Energy Rev 107:535–553. https://doi.org/10.1016/j.rser.2019.03.025

Karri RR, Jayakumar N, Sahu J (2017) Modelling of fluidised-bed reactor by differential evolution optimization for phenol removal using coconut shells based activated carbon. J Mol Liq 231:249–262. https://doi.org/10.1016/j.molliq.2017.02.003

Khalili NR, Campbell M, Sandi G, Golaś J (2000) Production of micro- and mesoporous activated carbon from paper mill sludge. Carbon 38(14):1905–1915. https://doi.org/10.1016/s0008-6223(00)00043-9

Lankford PW, Eckenfelder WW (1990) Toxicity reduction in industrial effluents. Van Nostrand Reinhold, New York, USA

Le-Minh N, Sivret EC, Shammay A, Stuetz RM (2018) Factors affecting the adsorption of gaseous environmental odors by activated carbon: a critical review. Crit Rev Environ Sci Technol 48(4):341–375. https://doi.org/10.1080/10643389.2018.1460984

Li Q, Qi Y, Gao C (2015) Chemical regeneration of spent powdered activated carbon used in decolorization of sodium salicylate for the pharmaceutical industry. J Clean Prod 86:424–431. https://doi.org/10.1016/j.jclepro.2014.08.008

Meidl JA (1997) Responding to changing conditions: how powdered activated carbon systems can provide the operational flexibility necessary to treat contaminated groundwater and industrial wastes. Carbon 35(9):1207–1216

Morgenroth E (2008) Biofilm reactors. In: Henze M, van Loosdrecht MCM, Ekama GA, Brdjanovic D (eds) Biological wastewater treatment, principles, modelling and design. IWA Publishing, pp 493–511

Mu'azu N, Jarrah N, Zubair M, Alagha O (2017) Removal of phenolic compounds from water using sewage sludge-based activated carbon adsorption: a review. Int J Environ Res Public Health 14(10):1094. https://doi.org/10.3390/ijerph14101094

Oladipo AA, Ifebajo AO, Nisar N, Ajayi OA (2017) High-performance magnetic chicken bone-based biochar for efficient removal of rhodamine-B dye and tetracycline: competitive sorption analysis. Water Sci Techno: J Int Assoc Water Pollut Res 76(2):373–385. https://doi.org/10.2166/wst.2017.209

Oladipo AA, Gazi M (2015) Microwaves initiated synthesis of activated carbon-based composite hydrogel for simultaneous removal of copper (II) ions and direct red 80 dye: a multi-component adsorption system. J Taiwan Inst Chem Eng 47:125–136. https://doi.org/10.1016/j.jtice.2014.09.027

Pam AA, Abdullah AH, Tan YP, Zainal Z (2018) Batch and fixed bed adsorption of Pb(II) from aqueous solution using EDTA modified activated carbon derived from palm kernel shell. BioResources 13(1). https://doi.org/10.15376/biores.13.1.1235-1250

Radic D, Stanojevic M, Obradovic M, Jovovic A (2017) Thermal analysis of physical and chemical changes occurring during regeneration of activated carbon. Therm Sci 21(2):1067–1081. https://doi.org/10.2298/tsci150720048r

Rafique RF, Min Z, Son G, Lee SH (2015) Removal of cadmium ion using micellar-enhanced ultrafiltration (MEUF) and activated carbon fiber (ACF) hybrid processes: adsorption isotherm study for micelle onto ACF. Desalin Water Treat 57(17):7780–7788. https://doi.org/10.1080/19443994.2015.1057538

Rafique RF, Lee S (2014) Micellar enhanced ultrafiltration (MEUF) and Activated carbon fiber (ACF) hybrid processes for the removal of cadmium from an aqueous solution. Korean Chem Eng Res 52(6):775–780. https://doi.org/10.9713/kcer.2014.52.6.775

Rafique RF, Chowdhury ZZ, Moon J, Lee S (2018) Application of micellar enhanced ultrafiltration (MEUF) and activated carbon fiber (ACF) hybrid processes for the removal of nickel from an aqueous solution. Int J Eng Technol 10:112–120

Razi MA, Al-Gheethi A, Al-Qaini M, Yousef A (2018) Efficiency of activated carbon from palm kernel shell for treatment of greywater. Arab J Basic Appl Sci 25(3):103–110. https://doi.org/10.1080/25765299.2018.1514142

Reza MS, Yun CS, Afroze S, Radenahmad N, Bakar MSA, Saidur R, Taweekun J, Azad AK (2020) Preparation of activated carbon from biomass and its' applications in water and gas purification, a review. Arab J Basic Appl Sci 27(1):208–238. https://doi.org/10.1080/25765299.2020.1766799

Regti A, Laamari MR, Stiriba S-E, El Haddad M (2017) Potential use of activated carbon derived from Persea species under alkaline conditions for removing cationic dye from wastewaters. J Assoc Arab Univ Basic Appl Sci 24:10–18. https://doi.org/10.1016/j.jaubas.2017.01.003

Sher MI, Arbuckle WB, Shen Z (2000) Oxygen uptake rate inhibition with PACTTM sludge. J Hazard Mater 73(2):129–142

Shi Q, Zhang J, Zhang C, Li C, Zhang B, Hu W, Xu J, Zhao R (2010) Preparation of activated carbon from cattail and its application for dyes removal. J Environ Sci 22(1):91–97. https://doi.org/10.1016/s1001-0742(09)60079-6

Silva M (2004) Preliminary feasibility study for the use of an adsorption/bio-regeneration system for molinate removal from effluents. Water Res. https://doi.org/10.1016/s0043-1354(04)00151-4

Snoeyink VL, Summers RS (1999) Adsorption of organic compounds. In: Letterman RD (ed) Water quality and treatment, 5th edn. McGraw-Hill, New York, NY

Suthersan SS (1999) Pump and treat systems. In: Suthersan SS, Raton B (eds) Remediation engineering: design concepts. CRC Press, LLC

Sun N, Okoye C, Niu CH, Wang H (2007) Adsorption of water and ethanol by biomaterials. Int J Green Energy 4(6):623–634. https://doi.org/10.1080/15435070701665396

Tan K, Hameed B (2017) Insight into the adsorption kinetics models for the removal of contaminants from aqueous solutions. J Taiwan Inst Chem Eng 74:25–48. https://doi.org/10.1016/j.jtice.2017.01.024

Thomas WJ, Crittenden B (1998) Adsorption technology and design. Elsevier Science and Technology Books. ISBN 0750619597

Veil JA (2011) Produced water management options and technologies. Produced water: Environmental risks and advances in mitigation technologies. Springer, New York, pp 537–571

Vinitnantharat S, Baral A, Ishibashi Y, Ha SR (2001) Quantitative bioregeneration of granular activated carbon loaded with phenol and 2,4-dichlorophenol. Environ Technol 22(3):339–344. https://doi.org/10.1080/09593332208618288

Worch E (2008) Fixed-bed adsorption in drinking water treatment: a critical review on models and parameter estimation. J Water Supply: Res Technol AQUA 57(3):171–183

Worch E (2009) Adsorptionsverfahren in der Trinkwasseraufbereitung. Wien Mitt 212:207–230

WWAP/UN-Water (2018) The United Nations world water development report 2018: nature-based solutions for water. 7, Place de Fontenoy, 75352 Paris 07 SP, France

Yonge DR, Keinath TM, Poznanska K, Jiang ZP (1985) Single-solute irreversible adsorption on granular activated carbon. Environ Sci Technol 19(8):690–694. https://doi.org/10.1021/es00138a006

Yonge DR, Keinath TM (1986) The effects of non-ideal competition on multi-component adsorption equilibria. J Water Pollut Control Fed 58:77–81

Yousef R, Qiblawey H, El-Naas MH (2020) Adsorption as a process for produced water treatment: a review. Processes 8(12):1657. https://doi.org/10.3390/pr8121657

Zbair M, Ainassaari K, Drif A, Ojala S, Bottlinger M, Pirilä M, Keiski RL, Bensitel M, Brahmi R (2017) Toward new benchmark adsorbents: preparation and characterization of activated carbon from argan nut shell for bisphenol A removal. Environ Sci Pollut Res 25(2):1869–1882. https://doi.org/10.1007/s11356-017-0634-6

Chapter 12
Thermochemical Conversion of Biomass Waste to Amorphous Phase Carbon for Treating Industrial Waste Water

Shajalal Md Shibly, Zaira Zaman Chowdhury, Abu Nasser Mohammad Faisal, Ahmed Elsayid Ali, Arnab Barua, Rahman F. Rafique, Rabia Ikram, Rafie Bin Johan, and Seeram Ramakrishnan

Abstract The Economic growth and industrial development of a country is leading to enhanced usage of chemicals and disposal of industrial and agricultural wastes to aquatic environment. This causes eco-toxicological hazards for environment. The ratio between fresh water to saline water in our planet is very low. Furthermore the pollution of fresh water is making the situation more critical. Water pollution needs to be addressed carefully to find out eco-friendly approach to purify the contaminated stream. Functionalized ligno-cellulosic fibre and its' derivatives can eliminate pollutant using biosorption mechanism, Biochar and activated carbon which is porous and amorphous can be considered as one of the best option to treat waste water. Activated carbon has suitable porosities with enlarged surface area and functional groups which helps them to entrap organic and inorganic pollutants from water. Activated carbon can be produced from low cost lingo-cellulosic waste using different types of thermochemical process to ensure targeted contaminants removal from waste water. The first part of the chapter overviews the recent advances in thermochemical processes; used to produce functionalized lingo-cellulosic fiber, biochar and activated carbon for water treatmentwhich can strengthen sustainable circular economy. The recycling and regeneration strategies with challenges and future prospects of these sorbents are illustrated in the concluding part of the chapter. Thus the ecosystem can be

---

S. M. Shibly · Z. Z. Chowdhury (✉) · A. N. M. Faisal · A. E. Ali · A. Barua · R. B. Johan
Nanotechnology and Catalysis Research Centre (NANOCAT), University of Malaya, 50603 Kuala Lumpur, Malaysia
e-mail: dr.zaira.chowdhury@um.edu.my

R. F. Rafique
John A. Reif, Jr. Department of Civil and Environmental Engineering, New Jersey Institute of Technology (NJIT), 323 MLK Blvd., New York, NJ 07102, USA

R. Ikram
Department of Chemical Engineering, University of Malaya, Kuala Lumpur, Malaysia

S. Ramakrishnan
Department of Mechanical Engineering, Centre for Nanofibers and Nanotechnology, National University of Singapore, Blk E3 #05-12, 2 Engineering Drive 3, Asia 117581, Singapore

S. Roy et al. (eds.), *Advanced Industrial Wastewater Treatment and Reclamation of Water*, Environmental Science and Engineering,
https://doi.org/10.1007/978-3-030-83811-9_12

protected by usage of lingo-cellulosic waste in multidimensional direction by purifying the waste water which can aid in solid biomass waste management up to a greater extent.

Keywords Activated carbon · Biosorption · Ligno-cellulosic waste · Thermochemical conversion · Water treatment · Aqueous pollutants

12.1 Introduction

The lack of comprehensive, advanced purification technologies for elimination of emerging toxins from waste water, aqueous pollution has become an inherent aspect; leading to environmental problem and economic crisis in developing countries. Thus, investigators have concentrated their efforts in current decades on discovering cost-effective strategies for alleviating environmental problems produced by untreated industrial effluents (Othmani et al. 2019). Numerous contaminants, such as carcinogenic metallic cations, anions, drugs, hormones, dyes and others salts have impacted negatively on human life and aquatic organism (John et al. 2018). Elimination of such hazardous chemicals is a critical requirement for environmental protection (Abdelwahab and Amin 2013). Over the last decade, researchers have identified numerous effective ways, including oxidation process, electrocoagulation (Othmani et al. 2017), reverse osmosis (Rajan et al. 2014), and anodic oxidation (Elaissaoui et al. 2016). However, the system's effectiveness cannot be determined solely by its ability to reduce contamination from water supply; it must also fulfil certain criteria such as affordable rates and ease of processing (Bhatnagar et al. 2015). Adsorption has a high capacity for pollutant removal from aqueous solutions (Gautam et al. 2014; Selmi et al. 2018a), enabling significant runoff volume reduction. For several millennia, commercial activated carbon has been the predominant sorbent utilised in water treatment (Liu et al. 2019). Indeed, adsorption on activated carbon has a number of advantages: it enables the removal of a broad variety of pollutants, including various types of organic and inorganic pollutants as well as can alter the composition-mainly taste and odour of water. Due to its relatively porous texture and enlarged surface area anchored with functional chemical groups, activated carbons are widely employed for contaminant removal (Bello 2013; Akinpelu et al. 2021; Chowdhury et al. 2016, 2018). Actually, the adsorption capacity of adsorbents is determined by a variety of parameters, including their textural qualities, which include porosity, surface area, pore size, and surface chemistry. These variables all play a significant role in pollutant adsorption effectiveness (Selmi et al. 2018b).

Nevertheless, parameters such as thermodynamic variables, agitation speed, solution pH, ionic strength, sorbent and sorbate characteristics, and contact time can all have an effect on adsorption capacity (Gallos et al. 2017). Activated carbon has demonstrated exceptional potential to eliminate various types of pollutants, validating its effectiveness as a promising adsorbent. Adsorption efficiencies onto ligno-cellulosic substrate are comparable to commercial activated carbon and are affordable. Recent statistics showed that, the annual production of ligno-cellulosic fibres is predicted to be between 10 and 50 billion tonnes [20]. The search for innovative activated carbon focusing on multiple biomass could lead to alternate approaches. Biosorption has stimulated a lot of interest in the filtration of industrial wastewaters (Kesraoui et al. 2015). It refers to the exploitation of biological samples for adsorption-based waste removal from aqueous effluents. Ligno-cellulosic biomass are organic skeletons composed primarily of polysaccharides. Polysaccharides are carbohydrate polymers of cellulose, hemicelluloses, lignin, and pectin with the general crude formula $Cx(H_2O)y$. Biosorbents are typically obtained from farms and industrial sites, particularly in the food, timber, and fisheries.

The potential and critical aspects for adsorptive removal of pollutants by ligno-cellulosic fibres and their multiple derivatives like biochar and activated carbon for purification of wastewater is discussed in this article. The surface chemistry of lignocellulosic fibre responsible for their biosorption capacity is also illustrated. The adsorption capabilities and the effect of operating conditions on process efficiency are also described. Future trends in the utilisation of ligno-cellulosic materials to enhance the purification process will be governed by the extent of forthcoming investigations.

12.2 Thermochemical Conversion of Ligno-Cellulosic Fibers: Critical Aspects and Effective Strategy for Strengthening the Sustainable Development of Green/Circular Economy

This survey is focused on the concept of a cyclical roll over of economy and the usage of biochar/AC to provide a practical solution for its effective management. During the pyrolysis and carbonization technique, the granules of organic compounds eventually breakdown into simpler molecules, which are expelled from the operation stream as fumes, condensable vapours (oils), and burnt ash. The quantity and quality of each finished product is dependent on the time, temperature, rate of heating, and mass and type of the precursors, as well as the furnace design and layout (Chowdhury et al. 2014; Yaashikaa et al. 2019, 2020). The thermo-chemical approaches used to produce biochar/AC, especially in rural communities, aid in the growth of that region and also assist medium and small enterprises in supplying sufficient energy, hence raising rural incomes and giving remedies for farming waste management. This allows the connection between small, micro supply chains to larger systems, for establishing closed loop concepts in which waste from one activity can be used as an intake for another, resulting in positive socioeconomic, and ecological implications in a circular, green economy.

Complementary connections between various biochar/AC generation and recycling technologies are needed for the generation of prospective customers. By utilising pollutants from one agricultural firm can shed light on removal of toxic pollutant. Subsequently after that, the spent adsorbent can be used by incorporating this soil amendment. Thus, a circular economy can be established, allowing the new implementation of process, as well as the possibility for the creation of new enterprises. Methodologies that bridge the gap between simplicity of usage, energy efficiency, and restrictive separations could be integrated into the nearby network for the practicable synthesis of biochar/AC (Chen et al. 2011). Ecological advantages include the following: Reduced greenhouse gas emissions; Economic advantage from cost savings associated with garbage disposal. This circular economy approach minimises wastes using a variety of procedures, thereby enhancing their potential. It demonstrates an opportunity for the development of a Circular Economy (CE) by combining novel methodology and advantageous tactical order to address the use of agricultural residues, by-products, and co-products. The suggested scheme might be a well-managed rural energy platform embedded in regional farming networks and compliant with circular economy principles. However, both anaerobic digestion and gasification frameworks with plantations supplying yield and business sector delivering food waste can generate electricity and fertiliser which can be used by the surrounding neighbourhood, rather than responding to social benefits associated with the transition from a straightforward to a circular economy (Dong et al. 2018; Dong et al. 2011; Beesley and Marmiroli 2011). Life cycle analysis (LCA) is a widely used and defined ecological evaluation technique for determining an item's sustainability impact. The findings indicated that the efficiency of pyrolysis can vary depending on a variety of factors, namely (a) agricultural residues, (b) Carbonization/pyrolysis technique, (c) co-products output, and (d) peripheral invention detection. Furthermore, the various life-cycle steps of bio-oil synthesis by pyrolysis have been divided into 5 stages: I feedstock preparation and drying, (ii) Reactor, (iii) extinguishing condensate, (iv) char recovery, and (v) heating cycle. The rationale for this concept was that a multi-phase paradigm would be usually beneficial for increasing productivity, upgrading activities, and sharing environmental loads rationally in order to achieve ecological benefits (Das et al. 2021a, 2021b) .

12.3 Source, Properties and Potential of Using Ligno-Cellulosic Fiber

Agricultural wastes plant residues are plentiful, renewable resources which are distributed on essentially every region. Their unique biochemical features, together with their minimal operating costs, promote the utilisation of such natural resources in different sector. The ligno-cellulosic residues can be used to resolve ongoing environmental issues effectively. These fibres have a high degree of variation and heterogeneity in their structures. Their analysis and modelling aid in the selection of preconditioning techniques that are most compatible with the subsequent modification or thermochemical transformation methods used to obtain the desired products, carbonaceous materials, energy, and bio-products. The major components of ligno-cellulosic fibres include cellulose, hemicellulose, lignin, and extractable compounds (Chowdhury et al. 2013; Fathy and El-Kady 2013). Cellulose and hemicellulose are composed of linear chain carbohydrates,

Cellulose is a homopolymer composed of D-glucose units that are joined together by -(1–4) links. It constitutes the bulk portion in plant cells. Cellulosic chain contains lignin which provides them mechanical and chemical stability. Hemicellulose polymers are crosslinked and amorphous in structure, whereas cellulose polymers are linear and crystalline. Lignin serve as an adhesive between the cellulose fibrils and the hemicellulose matrix. Hemicelluloses are connected using hydrogen bonds with cellulose fibrils. The aromatic part of lignocellulosic fibres is composed of lignin. Its structure is complicated and varied, and it is entangled with other cellulose and hemicellulose polymers. These three types of macromolecules, in combination with lignin, form the cell walls of the plant species. Lignin gives the plant fibre its stiffness (Hokkanen et al. 2016). The extractable materials composed of waxes, different fats, resins, tannins, starches, and colours, and other substances (Haldar and Purkait 2020).

Ligno-cellulose fiber contains crystalline and amorphous region. Hydrolysis proces using acid can yield microcrystalline cellulose having crystallinity between 5 and 80%. (Daneshvar et al. 2017). Similarly, Nanocrystalline cellulose can be synthesised using acid hydrolysis, and the critical steps for its extraction are identified as separation of acid molecules after the hydrolysis process and ensure uniform dispersion using mechanical dispersion methods (Brinchi et al. 2013). However, nanocrystalline cellulose generated using Ammonium-persulfate treatment demonstrated an outstanding alternative technique due to its consistent size (Leung et al. 2010). Apart from Microcrystalline cellulose, Nanocrystalline cellulose is shown to be advantageous due to the simplicity with which the surface can be modified by hydroxyl groups (Lam et al. 2012). Nanocrystalline cellulose can be functionalized with Au, Pt, Ag, Pd, and Cu and was observed to have a number of beneficial uses, including enzyme immobilisation, drug delivery, medicinal applications, and bio-sensors (Lam et al. 2012).

12.4 Ligno-Cellulosic Biomass-A Potential Candidate for Biosorption

Adsorption is a physical process in which the ions or molecules referred as (gas or liquid) adsorbate are firmly affixed over the solid surface termed as adsorbent in a somewhat or less reversible manner. Adsorption is classified into two types, as illustrated in Table 12.1. (a) Physisorption, in which the adsorbate and adsorbent atoms are attached using inter-molecular attractive forces (Hydrogen bonding,

Table 12.1 Comparative discussion regarding physical and chemical adsorption

Properties	Physical adsorption	Chemical adsorption
Temperature	With increasing temperature decreases	With increasing temperature increases
Sites for adsorption	Non-specific	Specific
Layer of coverage	Multi-layer	Mono-Layer
Nature of sorption	Reversible	Irreversible
Desorption	Easier with high temperature and pressure	Difficult as chemical interaction

Van Der Waals) or electrostatic attraction (Vo et al. 2019). This types of physical bonding is relatively weaker, and the adsorbed molecules are arranged in many layers. (b) Chemisorption, in which a strong, chemical reaction between the adsorbate and the adsorbent is initiated (covalent bonds or purely ionic bonds). These molecules attach with the surface by chemical bonds or ion-exchange interactions or can form surface complexes. The capacity of lignocellulosic fibres to absorb pollutants and their efficiency are reliant on their chemical structure, the biosorption influencing parameters, the surface chemistry, the chemical and physical characteristics, and the affinity between active available sites over the surfaces of the biosorbent and the pollutants (Chojnacka 2010). The energy of the adsorption mechanism can be used to determine the sort of adsorption procedure being used throughout the treatment process. The energy of adsorption can be calculated using the Dubinin-Radushkevich isotherm. If the energy value is 206 E8 KJ/mol, the adsorption process is classified as physisorption. However, when the energy value is 8E12 KJ/mol, the adsorption mechanism is chemisorption controlled by the ionic exchange process (Abdolali et al. 2014). Their attraction is stronger, and the adsorbed molecules are arranged in mono layer (Ali et al. 2012). Biosorption, as a result of these findings, can be regarded a potential strategy for eliminating a variety of harmful contaminants. The majority of contaminant removal percentages are greater than 50%, indicating the efficacy of the biosorbent utilised for water decontamination. A basic understanding of the adsorption mechanisms is critical for evaluating the aspects that can affect the sorption process. Basically, adsorption process is a complex phenomenon and it takes place via complexation, ion exchange, adsorption onto the surface and pores, micro precipitation, and at the final stage to some extent absorption (Hokkanen et al. 2016). Adsorption is governed by either physical or chemical forces.

During the adsorption process, four distinct mechanisms takes place: Electrostatic attraction I; hydrogen bonding is mechanism II; electron donor–acceptor mechanism III; and iV- interaction is mechanism (Tran et al. 2017). The adsorption of anionic dyes on a variety of cellulosic biomasses revealed that hydrogen bonding plays a significant role in the dyes' adsorption. The biosorption mechanism is also greatly regulated by the lingo-cellulosic fiber's chemical composition, particularly for the –OH groups found in cellulose and hemicellulose, as well as the aromatic structures found in lignin.

12.5 Physiochemical Properties of Sorbent Affecting Adsorption

Biosorption onto Ligno-cellulosic residues or adsorption onto its derivatives like biochar or activated carbon is not a consistent phenomenon but is somewhat depending on the multiple parameters enlisted below (Cheremisinoff 2002).

12.5.1 Surface Area

In general, the magnitude of sorption should be proportionate with the particular surface area. The precise/active surface area of a substance is the fraction of its overall surface area which is capable of binding during sorption leading to removal process. The more densely fragmented as well as porous sorbent, can provide more adsorption amount of pollutant per unit mass of it. External surfaces feature bulges or apertures with a width bigger than their depth, while interior surfaces have cracks and cavities-often referred as pores; with a depth bigger than their width.

12.5.2 Physiochemical Properties of the Adsorbate/Pollutants

12.5.2.1 Molecular Weight and Solubility

Increasing the weight of the molecules and the presence of specific functional chemical groups such as halogens and double or triple bonds increases the tendency to adsorb (Cheremisinoff 2002). Adsorption of larger molecules onto AC is superior to the smaller ones. The solubility of the solution is also critical for adsorption. The extent to which a solute adsorbs is inversely proportional to its solubility in the liquid from which the sorption takes place. High solubility indicates that the solute–solvent bonding is strong enough than the solute–sorbent attraction forces (Hung et al. 2005).

12.5.2.2 Polarity

Polar type pollutant solutes are more readily adsorbed by polar sorbents, while nonpolar solutes are more quickly adsorbed by nonpolar sorbents. Ac is more effective in adsorbing nonpolar molecules than the polar ones. Increased solubility indicates a stronger affinity between the liquid, which acts in opposition to the carbon's attraction.

As a result, any change which can enhance the solubility may be coupled with a decrease in absorbability. As a result, polar groups (which have an attraction for water typically inhibit sorption from the aqueous phase.

12.5.2.3 Dissociation Constant

The dissociation constant controls the degree of ionisation. Adsorption of molecular and ionic forms is significantly different, with the former being far better absorbed than the later on. Increased dissociation constant leads to the increased concentration of the ionic state. This is not favourable for adsorption onto AC.

12.5.2.4 Substituted Groups

Organic compounds' sorption aptitude is also affected by the existence of substituent groups (Table 12.2). The addition of a 2nd or 3rd substitution frequently enhances the effects of the previous one. The strength of a substituted group is determined by its orientation, for instance, ortho, meta, or para (Hung et al. 2005).

In general, branching in chains favours the adsorption than the straight chains. On the contrary, when the chain length increases, the sorption capability increases (Hung et al. 2005). The spatial configurations groups in a compound have an effect

Table 12.2 Effect of substituted group in organic compound for determining its sorption aptitude (Hung et al. 2005)

Chemical substituted group with organic compounds	Sorption aptitude
NO_2 (Nitro)	Enhances the sorption aptitude
R- SO_3H (Sulfonic)	Adjustable influence based on host compounds
Double bonds	Adjustable influence based on host compounds
Halogenated group	Adjustable influence based on host compounds
NH_2 (Amino)	Reduce the sorption aptitude
$C = O$ (Carbonyl)	Adjustable influence based on host compounds
– OH (Hydroxyl)	Reduce the sorption aptitude

on its adsorbtion aptitude. In general, aromatic molecules tend to adsorb more than the aliphatic ones of same molecular sizes.

12.5.3 pH

At higher pH levels, organic molecules tend to form negative anions; at lower pH, they produce positive cataions; and at intermediary pH levels, they produce neutral compounds. At neutral circumstances, the sorption of the majority of organic compounds is greater. In general, the adsorption of organic contaminants by AC in the aqueous phase increases as the pH decreases. This happen as a consequence of the neutralisation of negatively charged ions at the carbon's surface at lower pH levels. Negative charges are neutralised, which minimizes the diffusion resistance and results in much more effective sorption sites. The magnitude of this impact varies according on the sort and method of activation of AC. pH variations may also occur as a result of acidic or basic surface groups on AC.

12.5.4 Temperature

Adsorption processes are exothermic. Thus, the rate of sorption should increase as the temperature decreases. However, increasing the heating rate, also increases the speed of diffusion of solute across the solution towards the sorption sites, resulting in higher adsorption. The function of temperature plays a significant influence in the sorption of solutes vs gases. Increased temperature increases a gas's aptitude to evacuate from the surface, hence decreasing the sorption. Nevertheless, in case of sorption from a liquid, temperature has a greater effect on solvent affinity (Hung et al. 2005).

12.5.5 Pore Size Distribution

Micropores are classified by the IUPAC as pores having width lesser than 2 nm, mesopores are pores having diameter of between 2 and 50 nm, and macropores are pores which have diameter larger than 50 nm. The classification outlined above is broadly acknowledged in the literature. Ultra-micropores, on the other hand, are pores with a width of lesser than 0.7 nm (). The sorption mechanism on the surface of macropores is identical to that on the smooth surfaces. Macroporous adsorbents have a very narrow active surface area. As a result, sorption on this pores are frequently overlooked. Monolayer and multilayer sorption occur sequentially inside mesopores, and adsorption takes place via the capillary condensation of the pollutant. Mesopores are fundamentally defined by their explicit surface area, volume and dimension of the pores. Mesopores and macropores are critical for pollutant molecule movement through micropores. Given that the majority of the total active surface area is contained within micropores; macropores provide a negligible amount towards the overall surface area. Thus, the involvement of micropores for the sorption should be greater (Cheremisinoff 2002).

12.5.6 Surface Functional Groups

AC synthesised by chemical activation methods have a lesser hydrophobic sites. It has more negative charge and porous surface. Chemical processing upsurges the quantity of acidic chemical groups on AC, but physical activation decreases the amount of acidic surface groups (Cheremisinoff 2002). Thermally synthesised AC has greater sorption capacity than chemically treated AC (Aktaş and Çeçen 2005). This is undoubtedly due to the activation process, which can result in somewhat variable surface morphology. For chemical activation, oxygenation of the carbon surface diminishes its attraction for simple aromatic molecules. Intense oxidation of AC reduces the adsorption of nitrobenzene, phenol, benzene, and benzenesulfonate significantly (De Jonge et al. 1996). The development of functional chemical groups is temperature dependent. Thus, for instance, lactonic and phenolic functional groups are formed at temperatures as high as 400 °C (De Jonge et al. 1996). For pH values less than PZC, the lignocellulosic fibres' surface is positively charged, allowing them to absorb the negatively charged anionic contaminant.

For pH values less than PZC, the lignocellulosic fibres' surface is positively charged, allowing them to absorb the negatively charged anionic contaminant. The surface of the lignocellulosic fibres is negatively charged if pH values greater than PZC, and thus it is capable of adsorbing the positively charged contaminant (Othmani et al. 2019b). As a result, the acid–base characteristics of lignocellulosic fibres have a significant effect on their ability to absorb contaminants. Typically, lignocellulosic biomass is taken from plants and treated with chemicals or nanoparticles to increase the adsorptive as well as catalytic effectiveness. Alternatively, by altering the extraction procedure, various forms of lignocellulose with distinct properties can be obtained. Lignocellulosic fibres can be converted to Hydrochar that is produced through hydrothermal carbonization (HTC).

Hydrothermal magnetization transformed 75% of the carbon material into magnetic sorbent. Magnatic HTC is capable of adsorbing both cationic and anionic dyes. This magnetised adsorbent exhibited excellent stability up to six cycles and exhibited pseudo-second-order reaction kinetics. Lignocellulose aerogels are a popular cellulose-based advanced material in the twenty-first century. Lignocellulose aerogels are produced by creating cellulose gels from biomass and drying them to produce aerogel.

12.6 Derivatives of Ligno-Cellulosic Biomass to Treat Industrial Effluents

12.6.1 Biochar

Lignocellulosic fibre can be transformed into biochar and at higher temperature to activated carbon, using pyrolysis. The biochar thus obtained can be incorporated with nanoparticles (NPs) to improve its removal efficiencies. The coupling of NPs with biochar is primarily dependent on the physicochemical features of the biochar. pH and functional groups of biochar is inter-related. As the temperature of the pyrolysis process is enhanced; the ash residues also enhanced to rise the pH of the biochar. Additional functional groups like carbonates and hydroxides contribute to the increase level of pH. Furthermore, it was revealed earlier that biochar can exhibit the buffering capability, which varied according to the starting precursors. Biochar made from different biomass source had different buffering pH range (Li et al. 2017). This pH variation results in the alteration of surface charge of the biochar, which has a major effect on metallic cations sorption. Earlier metallic ferrous ion was doped over biochar and the doping process was facilitated by chelation with oxygen-rich functional groups (Yuan et al. 2011).

The difficulties associated with nanoparticle usage is recovering the spent NPs from the waste effluents. This created a significant secondary pollution in terms of wastewater treatment. To address this issue, magnetic NPs were used. Compared to indigenous biochar, biochar treated with Fe_3O_4 exhibited improved crystal violet (CV) sorption (Sun et al. 2015). Nano-scale zero-valent iron is utilised in combination with biochar matrix for optimum pollutant degradation due to its high interaction efficiency and simplicity of separation through magnetic field (Yan et al. 2015; Li et al. 2020). Earlier, Carbon Quantum Dots using biochar as a carbon source was synthesizede, which delayed recombination of electron–hole in nanosheets of Carbon Quantum Dots/$Bi_4Ti_3O_{12}$. The plasma resonance of nanoparticles increased the photocatalytic performance of these fabricated nano-sheets (Wang et al. 2019).

12.6.1.1 Nanoparticles/Biochar Composites

Conversion of biochar into its nano form has attracted considerable interest in recently due to its special environmental properties. The amorphous texture of biochar contains condensed aromatic ring and the crystalline structure contains nonpolar and polar surface properties (Das et al. 2011; Das and Mukherjee 2012). Modification of Charcoal's surface with qualities such as hydrophilicity/hydrophobicity, volume of pores, surface active area, and charge of the surface can be enhanced for increased sorption of organic contaminants (Das and Das 2020). Numerous external entities have been utilised to modify biochar, including nano-zero-valent iron, zeolite, nano cluster metal oxides, and nano-sized silica. A composite hydrogel composed of acrylamide, chicken charcoal (AAm), and carbon black can be used as a prototype for the synthesis of silver NPs (Choppala et al. 2013). Combining nano dimensional silica crystals or/and zeolite mixed with biochar improves the physicochemical characteristics of the biochar, as well as the ability of the biochar to address inorganic and organic pollutants in soil and water systems (El-Naggar et al. 2018). The accumulationpotential and oxidation state of nano zero valent iron (nZVI) complicate the procedure of application (Das and Ghosh 2020). However, the coupled biochar-nZVI composite has a broader application and a reduced aggregation rate. Thus, biochar modification using nano ZVI, zeolite, and/or silica was able to alter the functional chemical groups of biochar, resulting in increased covalent bond formation, hydrogen bond formation, and π-π acceptor–donor interactions, and eventually increased the sorption (Gudade et al. 2016; Rajput et al. 2018). As a sustainable green process, nano ZVI-DBC may be used to remove chlortetracycline from aquatic ecosystems (Das et al. 2020a, b, c, d, e). For the elimination of nitrate, the biochar-nano zero-valent iron (nano ZVI-BC) composite was more efficient than other alternatives even when 100 mg/L chloride ion was present concurrently (Das 2019a; b; Lashari et al. 2015). In comparison to bulk biochar, nano-biochar (100 nm) has superior transportation into the ground water and even dispersion in natural soils (Lashari et al. 2015). However, biochar nanoparticles enhanced P leakage unacceptably in alkaline soils (Lehmann and Joseph 2009). Additionally, it has been found that adding biochar-carbon NPs combinations to soil dramatically increased the growth of wheat (Zhang et al. 2020).

12.6.1.2 Biochar/Hydrogel Composits

Hydrogel is a type of superior quality absorbent material that has the capacity to absorb and retain 1000 times its own weight of moisture within its cross-linked polymer network. It is compatible with a variety of polymers and can be used to develop biochar/hydrogel hybrids. Acrylamide is commonly used monomer in hydrogel polymerization, with a marketplace price of USD 1.0/kg. However, the capacity of poly-acrylamide hydrogel to entrap certain heavy metals is significantly less than that of inexpensive biochar (Das 2015a; b). As a result, the coupling of biochar and hydrogel may be a viable technology (Das and Mukherjee 2011). It is an alternate sorbent for toxic gas and metallic cations entrapment due to its enlarged surface area and porous structure It enhances the soil's cation exchange capability allowing harmful gas and metals to be retained. Biochar-hydrogel composites demonstrated significant possibilities for organic pollutant removal from air and wastewaters.

When coal fly ash is combined with biochar derived from rice husk, a hydrogel capable of removing cations can be generated (Gopi et al. 2020; Das et al. 2018a, b). As a hydrophilic, tunable, and miscible compound, biochar-hydrogel is a material with outstanding physicochemical characteristics (Das et al. 2018c). Due to its advantageous binding sites and increased hydrophilicity, the biochar-hydrogel composite has benefits over empty AAm-hydrogel. Mukherjee et al. (2020) reported that a customized composite of rice husk based hydrogel/biochar had a 99 percent and 35.75 mg/g zinc removal capacity (Mukharjee et al. 2020). It is advantageous to use a hydrogel matrix as a supportive component for the synthesis of Ag-loaded coal because hydrogels can form a three-dimensional complex of stretchable hydrophilic cross linked polymer strands for trapping. Hydrogel composite of poly(acrylamide) and poultry biochars (p(AAm)-CB) utilised to remove phenol from water (Foereid 2015). Biochars from poultry dung can be inserted into hydrogel frameworks. The resulting hybrid also has the capacity to remove harmful organic pollutants (Roy et al. 2018).

Singh et al. (2018) synthesised, characterised, and used a biochar-hydrogel matrix to remove arsenic from effluent (Singh et al. 2018). They discovered that the composite has the highest monolayer adsorption capacity for arsenic and it can be easily separated from the aqueous phase. The use of biochar-hydrogel composites may be a lucrative method for removing heavy metals from polluted water.

12.6.2 Activated Carbon

12.6.2.1 Thermochemical Conversion of Ligno-Cellulosic Waste for AC Preparation

Two fundamental procedures are followed for the production of AC. The first stage is known as carbonization, whereas the second phase is denoted as activation. To generate biochar, pyrolysis/gasification at an elevated temperatures under inert atmosphere is used (Odetoye et al. 2019). The composition of carbon in biochar will be increased in this stage by eliminating volatile materials via heat treatment (Radenahmad et al. 2020). In this step, the time, temperature, rate of heating rate, flow rate of nitrogen gas, are all critical factors. Due to the limited adsorption capacity of the obtained biochar, an activation step is required to increase the volume fraction, diameter of the pore and surface area (Yang et al. 2019). During the earliest stages of activation, the disordered carbon was removed, exposing the lignin to activating chemicals and forming the micro-porous texture. Overall the Thermochemical conversion process of biomass can be summarized by Fig. 12.1.

Finally, by burning the sidewalls between the channels, the pre-existing pores are enlarged to a great size. This increases the amount of intermediary pores and macropores while reducing the number of micropores. Based on the degree of activation, it can be a technique performed before or afterwards the carbonization with the purpose of removing the accumulated tarry compounds from biochar (Ukanwa et al. 2019).

Recently hydrothermal carbonization is used to pretreat the biomass under pressure at 200–300 °C in presence of water to obtaining hydrochar. Hydrochar is considered as a novel precursors for AC production. Torrefacation is carried out at 200–300 °C under inert atmosphere to yield char with poor surface area which needs

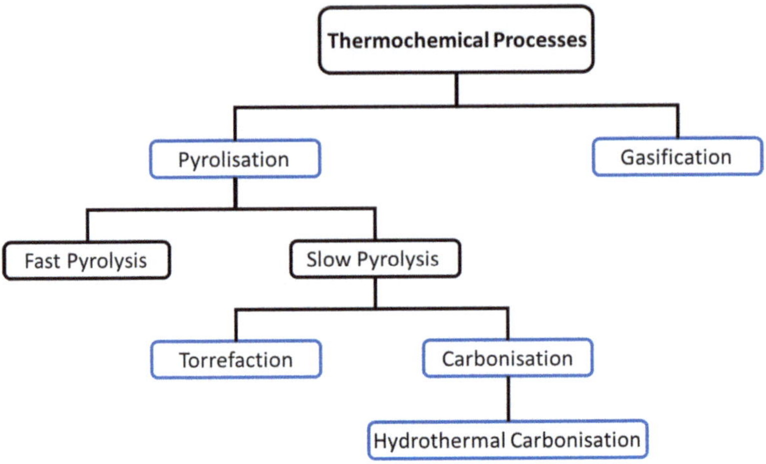

Fig. 12.1 Basic steps for thermochemical conversion of biomass

further activation. The advantages of Hydrothermal Carbonization and Torrefacation process are shown in Figs. 12.2 and 12.3.

Physical Activation

Activation of lignocellulosic fibres may be a one- or two-step process in case of physical adsorption process (Ao et al. 2018). Typically, it is a two-step process that begins with carbonization of dried specimens at 400–700 °C to form biochar and is accompanied by activation. The Activation process is carried out using an oxidising gas such as air, steam, argon, carbon di oxide (CO_2), and sometimes their blends at an elevated temperatures of about 800–1100 °C (Menya et al. 2018; Rashidi and Yusup 2017). Carbonized biochar usually has a surface area of lower than 300 m^2/gm (Ao et al. 2018).This reduced surface area and sorption capabilities are attributed to micro or meso-pores that have been blocked by tar compounds that can be eliminated through activation (Yahya et al. 2018).

Meanwhile, new crevices are observed on the surface of biochar, increasing the surface area and pore volume of the AC (Yahya et al. 2018). However, physical activation typically results in a lesser carbon yields of AC than chemical activation does, attributed to greater activating temperature and duration of the procedure (Rashidi and Yusup 2017).

Carbon di-oxide activation prioritises the emergence of new pores by expanding the existing pores, whereas steam activation favours the creation of mesopores and macro-pores from the start of the activating process at a constant rate of heating. This results in a more even distribution of pores on the final ACs (Gonzalez-Garcia 2018). Even with a higher rate of activation, this would result in a greater proportion of mesoporous and microporous texture for CO_2 and steam activation, correspondingly

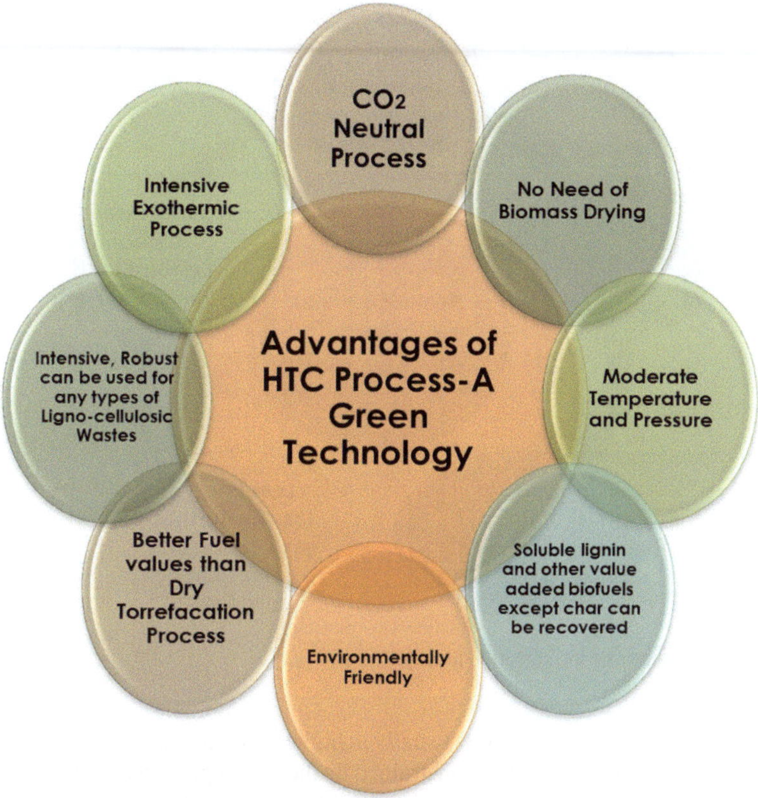

Fig. 12.2 Advantages of hydrothermal carbonization process

(Cao et al. 2006). In some circumstances, CO_2 activation seems to be the ideal method because it is cleaner, the gas is easier to manage, and the activation process is controlled even at higher temperatures of up to 800 °C due to the slower rate of reaction.

Chemical Activation

Activation using chemicals can be accomplished in a single phase or in two phases. Both procedures are comparable, with the exception of the carbonization step (Gonzalez-Garcia 2018). Carbonization does not occur in the single procedure and the dried substance is reactivated in the main phase by reacting with drying agents such as $ZnCl_2$, KOH, NaOH, and H_3PO_4. In double steps, char is produced at 400–600 °C first.

Various precursors respond variably based on the sort of dehydration agent used, resulting versatile morphological features with yields. However, base impregnation

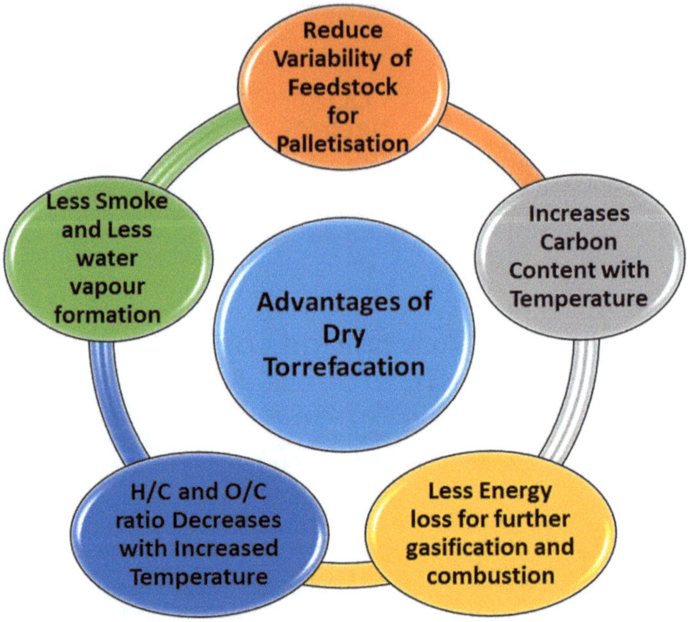

Fig. 12.3 Advantages of torrefacation process

method for activation of AC results usually in higher surface area of 2000 m^2/gm. The use of bases, on the other hand, is constrained by the sort of precursors. Base activation technique is more typically used in a double stage protocols because they function more successfully with the char (Cao et al. 2006). Char generated by carbonization of starting materials comprises a specific amount of pores that allow for the transport of drying agents and their subsequent reaction with carbons, resulting in the formation of ACs with enlarged surface area and porosity (Cao et al. 2006). Penetration using bases might have difficulty due to the low porosity of raw resources (Chowdhury et al. 2013). Impregnation by itself is inadequate to generate porous AC, requiring a two-stage chemical activation process (Isoda et al. 2014). Furthermore, biochars are believed to be substantially more responsive to activation using KOH, which may aid in the production of ACs with higher specific surface characteristics.

Chemical activation is preferred to physical activation. As it is more efficientfor developing the enlarged surface area, micro-pores, and increasing the carbon yield, as well as being commercially feasible due to the reduced temperature of activation and faster time for activation (Din et al. 2017). However, it has the disadvantage of leaving chemical compounds on char, which requires a subsequent cleaning stage to eradicate it completely (Din et al. 2017) or accompanied by physical activation of the char residues (Rashidi and Yusup 2017). Figure 12.4 summarizes different types of activation process.

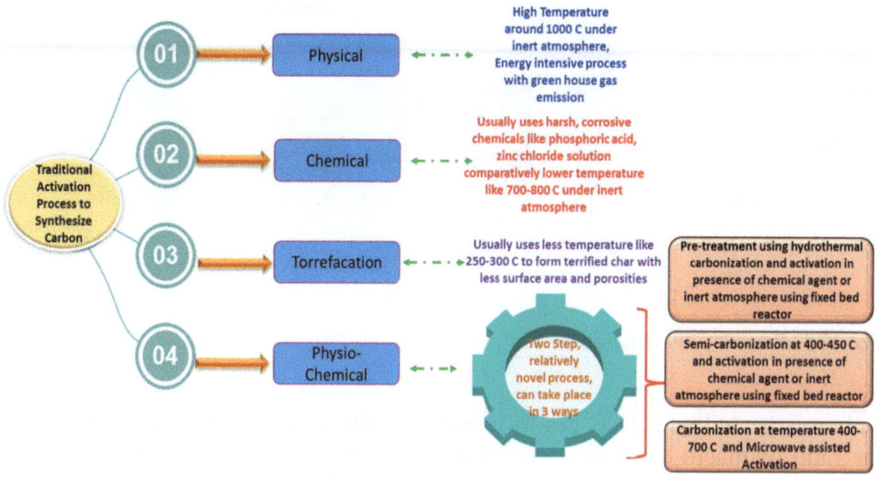

Fig. 12.4 Different types of activation technique

Physiochemical Activation

With the exception of chemical and physical activation, physiochemical activation can be used concurrently (Chowdhury et al. 2013). Two approaches for manufacturing ACs with physio-chemical activation are possible: I chemical treatments before to carbonization stated as pre-carbonization and (ii) chemical treatment following carbonization known as post-carbonization (Rashidi and Yusup 2017).

The first method involves carbonization of the substrates followed by mixing with biochar and then thermal pyrolysis under the flow of oxidising gas or combustion in an inert, oxygen deficient environment. The process is carried out at temperature of 600–850 °C (Din et al. 2017). The second approach involves chemical activation of the substrates and then heat treatment in the presence of oxidising gas (Mohd Din et al. 2009). The order of chemical activation in this process has an effect on the grade and morphological features of the produced ACs (Lee et al. 2014).

Numerous research analyzing physio-chemical and chemical activation techniques concluded that, double-stage activation results in carbon with distinct yet superior texture and quality attributes. ACs generated via physiochemical activation can yield carbon with adequate and consistentlydispersed pore configurations (Arami-Niya et al. 2012). They reported in another investigation that, the surface areas varied between 1035–1653 m^2/gm and 554–1213 m^2/gm for phosphoric acid and carbon di oxide activation and alone carbon di oxide activation, respectively, even though solitary activation using CO_2 had two-hour lengthier retention time (7 h) than the activation using two steps (5 h) (Arami-Niya et al. 2011).Additionally, the surface area of AC produced solely by CO_2 activation was found to be 632 m^2/gm, and the pore volume determined was 0.26 cm^3/g (Lee et al. 2014).

Furthermore, it was stated that, when carbons are soaked with KOH, merely micropores emerge onto the surface (Tseng et al. 2006). Meanwhile, when soaked

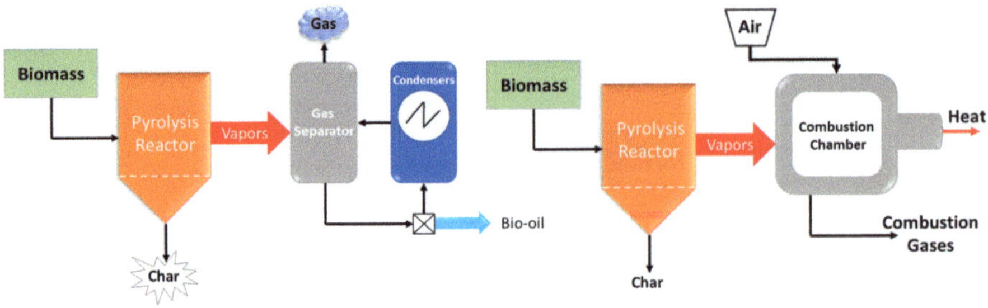

Fig. 12.5 Basic lay-out of pyrolysis unit

carbon is heated with CO_2 gas to initiate the physio-chemical activation process, the ratio of macro to meso pores changes, resulting in an enhancement in mass transfer and porosity for increased sorption capability (Danish and Ahmad 2018). Basic Pyrolysis set up unit is illustrated by Fig. 12.5.

Microwave Assisted Activation

Microwave (MW) assisted pyrolysis has emerged as a viable alternative to conventional methods for producing AC, owing to its unique characteristics. The process include preferential, rapid, even, heating, where indirect engagement between the heating framework and starting resources, takes place (Hoseinzadeh Hesas et al. 2013). The key operational limits for microwave-assisted activation include procedure setup, MW irradiation intensity activation time, precursor qualities, and correlations between the MW and chemical compounds (Ao et al. 2018) (Fig. 12.6). By virtue of dipole polarization and ionic conduction, energy is easily turned into heat within the macromolecules during heating in MW. When a high-frequency voltage is applied; particles having induced or permanent dipole-moment realign in the opposite direction of the delivered force (Alslaibi et al. 2013). As a result, a huge temperature difference exists between the inner and outer surfaces of the sample molecule, which renders it more operational and effective (Pathak and Mandavgane 2015). MW assisted activation is a procedure that combines physical and/or chemical activation to create higher-quality AC in a one-step or two-step (Abbas and Ahmed 2016) process.

Thermal decomposition via carbonization and activation are the primary processes for microwaves activation in a single phase using a heater. The advantage of this process is that it enables the AC to maintain a steady configuration by a simple approach. Microwave activation is a two-step process that involves carbonising and activating biochar using MW (Li et al. 2008). The BET surface area produced by this system is dependent on the cellulosic biomass, the pyrolysis temperatures, the irradiation power, the cycle time, and the chemical additives (Ao et al. 2018). Microwave

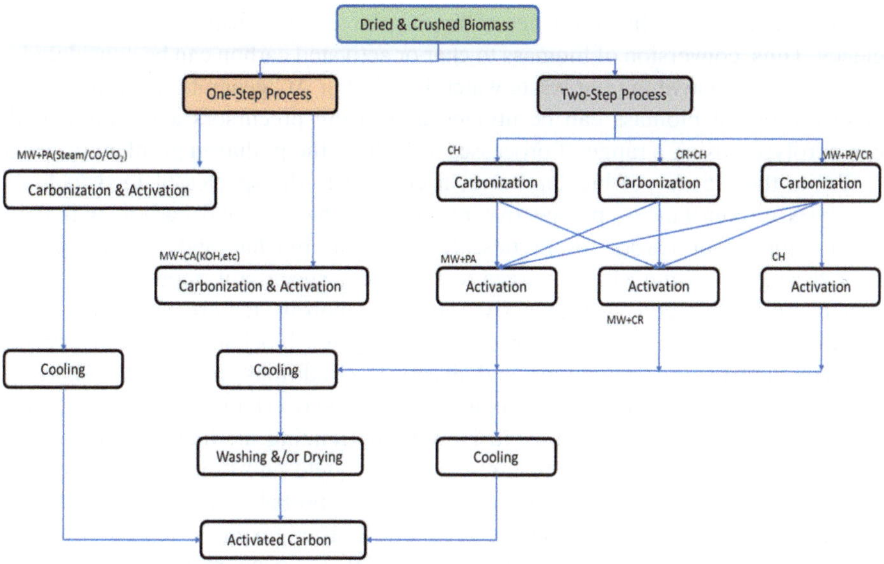

Fig. 12.6 Microwave assisted pyrolysis process

activation is a flexible technique for creating ACs with a more uniform textural characteristics.

12.7 Conclusion

This chapter focuses pertinent insight on application of various types of ligno-cellulosic biomass employed for water purification process. It has the greatest implications in comparison to the other alternatives as it can be used as biosorbent or converted to char and activated carbon for environmental application. The utilisation of such substances in waste effluent detoxification has been found to minimize the environmental footprint of these polluted fluids, that can contain a wide spectrum of harmful residues. Additionally, these ligno-cellulosic biomass can always be utilised as manure for agriculture and transformed into bio-fertilizers after being exploited to remove toxins from waste stream. When combined with various ion-exchange resins or converted to carbonaceous materials; this organic substrate can provide increased selectivity for pollutants. Owing to their easy availability, reduced cost and high sorption capacities, ligno-cellulosic fibres are mostly used to purify waste water containing metallic ions, inorganic ions, phenols, dyes, drugs, herbicides, pesticides and others present in waste stream. However, biosorption using the biomass directly has some disadvantages. Biosorption process is extremely relatively slow and pH variations reduce biosorption efficiency. It is biodegradable and the structure and surface characteristics of biomass have a significant impact on its' efficiency. Thus, conversion of biomass to char or activated carbon can be considered as promising alternative to treat waste water. Biochar or AC generation exhibits that, a diverse array of biomass can be utilized as starting precursors and subsequently can pyrolysed using a range of processes to address the pollution problem of water.

While the efficacy of biochar varies according to the species of feedstock and pyrolysis circumstances, prospective research on biochar and carbon will focus on fine-tuning of carbon properties. As a result, biochar and AC along with their derivatives/composites emerges to be a very successful approach for pollution control. Enlarged surface active area and porous texture are required for several biochar/AC activities. However, other type's biochar/AC features, which are regulated by feedstock type and pyrolysis conditions along with additional modification/functionalization strategies must not be overlooked. The efficiency of biochar/AC can be maximised exclusively by optimizing the link between surface area as well as other relevant features including yield and functional groups. For example, elevated temperature during pyrolysis are probable to lead in fewer functional groups, although large surface area with porous texture often requires very high temperature. Additionally, yield of biochar/AC is a benchmark for productivity, which varies in a manner, consistent with biochar/AC functionality. As a result, it is impracticable to enhance the surface area and porous texture of biochar/AC alone by raising the pyrolysis temperatures or by adjusting other operating conditions without taking into account regarding the yield and other inherent features of the biochar/AC. Additionally, certain procedures, such as doping/coating, can functionalize biochar/AC. Nevertheless, it is expected that additional holes will be clogged throughout the functionalization process, resulting worse porous structure. The solution to this challenge may reside in the advancement of coating/doping technologies. Additional research in this topic is necessary. Future study should concentrate on improving or inventing processing technologies that concurrently can produce biochar/AC with a functional groups and a porous texture.

Economical impact and reusability aspects of the carbon should be addressed for a broad range of environmental domain of sustainable applications. The relationship between various waste management and energy generating methods varies depending on the characteristics and different processes used to produce it. There are lot of social, economic, and ecological barriers which need to be overcome. Regardless of how the new framework is implemented in practise, the closed model outlines distinctions between the linear and circular models of waste management. Increased energy recovery is possible with this circular, green economy concept. This systematic review of this chapter summarised the current state-of-the-art info that might be used to identify special avenues for scientific entrepreneurship in the realm of carbon research.

References

Abbas AF, Ahmed MJ (2016) Mesoporous activated carbon from date stones (Phoenix dactylifera L.) by onestep microwave assisted K_2CO_3 pyrolysis. J Water Process Eng 9:201–207. https://doi.org/10.1016/j.jwpe.2016.01.004

Abdelwahab O, Amin N (2013) Adsorption of phenol from aqueous solutions by Luffa cylindrica fibers: kinetics, isotherm and thermodynamic studies. The Egypt J Aquatic Res 39(4):215–223. https://doi.org/10.1016/j.ejar.2013.12.011

Abdolali A, Guo W, Ngo H, Chen S, Nguyen N, Tung K (2014) Typical lignocellulosic wastes and by-products for biosorption process in water and wastewater treatment: a critical review. Biores Technol 160:57–66. https://doi.org/10.1016/j.biortech.2013.12.037

Akinpelu AA, Chowdhury ZZ, Shibly SM, Faisal ANM, Badruddin IA, Rahman MM, Amin MA, Sagadevan S, Akbarzadeh O, Khan TMY, Kamangar S, Khalid K, Saidur R, Johan MR (2021) Adsorption studies of volatile organic compound (Naphthalene) from aqueous effluents: chemical activation process using weak Lewis acid, equilibrium kinetics and isotherm modelling. Int J Mol

Aktaş Z, Çeçen F (2005) Effect of type of carbon activation on adsorption and its reversibility. J Chem Technol Biotechnol 81(1):94–101. https://doi.org/10.1002/jctb.1363

Ali I, Asim M, Khan TA (2012) Low cost adsorbents for the removal of organic pollutants from wastewater. J Environ Manage 113:170–183. https://doi.org/10.1016/j.jenvman.2012.08.028

Alslaibi TM, Abustan I, Ahmad MA, Foul AA (2013) A review: production of activated carbon from agricultural byproducts via conventional and microwave heating. J Chem Technol Biotechnol 88(7):1183–1190. https://doi.org/10.1002/jctb.4028

Ao W, Fu J, Mao X, Kang Q, Ran C, Liu Y, Zhang H, Gao Z, Li J, Liu G, Dai J (2018) Microwave assisted preparation of activated carbon from biomass: a review. Renew Sustain Energy Rev 92:958–979. https://doi.org/10.1016/j.rser.2018.04.051

Arami-Niya A, Daud WMAW, Mjalli FS (2011) Comparative study of the textural characteristics of oil palm shell activated carbon produced by chemical and physical activation for methane adsorption. Chem Eng Res Des 89(6):657–664.https://doi.org/10.1016/j.cherd.2010.10.003

Arami-Niya A, Wan Daud WMA., Mjalli S, Abnisa F, Shafeeyan MS (2012) Production of microporous palm shell based activated carbon for methane adsorption: modeling and optimization using response surface methodology. Chem Eng Res Des 90(6):776–784.https://doi.org/10.1016/j.cherd.2011.10.001

Beesley L, Marmiroli M (2011) The immobilisation and retention of soluble arsenic, cadmium and zinc by biochar. Environ Pollut 159:474–480

Bhatnagar A, Sillanpää M, Witek-Krowiak A (2015) Agricultural waste peels as versatile biomass for water purification—a review. Chem Eng J 270:244–271.https://doi.org/10.1016/j.cej.2015.01.135

Brinchi L, Cotana F, Fortunati E, Kenny J (2013) Production of nanocrystalline cellulose from lignocellulosic biomass: technology and applications. Carbohyd Polym 94(1):154–169. https://doi.org/10.1016/j.carbpol.2013.01.033

Cao Q, Xie K. Lv Y, Bao W (2006) Process effects on activated carbon with large specific surface area from corn cob. Bioresour Technol 97(1):110–115. https://doi.org/10.1016/j.biortech.2005.02.026

Chen XC, Chen GC, Chen LG, Chen YX, Lehmann JMB, McBride MB, Hay AG (2011) Adsorption of copper and zinc by biochars produced from pyrolysis of hardwood and corn straw in aqueous solution . Bioresour Technol 102:8877–8884

Cheremisinoff NP (2002) Handbook of water and wastewater treatment technologies. Butterworth-Heinemann, Woburn, MA, USA

Chojnacka K (2010) Biosorption and bioaccumulation—the prospects for practical applications. Environ Int 36(3):299–307. https://doi.org/10.1016/j.envint.2009.12.001

Choppala G, Bolan N, Kunhikrishnan A, Skinner W, Seshadri B (2013) Concomitant reduction and immobilization of chromium in relation to its bioavailability in soils. Environ Sci Pollut Res 22(12):8969–8978. https://doi.org/10.1007/s11356-013-1653-6

Chowdhury ZZ, Abd Hamid SB, Das R, Hasan MR, Zain SM, Khalid K, Uddin MN (2013) Preparation of carbonaceous adsorbents from lignocellulosic biomass and their use in removal of contaminants from aqueous solution. BioResources 8(4). https://doi.org/10.15376/biores.8.4.6523-6555

Chowdhury ZZ, Hamid SBA, Zain SM (2014) Evaluating design parameters for breakthrough curve analysis and kinetics of fixed bed columns for Cu(II) cations using lignocellulosic wastes. BioResources 10(1). https://doi.org/10.15376/biores.10.1.732-749

Chowdhury ZZ, Abd Hamid SB, Rahman MM, Rafique RF (2016) Catalytic activation and application of micro-spherical carbon derived from hydrothermal carbonization of lignocellulosic biomass: statistical analysis using Box-Behnken design. RSC Adv 6(104):102680–102694. https://doi.org/10.1039/c5ra26189a

Chowdhury Z, Krishnan B, Sagadevan S, Rafique R, Hamizi N, Abdul Wahab Y, Khan A, Johan R, Al-douri Y, Kazi S, Tawab Shah S (2018) Effect of temperature on the physical, electro-chemical and adsorption properties of carbon micro-spheres using hydrothermal carbonization process. Nanomaterials 8(8):597.https://doi.org/10.3390/nano8080597

Daneshvar E, Vazirzadeh A, Niazi A, Sillanpää M, Bhatnagar A (2017) A comparative study of methylene blue biosorption using different modified brown, red and green macroalgae—effect of pretreatment. Chem Eng J 307:435–446. https://doi.org/10.1016/j.cej.2016.08.093

Danish M, Ahmad T (2018) A review on utilization of wood biomass as a sustainable precursor for activated carbon production and application. Renew Sustain Energy Rev 87:1–21. https://doi.org/10.1016/j.rser.2018.02.003

Das SK, Mukherjee I, Das SK (2011) Dissipation of Flubendiamide in/on Okra [Abelmoschus esculenta (L.) Moench] Fruits. Bull Environ Contam Toxicol 88(3):381–384. https://doi.org/10.1007/s00128-011-0491-9

Das SK, Mukherjee I (2012) Effect of moisture and organic manure on persistence of flubendiamide in soil. Bull Environ Contam Toxicol 88(4):515–520. https://doi.org/10.1007/s00128-012-0551-9

Das SK, Das SK (2020) Influence of phosphorus and organic matter on microbial transformation of arsenic. Environ Technol Innov 19:100930.https://doi.org/10.1016/j.eti.2020.100930

Das SK, Ghosh GK (2020) Soil health management through low cost biochar technology. In: Singh J, Singh C (eds) Biochar applications in agriculture and environment management. Springer. https://doi.org/10.1007/978-3-030-40997-5_9

Das SK, Ghosh GK, Avasthe R (2020a) Application of biochar in agriculture and environment, and its safety issues. Biomass Convers Biorefinery.https://doi.org/10.1007/s13399-020-01013-4

Das SK, Ghosh GK, Avasthe R (2020b) Biochar application for environmental management and toxic pollutant remediation. Biomass Convers Biorefinery.https://doi.org/10.1007/s13399-020-01078-1

Das SK, Ghosh GK, Avasthe R (2020c) Ecotoxicological responses of weed biochar on seed germination and seedling growth in acidic soil. Environ Technol Innov 20:101074.https://doi.org/10.1016/j.eti.2020.101074

Das SK, Ghosh GK, Avasthe R (2020d) Evaluating biomas-derived biochar on seed germination and early seedling growth of maize and black gram. Biomass Convers Biorefinery.https://doi.org/10.1007/s13399-020-00887-8

Das SK, Ghosh GK, Avasthe R (2020e) Valorizing biomass to engineered biochar and its impact on soil, plant, water, and microbial dynamics: a review. Biomass Convers Biorefinery.https://doi.org/10.1007/s13399-020-00836-5

Das SK, Ghosh GK, Avasthe R, Sinha K (2021a) Compositional heterogeneity of different biochar: effect of pyrolysis temperature and feedstocks. J Environ Manage 278:111501.https://doi.org/10.1016/j.jenvman.2020.111501

Das SK, Ghosh GK, Avasthe R, Sinha K (2021b) Morpho-mineralogical exploration of crop, weed and tree derived biochar. J Hazard Mater 407:124370.https://doi.org/10.1016/j.jhazmat.2020.124370

Das SK (2015a) Acid sulphate soil: management strategy for soil health and productivity. Popular Kheti 3(2):2–7

Das SK (2015b) Integrated nematode management in chickpea against Meloidogyne incognita-a view point. J Agric Res 5(5):145–149
Das SK (2019a) Qualitative evaluation of fodder trees and grasses in hill region. J Krishi Vigyan 7(2):276.https://doi.org/10.5958/2349-4433.2019.00055.2
Das SK (2019b) Soil carbon sequestration strategies under organic production system: a policy decision. Agrica 8(1):1.https://doi.org/10.5958/2394-448x.2019.00001.4
Das SK, Mukherjee I (2011) Effect of light and pH on persistence of flubendiamide. Bull Environ Contam Toxicol 87:292–296
Das SK, Avasthe RK, Singh M, Roy A (2018a) Managing soil fertility under organic production system through integrated approach. Green Farming 9(3):449–454
Das SK, Avasthe RK, Singh M, Yadav A (2018b) Soil health improvement using biochar application in Sikkim: a success story. Innov Farming 3(1):48–50
Das SK, Ghosh GK, Avasthe RK (2018c) Preparation and characterization of biochar for their application as a soil amendment. Indian J Hill Farming 31(1):141–145
De Jonge RJ, Breure AM, van Andel JG (1996) Reversibility of adsorption of aromatic compounds onto powdered activated carbon (PAC). Water Res 30(4):883–892. https://doi.org/10.1016/0043-1354(95)00248-0
Din MI, Ashraf S, Intisar A (2017) Comparative study of different activation treatments for the preparation of activated carbon: a mini-review. Sci Prog 100(3):299–312. https://doi.org/10.3184/003685017x14967570531606
Dong CD, Chen CW, Kao CM, Chien CC, Hung CM (2018) Wood-biochar-supported magnetite nanoparticles for remediation of PAH-contaminated estuary sediment. Catalysts 8(2):73.https://doi.org/10.3390/catal8020073
Dong XL, Ma L, Li YC (2011) Characteristics and mechanisms of hexavalent chromium removal by biochar from sugar beet tailing. J Hazard Mater 190:909–915
Elaissaoui I, Akrout H, Grassini S, Fulginiti D, Bousselmi L (2016) Role of SiO interlayer in the electrochemical degradation of Amaranth dye using SS/PbO anodes. Mater Des 110:633–643. https://doi.org/10.1016/j.matdes.2016.08.044
El-Naggar A, Shaheen SM, Ok YS, Rinklebe J (2018) Biochar affects the dissolved and colloidal concentrations of Cd, Cu, Ni, and Zn and their phytoavailability and potential mobility in a mining soil under dynamic redox-conditions. Sci Total Environ 624:1059–1071.https://doi.org/10.1016/j.scitotenv.2017.12.190
Fathy A, El-Kady O (2013) Thermal expansion and thermal conductivity characteristics of Cu–Al_2O_3 nanocomposites. Mater Des 46:355–359. https://doi.org/10.1016/j.matdes.2012.10.042
Foereid B (2015) Biochar in nutrient recycling-the effect and its use in wastewater treatment. Soil Sci 5:39–44
Gallos A, Paës G, Allais F, Beaugrand J (2017) Lignocellulosic fibers: a critical review of the extrusion process for enhancement of the properties of natural fiber composites. RSC Adv 7(55):34638–34654. https://doi.org/10.1039/c7ra05240e
Gautam RK, Mudhoo A, Lofrano G, Chattopadhyaya MC (2014) Biomass-derived biosorbents for metal ions sequestration: adsorbent modification and activation methods and adsorbent regeneration. J Environ Chem Eng 2(1):239–259. https://doi.org/10.1016/j.jece.2013.12.019
González-García P (2018) Activated carbon from lignocellulosics precursors: a review of the synthesis methods, characterization techniques and applications. Renew Sustain Energy Rev 82:1393–1414.https://doi.org/10.1016/j.rser.2017.04.117
Gopi R, Avasthe RK, Kalita H, Yadav A, Das SK, Rai D (2020) Eco-friendly management of tomato late blight using botanicals, bio-control agents, compost tea and copper fungicides. Indian J Agric Sci 90(1):35–39
Gudade B, Babu S, Bora S, Dhanapal K, Singh R (2016) Effect of boron on growth, nutrition and fertility status of large cardamom in Sikkim Himalaya, India. J Appl Nat Sci 8(2):822–825. https://doi.org/10.31018/jans.v8i2.879

Haldar D, Purkait MK (2020) Micro and nanocrystalline cellulose derivatives of lignocellulosic biomass: a review on synthesis, applications and advancements. Carbohydr Polym 250:116937.https://doi.org/10.1016/j.carbpol.2020.116937

Hokkanen S, Bhatnagar A, Sillanpää M (2016) A review on modification methods to cellulose-based adsorbents to improve adsorption capacity. Water Res 91:156–173. https://doi.org/10.1016/j.watres.2016.01.008

Hoseinzadeh Hesas R, Wan Daud WMA, Sahu JN, Arami-Niya A (2013) The effects of a microwave heating method on the production of activated carbon from agricultural waste: a review. J Anal Appl Pyrol 100:1–11.https://doi.org/10.1016/j.jaap.2012.12.019

Hung Y, Lo HH, Wang LK, Taricska JR, Li KH (2005) Granular activated carbon adsorption. In: Wang LK, Hung Y, Shammas NH (eds) Handbook of environmental engineering—vol. 3: physicochemical treatment processes. The Humana Press Inc., Totowa, New Jersey, USA

Isoda N, Rodrigues R, Silva A, Gonçalves M, Mandelli D, Figueiredo FCA, Carvalho WA (2014) Optimization of preparation conditions of activated carbon from agriculture waste utilizing factorial design. Powder Technol 256:175–181. https://doi.org/10.1016/j.powtec.2014.02.029

John Y, David VE, Mmereki D (2018) A comparative study on removal of hazardous anions from water by adsorption: a review. Int J Chem Eng 1–21. https://doi.org/10.1155/2018/3975948

Kesraoui A, Moussa A, Ali GB, Seffen M (2015) Biosorption of alpacide blue from aqueous solution by lignocellulosic biomass: luffa cylindrica fibers. Environ Sci Pollut Res 23(16):15832–15840. https://doi.org/10.1007/s11356-015-5262-4

Lam E, Male KB, Chong JH, Leung AC, Luong JH (2012) Applications of functionalized and nanoparticle-modified nanocrystalline cellulose. Trends Biotechnol 30(5):283–290. https://doi.org/10.1016/j.tibtech.2012.02.001

Lashari MS, Ye Y, Ji H, Li L, Kibue GW, Lu H, Zheng J, Pan G (2015) Biochar-manure compost in conjunction with pyroligneous solution alleviated salt stress and improved leaf bioactivity of maize in a saline soil from central China: a 2-year field experiment. J Sci Food Agric 95(6):1321–1327

Lehmann J, Joseph S (2009) Biochar for environmental management: an introduction. In: Lehmann J, Joseph S (eds) Biochar for environmental management: science and technology. Earthscan London, pp 1–12

Lee T, Zubir ZA, Jamil FM, Matsumoto A, Yeoh FY (2014) Combustion and pyrolysis of activated carbon fibre from oil palm empty fruit bunch fibre assisted through chemical activation with acid treatment. J Anal Appl Pyrol 110:408–418. https://doi.org/10.1016/j.jaap.2014.10.010

Leung ACW, Hrapovic S, Lam E, Liu Y, Male KB, Mahmoud KA, Luong JHT (2010) Characteristics and properties of carboxylated cellulose nanocrystals prepared from a novel one-step procedure. Small 7(3):302–305. https://doi.org/10.1002/smll.20100171

Li H, Dong X, da Silva EB, de Oliveira LM, Chen Y, Ma LQ (2017) Mechanisms of metal sorption by biochars: biochar characteristics and modifications. Chemosphere 178:466–478. https://doi.org/10.1016/j.chemosphere.2017.03.072

Li Z, Sun Y, Yang Y, Han Y, Wang T, Chen J, Tsang DC (2020) Biochar-supported nanoscale zero-valent iron as an efficient catalyst for organic degradation in groundwater. J Hazard Mater 383:121240. https://doi.org/10.1016/j.jhazmat.2019.121240

Li W, Zhang LB, Peng JH, Li N, Zhu XY (2008) Preparation of high surface area activated carbons from tobacco stems with K_2CO_3 activation using microwave radiation. Ind Crops Prod 27(3): 341–347.https://doi.org/10.1016/j.indcrop.2007.11.011

Liu X, Lai D, Wang Y (2019) Performance of Pb(II) removal by an activated carbon supported nanoscale zero-valent iron composite at ultralow iron content. J Hazard Mater 361:37–48.https://doi.org/10.1016/j.jhazmat.2018.08.082

Menya E, Olupot P, Storz H, Lubwama M, Kiros Y (2018) Production and performance of activated carbon from rice husks for removal of natural organic matter from water: a review. Chem Eng Res Des 129:271–296. https://doi.org/10.1016/j.cherd.2017.11.008

Mohd Din AT, Hameed B, Ahmad AL (2009) Batch adsorption of phenol onto physiochemical-activated coconut shell. J Hazard Mater 161(2–3):1522–1529. https://doi.org/10.1016/j.jhazmat.2008.05.009

Mukherjee I, Das SK, Kumar A, Shukla L (2020) Sludge amendment affects the persistence, carbon mineralization and enzyme activity of atrazine and bifenthrin. Bull Environ Contam Toxicol 105(2):291–298

Odetoye T, Abu Bakar M, Titiloye J (2019) Pyrolysis and characterization of Jatropha curcas shell and seed coat. Niger J Technol Dev 16(2):71.https://doi.org/10.4314/njtd.v16i2.4

Othmani A, Kesraoui A, Seffen M (2017) The alternating and direct current effect on the elimination of cationic and anionic dye from aqueous solutions by electrocoagulation and coagulation flocculation. Euro-Mediterranean J Environ Integr 2(1). https://doi.org/10.1007/s41207-017-0016-y

Othmani A, Kesraoui A, Boada R, Seffen M, Valiente M (2019) Textile wastewater purification using an elaborated biosorbent hybrid material (Luffa–Cylindrica–Zinc Oxide) assisted by alternating current. Water 11(7):1326.https://doi.org/10.3390/w11071326

Othmani A, Kesraoui A, Hanene A, Elaissaoui I, Seffen M (2020) Coupling anodic oxidation, biosorption and alternating current as alternative for wastewater purification. Chemosphere 249:126480.https://doi.org/10.1016/j.chemosphere.2020.126480

Pathak PD, Mandavgane SA (2015) Preparation and characterization of raw and carbon from banana peel by microwave activation: application in citric acid adsorption. J Environ Chem Eng 3(4):2435–2447. https://doi.org/10.1016/j.jece.2015.08.023

Radenahmad N, Azad AT, Saghir M, Taweekun J, Bakar MSA, Reza MS, Azad AK (2020) A review on biomass derived syngas for SOFC based combined heat and power application. Renew Sustain Energy Rev 119:109560.https://doi.org/10.1016/j.rser.2019.109560

Rajan PA, Sp A, Balaji S (2014) Treatment study of dyeing industry effluents using reverse osmosis technology. Res J Recent Sci 3:58–61

Rajput BS, Bhardwaj DR, Pala NA (2018) Soil organic carbon (SOC) density under different agroforestry systems along an elevation gradient in north-western Himalaya. Agrofor Syst 89(3):525–536. https://doi.org/10.1007/s10457-015-9788-8

Rashidi NA, Yusup S (2017) A review on recent technological advancement in the activated carbon production from oil palm wastes. Chem Eng J Biochem Eng J 314:277–290.https://doi.org/10.1016/j.cej.2016.11.059

Roy A, Das A, Das SK, Datta M, Datta J, Tripathi AK, Singh NU (2018) Impact analysis of National Agricultural Innovation Project (NAIP): a paradigm shift in income and consumption in Tripura. Green Farming 9(3):559–564

Selmi T, Sanchez-Sanchez A, Gadonneix P, Jagiello J, Seffen M, Sammouda H, Celzard A, Fierro V (2018a) Tetracycline removal with activated carbons produced by hydrothermal carbonisation of Agave americana fibres and mimosa tannin. Ind Crops Prod 115:146–157.https://doi.org/10.1016/j.indcrop.2018.02.005

Selmi T, Seffen M, Sammouda H, Mathieu S, Jagiello J, Celzard A, Fierro V (2018b) Physical meaning of the parameters used in fractal kinetic and generalised adsorption models of Brouers–Sotolongo. Adsorption 24(1):11–27.https://doi.org/10.1007/s10450-017-9927-9

Singh NS, Mukherjee I, Das SK, Varghese E (2018) Leaching of clothianidin in two different Indian soils: effect of organic amendment. Bull Environ Contam Toxicol 100(4):553–559

Sun P, Hui C, Azim Khan R, Du J, Zhang Q, Zhao YH (2015) Efficient removal of crystal violet using Fe3O4-coated biochar: the role of the Fe_3O_4 nanoparticles and modeling study their adsorption behavior. Sci Rep 5(1). https://doi.org/10.1038/srep12638

Tran HN, You SJ, Nguyen TV, Chao HP (2017) Insight into the adsorption mechanism of cationic dye onto biosorbents derived from agricultural wastes. Chem Eng Commun 204(9):1020–1036.https://doi.org/10.1080/00986445.2017.1336090

Ukanwa P, Sakrabani A, Mandavgane A (2019) A review of chemicals to produce activated carbon from agricultural waste biomass. Sustainability 11:6204. https://doi.org/10.3390/su11226204

Vo AT, Nguyen VP, Ouakouak A, Nieva A, Doma BT, Tran HN, Chao HP (2019) Efficient removal of Cr(VI) from water by biochar and activated carbon prepared through hydrothermal carbonization and pyrolysis: adsorption-coupled reduction mechanism. Water 11(6):1164.https://doi.org/10.3390/w11061164

Wang T, Liu X, Men Q, Ma C, Liu Y, Ma W, Liu Z, Wei M, Li C, Yan Y (2019) Surface plasmon resonance effect of Ag nanoparticles for improving the photocatalytic performance of biochar quantum-dot/Bi4Ti3O12 nanosheets. Chin J Catal 40(6):886–894. https://doi.org/10.1016/s1872-2067(19)63330-9

Yaashikaa PR, Senthil Kumar P, Varjani SJ, Saravanan A (2019) A review on photochemical, biochemical and electrochemical transformation of CO_2 into value-added products. J CO_2 Util 33:131–147

Yaashikaa PR, Senthil Kumar P, Sunita Varjani A, Saravanan A (2020) A critical review on the biochar production techniques, characterization, stability and applications for circular bioeconomy Biotechnol Rep 28:e00570. https://doi.org/10.1016/j.btre.2020.e00570

Yahya MA, Mansor MH, Zolkarnaini WAAW, Rusli NS, Aminuddin A, Mohamad K, Sabhan FA, Atik AA, Ozair LN (2018) A brief review on activated carbon derived from agriculture by-product. In: AIP conference proceedings, pp 030023–1–030023–8. https://doi.org/10.1063/1.5041244

Yan J, Han L, Gao W, Xue S, Chen M (2015) Biochar supported nanoscale zerovalent iron composite used as persulfate activator for removing trichloroethylene. Biores Technol 175:269–274. https://doi.org/10.1016/j.biortech.2014.10.103

Yang X, Wan Y, Zheng Y, He F, Yu Z, Huang J, Wang H, Ok YS, Jiang Y, Gao B (2019) Surface functional groups of carbon-based adsorbents and their roles in the removal of heavy metals from aqueous solutions: a critical review. Chem Eng J 366:608–621.https://doi.org/10.1016/j.cej.2019.02.119

Yuan JH, Xu RK, Zhang H (2011) The forms of alkalis in the biochar produced from crop residues at different temperatures. Biores Technol 102(3):3488–3497. https://doi.org/10.1016/j.biortech.2010.11.018

Zhang P, Huang P, Xu X, Sun H, Jiang B, Liao Y (2020) Spectroscopic and molecular characterization of biochar-derived dissolved organic matter and the associations with soil microbial responses. Sci Total Environ 708

Index

S. Roy et al. (eds.), *Advanced Industrial Wastewater Treatment and Reclamation of Water*, Environmental Science and Engineering,
https://doi.org/10.1007/978-3-030-83811-9

MIX
Papier aus verantwortungsvollen Quellen
Paper from responsible sources
FSC® C105338

If you have any concerns about our products, you can contact us on
ProductSafety@springernature.com

In case Publisher is established outside the EU, the EU authorized representative is:
Springer Nature Customer Service Center GmbH
Europaplatz 3, 69115 Heidelberg, Germany

Printed by Libri Plureos GmbH
in Hamburg, Germany